Array and Phased Array Antenna Basics

Array and Phased Array Antenna Basics

Hubregt J. Visser

Antenna Engineer, The Netherlands

John Wiley & Sons, Ltd

Telephone (+44) 1243 779777

Email (for orders and customer service enquiries): cs-books@wiley.co.uk

Visit our Home Page on www.wiley.com

Reprinted with corrections January 2006

Other Wiley Editorial Offices

John Wiley & Sons Inc., 111 River Street, Hoboken, NJ 07030, USA

Jossey-Bass, 989 Market Street, San Francisco, CA 94103-1741, USA

Wiley-VCH Verlag GmbH, Boschstr. 12, D-69469 Weinheim, Germany

John Wiley & Sons Australia Ltd, 42 McDougall Street, Milton, Queensland 4064, Australia

John Wiley & Sons (Asia) Pte Ltd, 2 Clementi Loop #02-01, Jin Xing Distripark, Singapore 129809

John Wiley & Sons Canada Ltd, 22 Worcester Road, Etobicoke, Ontario, Canada M9W 1L1

Wiley also publishes its books in a variety of electronic formats. Some content that appears in print may not be available in electronic books.

British Library Cataloguing in Publication Data

A catalogue record for this book is available from the British Library

ISBN-13 978-0-470-87117-1 (HB)
ISBN-10 0-470-87117-2 (HB)

Typeset by the author using LaTeX software.
Printed and bound in Great Britain by Antony Rowe Ltd, Chippenham, Wiltshire.
This book is printed on acid-free paper responsibly manufactured from sustainable forestry in which at least two trees are planted for each one used for paper production.

Contents

Preface

Array and phased array antennas are gaining in popularity. They seem no longer to be of interest to military (radar) systems only, but are encountered today in many civilian systems, like for example in mobile communications base stations. An array antenna is a group of individual radiators, positioned in such a way as to produce a maximum radiation into a forward direction. With a *phased* array antenna, we mean an array antenna, wherein we change the phases between the array antenna elements - added to the already existing phase differences due to the differences in position - to create maximum radiation in a desired direction.

One should be aware that for some people all array antennas are *phased* array antennas, due to the fact that the array antenna principle is also based on phase differences. In this book we adopt the term *phased* array antenna for *scanned beam array antennas.*

Over the decades many books have been written, dealing solely or partly with array and phased array antenna theory. The books by Mailloux [1] and Hansen [2], are two fine examples of the many excellent books available.

So, why another book on phased array antennas when there are already so many books on the topic available? The reason for this book lies in the fact that more and more engineers nowadays are getting involved in antenna and (phased) array antenna technology. Not all of them are thoroughly trained in electromagnetics or antennas and are easily scared away by texts that assume a basic understanding of electromagnetics or antennas or both, even if this fear is not realistic.

Often the question arises if it is possible to teach antenna technology without using equations. Although the answer to that question is negative, this book tries to create a compromise, providing an easy-to-read text, explaining antennas in general, based on historical development and physical characteristics rather than mathematics. Chapters 1, 2 and 3 are dealing in this way with, respectively, radiation, antennas and antenna parameters. The mathematics are introduced where and when necessary and then it will appear that the mathematics are not that complicated at all. The mathematical treatment starts in chapter 4, dealing with the broadside linear array antenna. The next chapter, chapter 5, is a 'how to' chapter that will provide the reader with detailed information on how to use the acquired knowledge in the design of a microstrip patch array antenna. Chapter 6 deals with linear endfire array antennas and will treat the Yagi–Uda array antenna in more detail. Chapter 7 is concerned with the linear phased array antenna and is followed again by a 'how to' chapter, chapter 8, wherein the design of a frequency scanned, slotted waveguide array antenna will be dealt with in detail. Chapter 9 then extends the acquired knowledge of linear array and phased array antennas to planar array and phased array antennas and in chapter 10 some special array antenna configurations will be discussed. These configurations are conformal and volume array antennas, sequentially rotated and fed arrays for the creation of circular polarisation and reactively loaded array antennas. Finally, in chapter 11, antenna measurement techniques in general and phased array antenna measurement techniques in particular, including the use of the scan element pattern (also known as active element pattern), will be discussed.

As an additional resource, this book is supported by a companion website on which instructors and lecturers can find electronic versions of the figures. Please go to ftp://ftp.wiley.co.uk/pub/books/visser

The aim of *Array and Phased Array Antenna Basics* is to provide an introduction to (phased) array antennas that will allow the reader to move onto specialist books on the topic with a greater understanding.

Therefore, topics like beam synthesis, phased array antenna errors, beam switching, digital beam forming, array thinning and adaptive array antennas will not be treated in this book.

REFERENCES

1. Robert J. Mailloux, *Phased Array Antenna Handbook*, Artech House, Boston, 1994.

2. R.C. Hansen, *Phased Array Antennas*, John Wiley & Sons, New York, 1998.

HUBREGT J. VISSER

Veldhoven, The Netherlands

Acknowledgments

I would like to express my gratitude to the people who indirectly contributed to this book. First to my colleagues, past and present, of the Netherlands Organisation for Applied Scientific Research (TNO), especially Gerard van der Spek, Will Keizer and Frank van den Boogaart who made it possible for me to work on phased array antennas. To Antoine Roederer from the European Space Research and Technology Centre (ESA-ESTEC) for his support of my phased array antenna work over the many years. To my colleagues of Eindhoven University of Technology, especially Anton Tijhuis, for introducing me into the fascinating world of teaching. Special thanks I owe to Ad Reniers for his innovative ways of constructing antennas for this book and for providing the measurement data of these antennas. I also would like to thank Sarah Hinton from John Wiley & Sons, Ltd, for her support and patience.

I am most grateful to Dianne, for her patience and understanding during the many evening and weekend hours of neglect that - as I understand now - inherently come with the writing of a book.

H.J.V.

Acknowledgments

Acronyms

AC	Alternating Current
A/D	Analog to Digital
AI	Aircraft Interception
AM	Amplitude Modulation
ASV	Air to Surface Vessel
AUT	Antenna Under Test
CATR	Compact Antenna Test Range
CH	Chain Home
CRT	Cathode Ray Tube
CW	Continuous Wave
DUT	Device Under Test
HF/DF	High Frequency/Direction Finding
HPBW	Half Power Beam Width
IFF	Identification Friend or Foe
LHCP	Left Hand Circular Polarisation
LORAN	Long Range Navigation
PCB	Printed Circuit Board
PPI	Plan Position Indicator

PRF	Pulse Repetition Frequency
RADAR	Radio Detection and Ranging
RAM	Radar Absorbing Material
RCS	Radar Cross Section
RDF	Radio Detection and Finding
RF	Radio Frequency
RHCP	Right Hand Circular Polarisation
SLL	Side Lobe Level
SGA	Standard Gain Antenna
UHF	Ultra High Frequency
VHF	Very High Frequency
VNA	Vector Network Analyser
VSWR	Voltage Standing Wave Ratio

1

Radiation

The key to understanding phased array antennas is to understand antennas and the key to understanding antennas is to understand electromagnetic radiation. In contrast to what is widely believed, one does not need to be a specialist in integro-differential equations and vector mathematics to grasp the mechanism of electromagnetic radiation. As in Faraday's time, a vast majority of educators prefers the rigor of a mathematical description to the insight of a physical understanding for explaining the mechanism of radiation. It is the author's belief though that the latter is needed first to form the basic understanding and once this understanding has been accomplished, the former may be used to develop this understanding and put it to practical use.

For the basic understanding of electromagnetic radiation one only needs an understanding of electricity and magnetism at a level as educated in secondary school. By following the historical developments in the field of electricity and magnetism, the interaction between the two - electromagnetism - and electromagnetic radiation follows naturally.

1.1 THE EARLY HISTORY OF ELECTRICITY AND MAGNETISM

Research in the field of electricity and magnetism goes back a long way. Hundreds of years BC, experiments dealing with these two phenomena have been described. However, for nearly two thousand years experiments have been concentrated mainly on static electricity. The absence of a source of continuous electrical energy posed a severe limitation in the progress of understanding the underlying physics of the observed electrical and magnetic phenomena. It lasted until the invention of the electric battery

Array and Phased Array Antenna Basics Hubregt J. Visser

in 1800 by *Alessandro Volta* (1745–1827) - a logical next step to the famous 'frog-experiments' of *Aloys Galvani* (1737–1798) nine years earlier - before research could be conducted in a reproducible way.

In 1819, the Danish professor *Johannis Ørsted* (1777–1851) observed the change in position of a compass needle, when brought into the vicinity of a current carrying wire, see figure 1.1.

Fig. 1.1 The Ørsted experiment: a current-carrying wire deflects a compass needle such that the needle positions itself perpendicularly to the wire.

This current originated from a voltaic pile. Although Ørsted did not fully understand that the compass needle was not directly influenced by the electric current, but rather indirectly through the induced magnetic field around the current, he did notice the importance of his observation. It opened up the possibility to find a relation between electricity and magnetism. A few months after the publication of his experiences, Ørsted introduced the term *electromagnetism*.

In that same year (1819), the French professor *André-Marie Ampère* (1775–1836) observed a reproduction of the Ørsted experiment at the Parisian Academy. Only a week afterwards he produced a document, giving a theoretical explanation of the experiment. He assumed - correctly - that an electrical current is capable of inducing a magnetic field, see figure 1.2.

It is this magnetic field that explains why parallel currents attract and anti-parallel currents repel. For two parallel currents, the compass needles, indicating the direction of the magnetic induction, are positioned such that they will attract one another, see figure 1.3. For two anti-parallel currents, the compass needles are positioned such that they will repel one another (opposite poles attract, equal poles repel), see figure 1.4.

If we bend a current-carrying wire into a loop as shown in figure 1.5, the magnetic inductions of all parts of the wire add up to form distinct poles at the top and bottom of the loop.

Ampère (rightfully) assumed that in solid matter, microscopic parts contain a circulating current and that in a magnet (like a piece of magnetised iron) all these

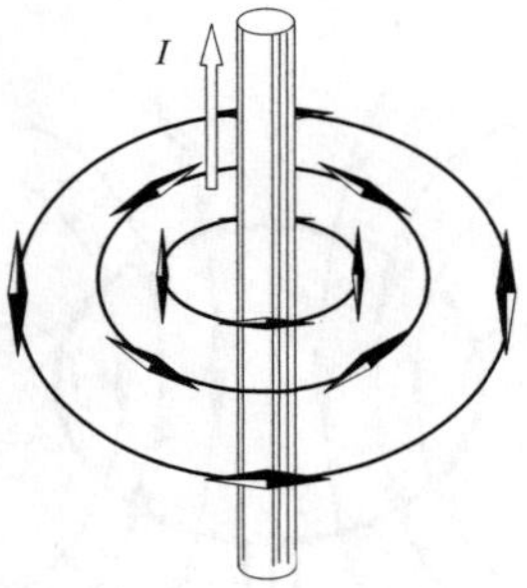

Fig. 1.2 The findings of Ampère: a current-carrying wire creates a magnetic induction around the wire. The direction of the magnetic field may be found by placing compass needles in the vicinity of the wire.

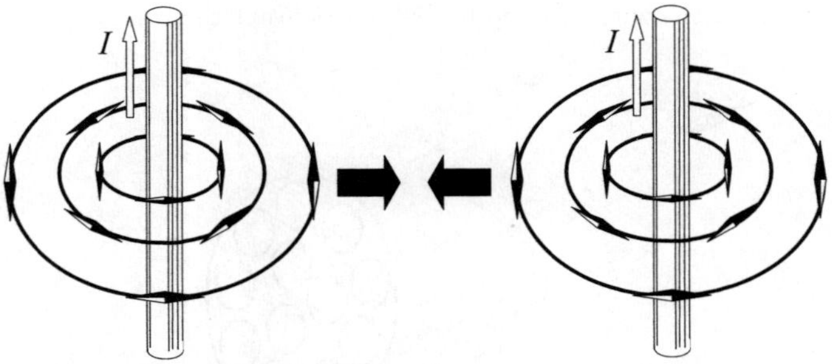

Fig. 1.3 The findings of Ampère: parallel currents attract. The compass needles indicate that the induced magnetic fields are such that the wires will attract. Opposite poles of the compass needles are placed next to each other.

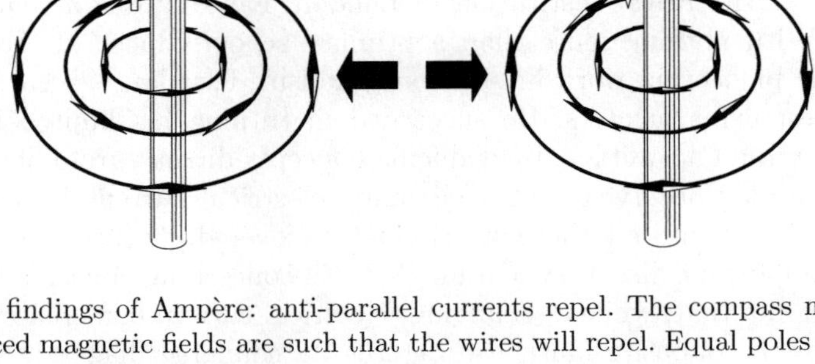

Fig. 1.4 The findings of Ampère: anti-parallel currents repel. The compass needles indicate that the induced magnetic fields are such that the wires will repel. Equal poles of the compass needles are placed next to each other.

microscopic current loops are lined up in the same direction, resulting in the forming of distinct macroscopic magnetic poles. This process is schematically depicted in figure 1.6.

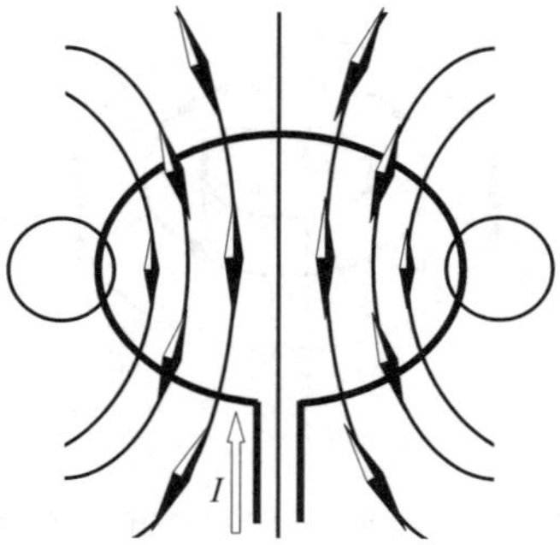

Fig. 1.5 The findings of Ampère: for a current-carrying wire bent into a loop, the magnetic induction of all parts of the wire adds up to distinct poles at the top and bottom of the loop.

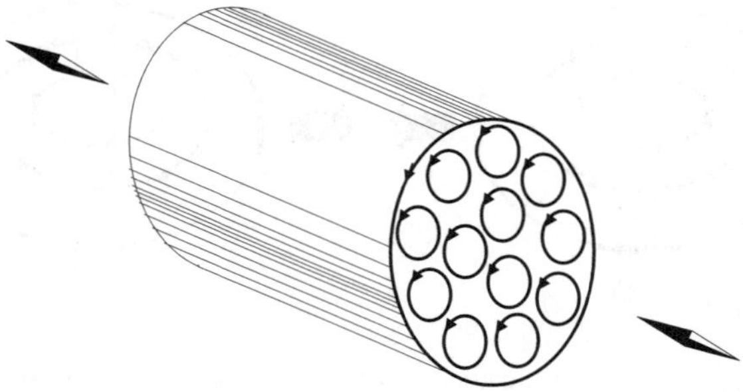

Fig. 1.6 The findings of Ampère: the forming of a macroscopic magnet by the lining up of microscopic current loops.

While Ørsted was conducting his experiments in Denmark, *Michael Faraday* (1791–1867) worked at the Royal Institution in London. Faraday was a remarkable 'self-made' man. With nothing more than a primary school education, he has become famous for his pioneering work in electromagnetism. Faraday is seen as one of the world's greatest experimenters. He succeeded in turning his limited knowledge of mathematics into an advantage, by deducing concepts directly from observations.

In 1831, Faraday observed that a changing electrical current in a coil, induced an electrical current into another coil. He had discovered *electromagnetic induction*. This was an important discovery. Faraday's ideas concerning conservation of energy had convinced him that, while a (changing) electrical current can create a (changing) magnetic field, the opposite must also be true: 'a changing magnetic field must be able to produce an electric field'. One of his experiments, following the discovery of the electromagnetic induction, showed that a *moving* magnet induced an electrical current, see figure 1.7.[1]

[1]In his first experiment, the changing electrical current applied to a coil took care of the changing magnetic induction, created by moving the magnet in his second experiment.

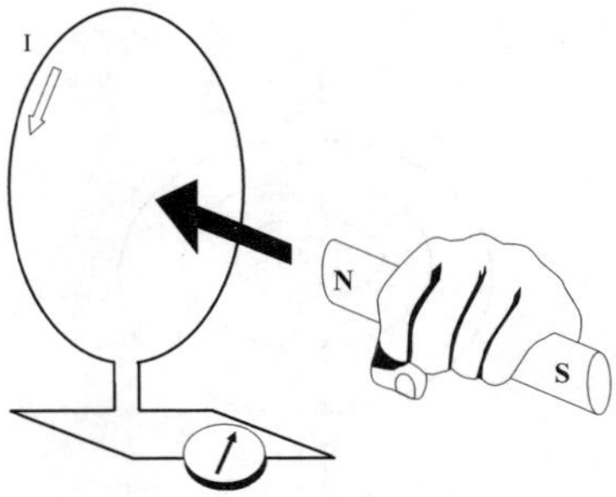

Fig. 1.7 The findings of Faraday: a *moving* magnet induces an electrical current in a loop.

The direction of the current in the loop is such that it opposes the change in magnetic induction. If the north pole of the magnet is approaching the loop, the current in the loop will be directed such that a north pole is formed in the direction of the magnet to repel this magnet. If the north pole of the magnet is moving away from the loop, the current in the loop will be directed such that a south pole is formed in the direction of the magnet to attract the magnet. If the magnet remains static, there will be no current in the loop.

To understand how a magnet influences a wire from some distance, Faraday visualised a 'magnetic field'. He saw this 'magnetic field' as magnetic force lines, laying closer together at places where the field is stronger. These magnetic field lines can be shown by placing compass needles in the vicinity of the magnet, as we have done thus far. These compass needles will direct themselves tangential to the magnetic field lines. More detail can be obtained by using iron filings that may be regarded as very small compass needles. In fact, this is how Faraday constructed his magnetic field lines. In his 1831 notes[2] [1] he wrote,

> By magnetic curves I mean lines of magnetic forces which would be depicted by iron filings.

So, instead of visualising the magnetic field by placing compass needles, as we did in figure 1.2 for a current-carrying wire, we may now draw the magnetic field lines as shown in figure 1.8.

The electrical current, induced in a loop or a (piece of) wire at some distance from the 'source', Faraday expressed in terms of the number of magnetic field lines cut by the loop or wire (flux).

Now, let's have a closer look at the first electromagnetic induction experiment of Faraday, see figure 1.9.

Obviously, we can transport the changing electric field in one coil to an isolated second coil. So, the question arises does this mean that we are dealing with electromagnetic radiation? The answer to that question is: 'No'!

If we take away the changing electric field in the bottom coil, by keeping the switch open or closed, the changing magnetic field vanishes everywhere and no current flows in the top coil. If we had had an electromagnetic radiating system, the changing

[2]Read at the Royal Society on 24 November 1831.

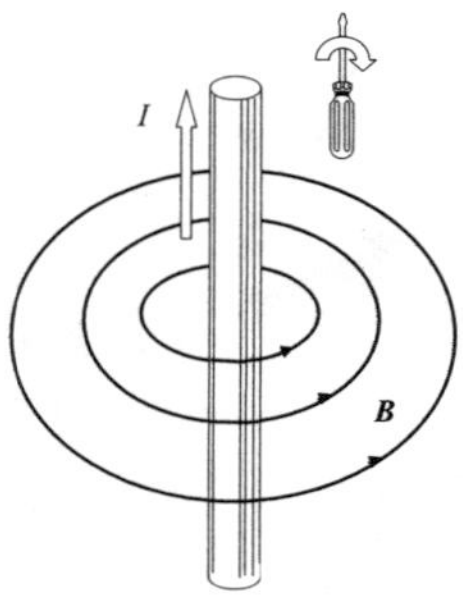

Fig. 1.8 The findings of Faraday: the magnetic field. The direction of the field relative to the direction of the current is dictated by a right-hand screw in the direction of the current. The separation between the field lines is inversely proportional to the magnitude of the field.

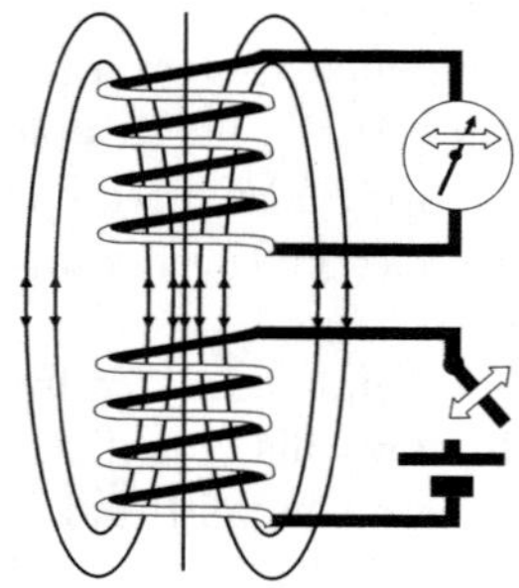

Fig. 1.9 The findings of Faraday: a *changing* electrical current in a coil, induces a changing electrical current into another coil.

magnetic field would have induced a changing electric field, regardless of the presence of the top coil. How the *radiation* mechanism works will be explained further on.

For the moment we will dwell on the bottom coil of figure 1.9 only. If we connect an alternating current (AC) source to the coil, the current through the coil changes continuously and because of that, the magnetic field lines change continuously too. Because of these changing field lines, a current is recreated in this coil. Energy is thus delivered to a magnetic field and this energy is returned to the circuit.

When the magnetic field is increasing, due to the current flowing through the coil, a voltage over the coil is being created. When the current has reached its maximum value and starts to decrease, the magnetic field strength decreases. The coil, however, opposes this change and therefore tries to maintain a voltage over the coil, such that the field remains static. *Therefore the current flowing through a coil lags the voltage over it.*

Thus, because of this energy cycling, the current and voltage of the coil are out of phase. For a radiating electromagnetic field to exist, the electric and magnetic field need to be in phase. The field now is purely a storage field; the energy is stored in the magnetic field surrounding the inductor. If we place a second coil in the vicinity of the first coil, see figure 1.9, we can intercept some of the changing magnetic field lines and thereby create an electric current in the second coil. With the second coil we can thus take energy from the storage field of the first coil.[3]

The field line concept may also be applied to electric fields. Electric field lines are then imaginary lines for which the tangent vector at a given point is directed such as to coincide with the direction in which a positively charged unit charge would accelerate. As with magnetic field lines, the separation between adjacent field lines is inversely proportional to the magnitude of the field. Tightly packed electric field lines indicate a strong field; sparsely packed electric field lines a weak one.

Faraday observed that magnetic field lines (made visible by employing iron filings) originate from one pole of a magnet and terminate on the other. So he imagined that the lines of force of an electrical field would originate on a positive charge and end on a negative charge. It appears that this is not completely true. Magnetised objects always form poles in pairs. Magnetic field lines originate from the north pole of the object and terminate on the south pole of the object. Electrically charged objects however may exist as monopole (positively charged or negatively charged). The field lines are always directed perpendicular to the surface of the charged object. The electric field lines of an isolated, positively charged monopole, start on the monopole and extend radially to infinity, those of an isolated, negatively charged monopole, start at infinity and convert radially on the monopole, see figure 1.10.

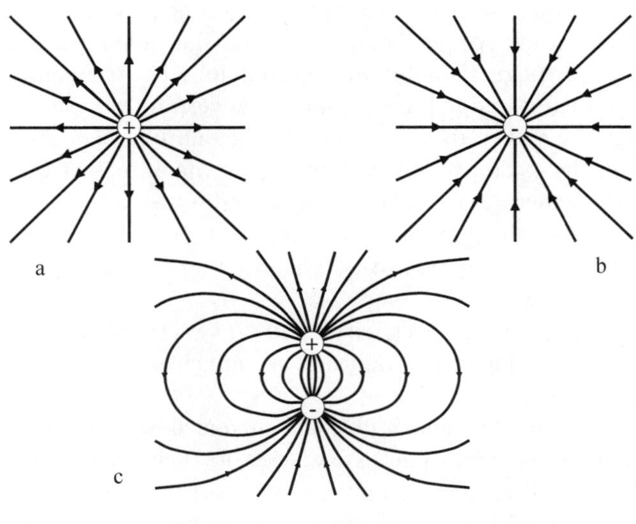

Fig. 1.10 Electric field lines. a: isolated, positively charged point source monopole. b: isolated, negatively charged point source monopole. c: system of one positively and one negatively charged point source.

It was Faraday's belief that the physical lines of force he envisaged were really present everywhere in space, i.e. were an attribute of this space. Even though we know now that this is not true, we do understand that a magnet or electric charge brought into empty space, modifies this space. It is this understanding that formed a breakthrough in the nineteenth century in explaining the action-at-a-distance

[3]Of course the battery has to replenish the energy taken from the storage field.

which occurs between (magnetic, electrically charged or gravitational) objects. It was Faraday's genius that perceived the concept of a field to explain how e.g. a charged object affects the surrounding space. When another charged object is brought into this space, it becomes affected by the field of the object already present as opposed to the object itself.

1.2 JAMES CLERK MAXWELL, THE UNION OF ELECTRICITY AND MAGNETISM

In the year that Michael Faraday discovered electromagnetic induction (1831), *James Clerk Maxwell* (1831–1879) was born in Edinburgh, Scotland. In contrast to Faraday, Maxwell received an academic education and evolved not into a experimental scientist but a brilliant thinker and mathematician.

Although Maxwell has performed monumental scientific work, like for example proving that the rings of Saturn are made up of dense particles (1859) and demonstrating the first ever colour photograph (1860)[4] [2], he is best known for what are currently called the *Maxwell equations* (1873).

The remarkable thing about the Maxwell equations is that he did not derive all of them, but rather saw the connection between Ampère's, Faraday's and Gauss's law. By extending Ampère's law with what he called a *displacement current* term, electricity and magnetism became united into electromagnetism. With this displacement current term added, the equations governing electricity and magnetism allow electromagnetic waves to exist, light being one out of a spectrum of waves. Maxwell predicted the existence of electromagnetic waves tens of years before he was proven right by the generation and reception of radio waves.

Before we move on to the mechanism of electromagnetic radiation, we will first pay some attention to electromagnetic waves. Therefore we will have to dwell a bit longer on the *displacement current*.

The equations governing electricity and magnetism before Maxwell were incomplete. This was evident in analysing a capacitor in a circuit supporting a changing current, see figure 1.11.

Although the capacitor prevents a physical current flowing through the plates, the circuit still supports a current. The explanation for this effect, Maxwell attributed to what he called the *displacement current*, which turns out to be the time rate of change of the electric field between the capacitor plates.

Let's assume that we look at the capacitor in figure 1.11 at the moment that the capacitor is fully charged. The charge on the right plate is Q, the charge on the opposite plate is the negative of that, -Q. Because of these charges, the current in the circuit will start flowing to the right. The electric field between the plates is directed to the left. Since a current is flowing to the right, the strength of the electric field is decreasing, so the direction of *change* of the electric field is to the right. Attached to this *changing* electric field is a surrounding magnetic field, its direction connected to

[4]And - as the rumour goes and denied by Maxwell - at Trinity College, Cambridge, inventing a method of throwing a cat out of a window in such a way that it cannot land on its feet.

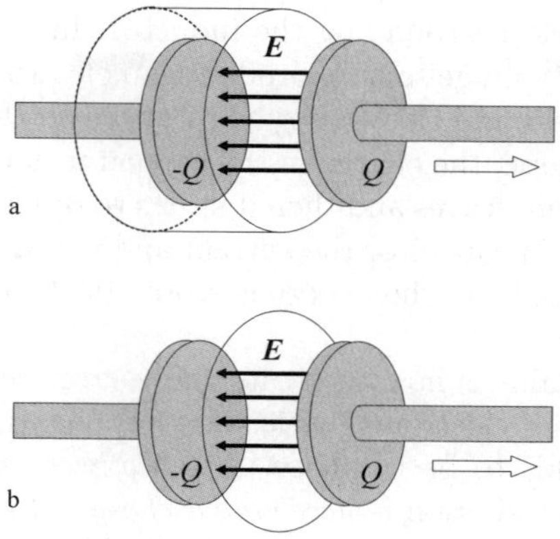

Fig. 1.11 Capacitor in a circuit supporting a changing current, i.e. before steady state. a. A current is flowing through the transparent surface. b. No current is flowing through the transparent surface.

the direction of *change* like a right-hand screw.[5] Since this magnetic field could have been due to a physical current - one cannot tell the source from the magnetic field alone - Maxwell named the source, the changing electric field: *displacement current.*

It is this specialty - that a changing electric field creates a (changing) magnetic field - that makes the existence of electromagnetic waves possible. It means that once you create the 'correct'[6] changing electric field, this field will create a changing magnetic field that in turn will create a changing electric field and so on. This expansion of the disturbance in space will continue, even when the source has ceased to exist. This is what we call *wave propagation.* This process is depicted in figure 1.12.

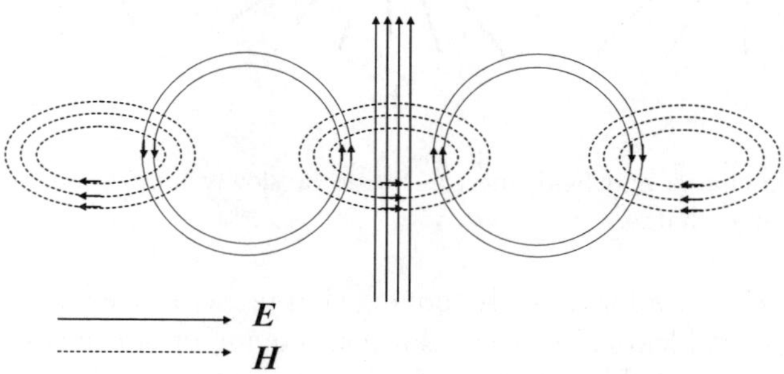

Fig. 1.12 The process of wave propagation. In reality the process is three dimensional and waves expand spherically.

We have seen that a coil (an inductor), when connected to an AC current source, produces an electromagnetic field. Energy is delivered to a magnetic field and this energy is returned to the circuit. The field is purely a storage field; the energy is

[5]Just like the direction of the magnetic field attached to the current in the circuit.
[6]We will get to the definition of 'correct' later on.

stored in the magnetic field surrounding the inductor. In a similar way a capacitor, when connected to an AC voltage source, produces an electromagnetic field. Energy is delivered to an electric field and this energy is returned to the circuit. When the AC cycle starts on its down slope, the charge in the capacitor holds the voltage (opposing the change) until the current leaves and then it starts to discharge, causing the voltage to lag the current. Also for a capacitor the current and voltage are out of phase. Again the field is purely a storage field; the energy is stored in the electric field surrounding the capacitor.

The question that remains is how to create the *correct* changing electric field, i.e. how to create the source of electromagnetic wave propagation; how to get changing electric and magnetic fields to be in phase. The Maxwell equations reveal that the source of electromagnetic radiation is *accelerated charge*. Rather than elaborating on the Maxwell equations, we will discard the mathematics and explain the radiation of accelerated charge by physical reasoning.

1.3 RADIATION BY ACCELERATED CHARGE

When looking at an electric charge, either positively or negatively charged, the electric field lines extend radially from this charge to infinity, see figure 1.13a.

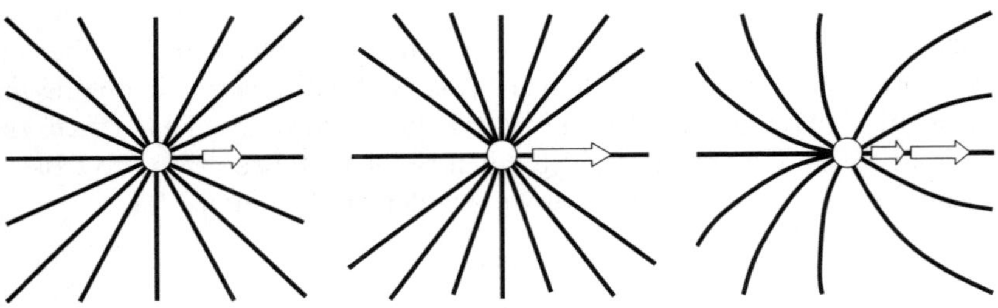

a b c

Fig. 1.13 Electric charge and field lines. a. Static or slowly moving charge. b. Fast moving charge. c. Accelerated charge.

The electric field lines for a slowly moving charge, i.e. having a velocity well below that of light, behave identical to that of a non-moving or static charge, figure 1.13a. This is a consequence of the *principle of relativity in the restricted sense* [3],

> If, relative to K, K' is a uniformly moving co-ordinate system devoid of rotation, then natural phenomena run their course with respect to K' according to exactly the same general laws as with respect to K.

In other words: 'An observer moving at the same speed as the charge still sees only static fields.'

When the charge is moving fast, i.e. at velocities approaching the velocity of light, the electric field lines tend to compress in the direction perpendicular to the direction the charge is moving, see figure 1.13b. This is also a consequence of the principle of relativity [3]. Only when a charge is accelerated, that is acted upon by an external force, can it radiate. The electric field lines will bend, see figure 1.13c, thus creating a transversal component - next to the radial components - in the electric field that propagates away from the charge at the speed of light. We will explain this by having a closer look at figure 1.13c.

Let's assume that a charged particle is uniformly moving along a horizontal line as depicted in figure 1.14. At a certain moment of time the particle is accelerated for a short period of time and, afterwards, it continues its uniform movement, see figure 1.14.

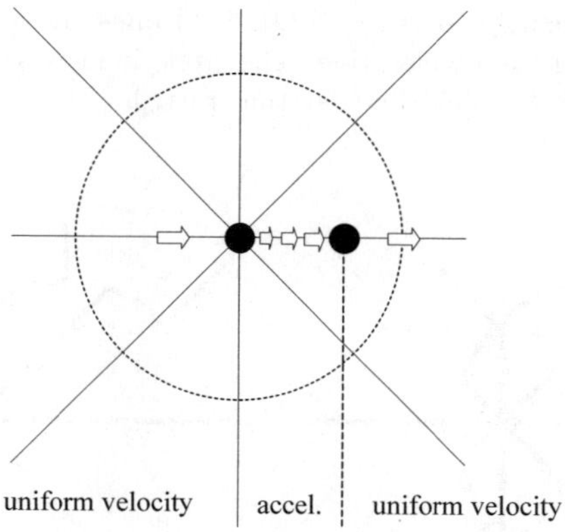

Fig. 1.14 A charged particle, uniformly moving along the horizontal axis is accelerated for a brief moment of time.

In the figure we have indicated the position of the particle at the start and finish of the acceleration. Some of the static electric field lines at the start position are shown, those at the finish position are left out for the sake of clarity. Also indicated in figure 1.14 is the position of an observer that has moved with the speed of light along a static electric field line from the particle, for the duration of the acceleration.

In figure 1.15 we repeat figure 1.14, but now also indicate the static field lines associated with the particle at the end of its acceleration.

If we now think of ourselves positioned anywhere on the 'observer circle' and accepting the fact that nothing can move faster than the velocity of light, we see that everywhere from the circle to infinity, the static field lines must follow those from the initial particle position. Everywhere inside the circle, the static field lines must follow those associated with the final particle position. Since electric field lines must be continuous, so-called *kinks* must exist at the observer position to make the field lines connect, see figure 1.16.

For a continuously accelerated charge, we would have found the field lines as shown in figure 1.13.

Now that we know how to construct the electric field lines of an accelerated, charged particle, we can take a look at these field lines for different times inside the time interval of acceleration. This has been done in figure 1.17 for subsequent moments within the acceleration time frame.

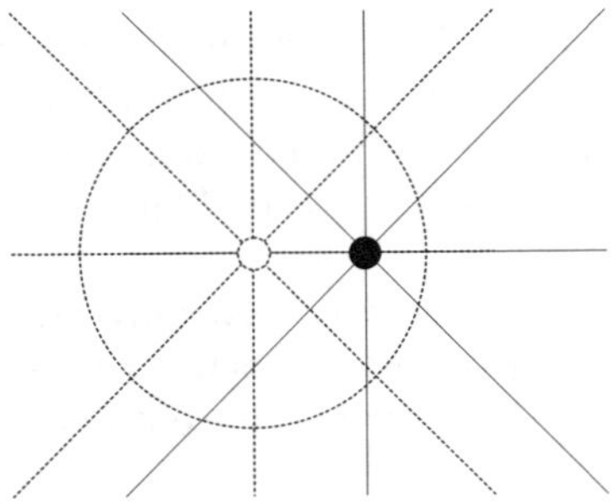

Fig. 1.15 An accelerated charged particle. Static field lines are shown for the particle at the beginning and at the end of the acceleration. The circle indicates the position of an observer that travelled outward with the field lines at the start, for the duration of the acceleration, with the speed of light.

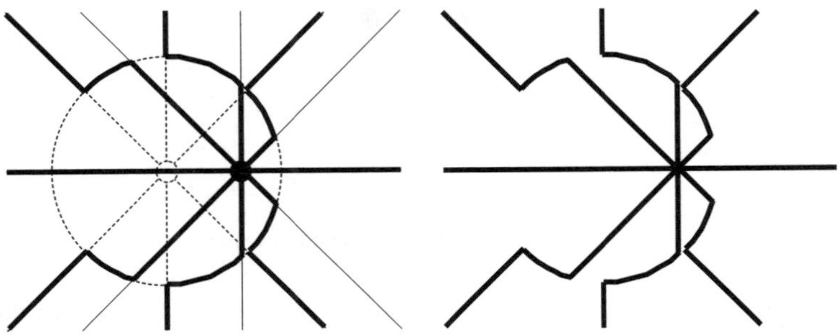

Fig. 1.16 The electric field lines of a shortly accelerated particle must form *kinks* in order to connect the field lines associated with the end and initial position of the particle to form continuous field lines.

When we take the disturbances, i.e. the transverse components of the electric field, taken at the subsequent moments and add them in one graph, see figure 1.17d, we see that these disturbances move out from the accelerated charge at the speed of light. Associated with this changing electric field is a changing magnetic field. Unlike the situation as described for the coil or the capacitor, the electric and magnetic fields are *in phase* now, due to the fact that they were produced by a single event, the acceleration of charge. The electric and magnetic field travel along in phase, their directions being perpendicular to one another, see also figure 1.12. This is what we call an *electromagnetic wave.*

In real life we will find accelerating and decelerating[7] charges in time-varying currents. Charge acceleration or deceleration in an electrically conducting, wire object may be found where the wire is curved, bent, discontinuous or terminated. All these origins of radiation are shown in figure 1.18 [4].

[7]Deceleration is a negative acceleration and will therefore be treated as acceleration.

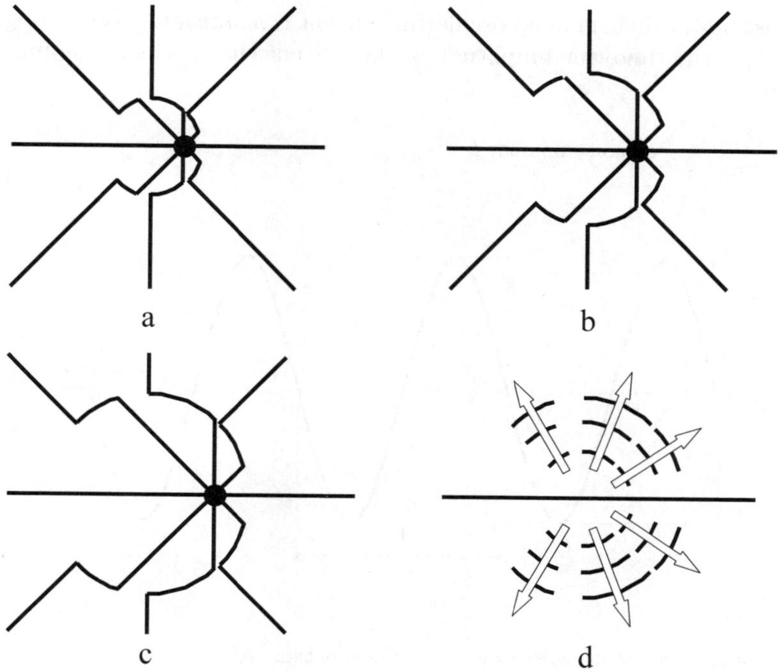

Fig. 1.17 The electric field lines of an accelerated, charged particle at subsequent instances of time. a. Time t_1. b. Time $t_2 > t_1$. c. Time $t_3 > t_2 > t_1$. c. Transverse electric field moving out at the speed of light.

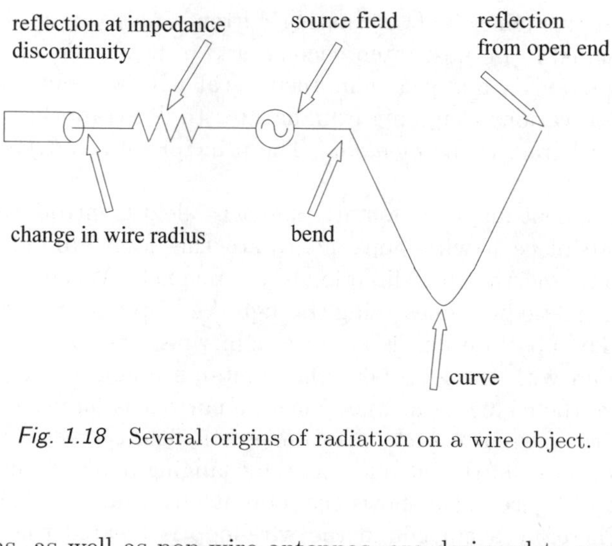

Fig. 1.18 Several origins of radiation on a wire object.

Wire antennas, as well as non-wire antennas, are designed to support oscillating currents. An oscillating current consists of charges accelerated back and forth. These oscillating currents thus create a regular disturbance or continuous radiation.

The most 'natural' form of an oscillating current is a *sinusoidal* oscillating current. As function of the time, the amplitude of the current changes as a sine function, see figure 1.19.

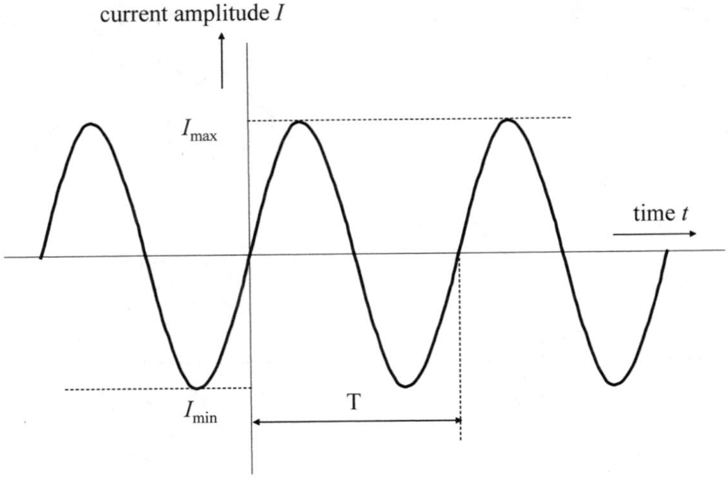

Fig. 1.19 Current changing with time as a sine function. After every period of time T, the changing of the amplitude with time repeats itself. T is called the period of the current-amplitude function.

We see that the current as function of time repeats itself every period T. The amplitude of the current therefore is a periodic function. We call T the *period* of the function.

We designated this sinusoidal alternating current (AC) the most 'natural', since it is a periodic signal form most often encountered in nature. We have to remember that the sine function is not just a mathematical frivolity, but rather a compact description of observations stemming from nature. To illustrate this, let's have a look at an invention of Faraday: the *dynamo*.[8] The principle of the dynamo is illustrated in figure 1.20.

With two permanent magnets, a static magnetic field is introduced. In this static magnetic field we place a wire-loop and rotate this loop around an axis, placed perpendicular to the north-south direction of the magnets. When the loop is rotated, the number of magnetic field lines going through the loop (this is called the magnetic *flux*), changes. The absolute flux is at maximum when the loop is directed vertical and is at minimum with the loop being horizontal, see figure 1.20a. Because of the changing flux, i.e. the changing magnetic field, a current is induced in the wire-loop. The direction of the current is such, that a magnetic field is created through the loop that opposes the source of the current, i.e. the changing magnetic field.

The snapshot of figure 1.20a shows the loop at an instant of time where the flux is decreasing. Therefore a current in the wire-loop is created that has a direction

[8] Although Faraday discovered the principle of the dynamo, he never foresaw the economical potential of it, being interested only in the physics of electricity and magnetism.

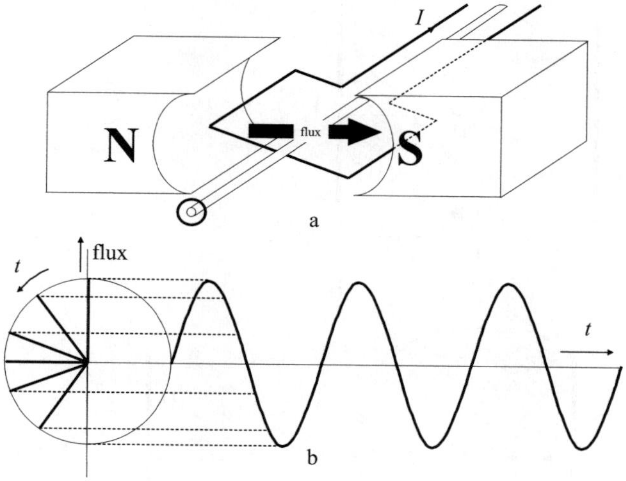

Fig. 1.20 Principle of the dynamo. a. A wire-loop is rotated in a static magnetic field and thus in the loop a changing magnetic field is created, inducing an electric current through the wire of the loop. b. The number of magnetic field lines, cut by the loop (flux) whilst the loop is rotating, changes with time as a sine function.

such that a north pole is created in the direction of the static north pole to oppose the rotation and thereby to oppose the decrease of flux. When the loop is making a complete turn, it subsequently forms north and south poles to oppose decreasing and increasing fluxes. The direction of the current therefore changes while the loop rotates. As Faraday already found (and as put into mathematical form by Ampère), the change in flux is proportional to the induced current.

In figure 1.20b we see, on the left, the position of the loop when the loop is rotating. When the loop is vertical, the flux is at maximum. When the loop is horizontal, the flux is zero. Therefore, if we associate with the loop position also the direction of maximum flux, i.e. we draw a circle with radius equal to the maximum flux, we can obtain the amplitude of the flux by projecting the instantaneous maximum flux onto the vertical axis. In the same figure we see how we can, from this figure, construct the amplitude of the flux (and thus the amplitude of the induced current) as a function of time. We see that this function is identical to the sine function as shown in figure 1.19.

The principle of the dynamo is still put into practice by power plants, where a rotational motion, initiated by steam or water, is converted to AC electrical power that is delivered to our homes. Also when AC electrical power is obtained from a direct current (DC) source, like a battery, the most natural form, obtained by using capacitors and inductors, is a sinusoidal one.

After this digression about oscillating currents, let's return to the nature of electromagnetic radiation by charges in oscillatory acceleration/deceleration.

To that purpose, consider two electrically conducting wires of infinitesimal length, folded 90 degrees, and connected to an AC source as described before. We will look

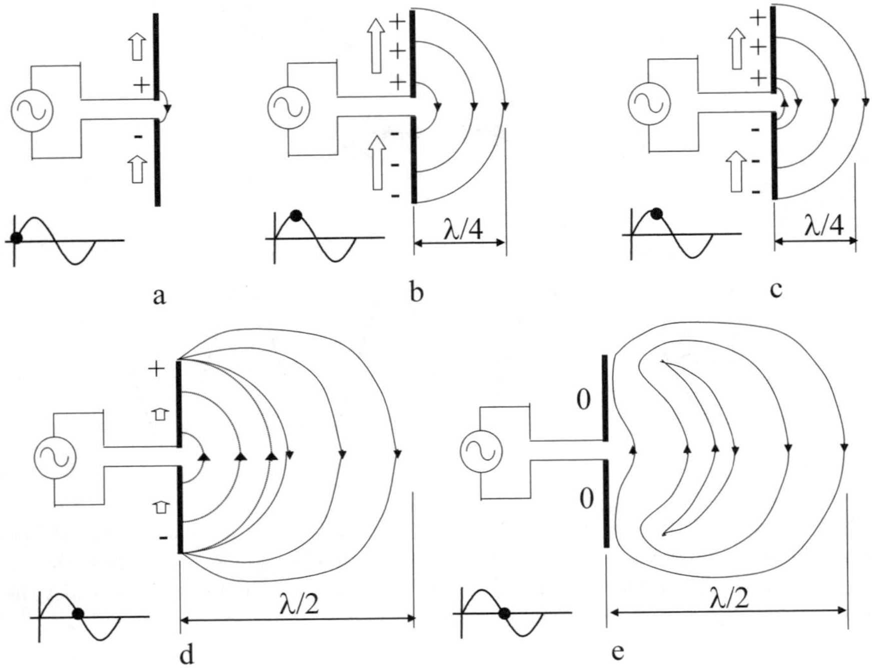

Fig. 1.21 Electromagnetic radiation by charges in oscillatory acceleration. a. $t = 0^+$. b. $t = \left(\frac{T}{4}\right)$. c. $t = \left(\frac{T}{4}\right)^+$. d. $t = \left(\frac{T}{2}\right)^-$. e. $t = \left(\frac{T}{2}\right)$.

at the electric field around this structure at different instances of time within one half of the period T, see figure 1.21.

Figure 1.21a shows the time $t = 0^+$, meaning: a short while after $t = 0$. The source has been turned on and charge is accelerated from the source to the wire ends. Because of the accelerating charges at the feed point, a transverse electric field component is travelling outward, in a direction perpendicular to the wire. Since field lines have to be continuous and end and start perpendicular to a charged body, the electric field line takes the form as shown in figure 1.21a. The direction and amplitude of the current is indicated by the white arrows next to the dipole. Underneath the dipole is the current shown as function of time; the time of the snapshot is indicated with a black dot. Figure 1.21b shows the infinitesimal antenna at time $t = \left(\frac{T}{4}\right)$, where T is the period, see figure 1.19. At this moment, the current has reached its maximum value, its change with time has become zero. The electric field lines are as shown in the figure. The transverse electric field component that was created at $t = 0^+$ has travelled a distance of a quarter of a wavelength.[9] New transverse electric field components have been created after the creation of this first one. Shortly after this moment, that is at $t = \left(\frac{T}{4}\right)^+$, the current has become less than the maximum value and the time change of the current has changed sign. This means that charges are

[9] A wavelength (λ) is the distance a 'disturbance' travels in the time frame of one period, moving at the speed of light.

accelerated into the opposite direction, leading to the creation of a new transversal electric field component, also directed into the opposite direction, see figure 1.21c. In figure 1.21d, we see the situation shortly before half a period has elapsed, that is at $t = \left(\frac{T}{2}\right)^{-}$. The current amplitude has become very small, the initial excess charges (positive in the upper half of the dipole, negative in the lower half) are only present at the tips and charge *acceleration* is directed downward. Additional, upward directed, transverse electric field components have been created, the first one has travelled nearly a quarter of a wavelength outward. At $t = \left(\frac{T}{2}\right)$, both halves of the dipole have become neutral, see figure 1.21e, no excess charge is present anymore and the current has become zero. The need for the electric field lines to be directed perpendicular to the conductors is not longer present and consequently, the field lines form closed loops and detach from the conductors.

We have shown this phenomenon in a plane and at one side of the antenna. Of course, in this plane, the phenomenon exists on both sides of the antenna. Furthermore, the phenomenon is three-dimensional and includes the existence of magnetic field lines that are positioned perpendicular to the electric field lines and form closed loops, as indicated in figure 1.12. For the sake of clarity we have not shown these field lines in figure 1.21. Figure 1.22, finally, shows the detachment of electric field lines from the infinitesimal (dipole) antenna at times $t = \left(\frac{T}{2}\right)^{+}$ and $t = T^{+}$ within a plane.

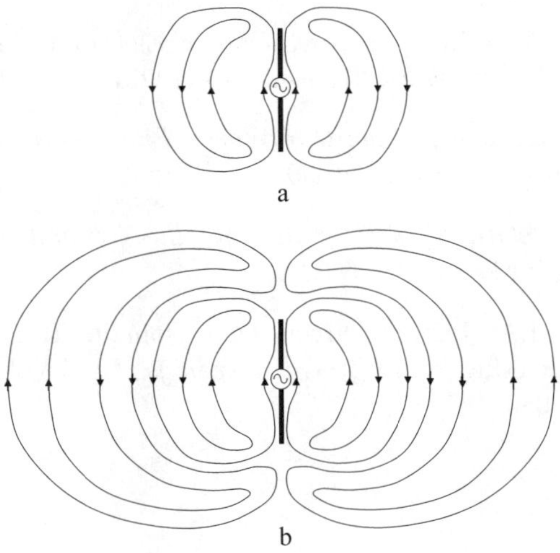

Fig. 1.22 Detachment of electric field lines from an infinitesimal dipole antenna. a. $t = \left(\frac{T}{2}\right)^{+}$. b. $t = T^{+}$.

However, the *radiating* electromagnetic fields are not the only electromagnetic fields present in the vicinity of our infinitesimal dipole antenna.

1.4 REACTIVE AND RADIATING ELECTROMAGNETIC FIELDS

We have seen that the current from the source transports charges to the ends of the dipole. The thus formed charge densities at the ends of the dipole will produce a voltage between the ends and electric field lines may be envisaged between both ends. Apart from the magnetic fields produced by the AC current flowing in the dipole wires, magnetic fields are also produced by virtue of the electric field between the ends increasing or decreasing. These particular electric and magnetic fields are not in phase. The magnetic field reaches its maximum when the electric field is changing the fastest and the electric field is at maximum when the magnetic field is changing the fastest (i.e. when the current is changing direction). Therefore the antenna also behaves as a capacitor.[10] However, the radiating part dominates the storage or reactive part and sufficiently far away from the antenna only the radiating part remains. What we mean by sufficiently far away will be explained in the next chapter.

Similarly, the capacitor and inductor are not purely capacitive and inductive, respectively. Since charge is being accelerated in the circuits containing the capacitor or inductor, electromagnetic radiation will occur. However, if the capacitor or inductor is designed well, this radiation will be of second-order importance and may be neglected for most real-life applications.

REFERENCES

1. Michael Faraday, Experimental Researches in Electricity, *The Royal Institution of Great Britain, Philosophical Transactions*, 122, pp. 125–162, 1832.

2. John Hamilton (ed.), *They Made Our World, Five Centuries of Great Scientists and Inventors*, Broadside Books, 1990.

3. Albert Einstein, *Relativity, the Special and the General Theory*, Three Rivers Press, New York, 1961.

4. E.K. Miller and J.A. Landt, Direct Time-Domain Techniques for Transient Radiation and Scattering from Wires, *Proceedings IEEE*, Vol. 68, No. 11, pp. 1396–1423, November 1980.

[10] If we take an infinitesimal small wire-loop, we can envisage radiating electromagnetic fields along the same line of reasoning as we did for the infinitesimal small dipole antenna. For the loop we will find that the antenna also behaves as an inductor; a one-turn coil. Also by the wire-loop antenna reactive fields are created.

2

Antennas

The key to understanding phased array antennas is to understand antennas. It suffices, however, to develop this understanding to a basic level where the antenna functioning is understood on the basis of physical reasoning. It won't hurt of course to develop a thorough understanding of antennas, based on electromagnetic theory, but for acquiring a basic knowledge of phased array antennas this is not considered as a prerequisite. Having accepted this, we may limit the education to explaining antenna functioning by physical reasoning. As in the previous chapter, following the course of history naturally brings us to an understanding of antennas.

2.1 THE EARLY HISTORY OF ANTENNAS

The graphical representation of the electric field lines, detaching themselves from the (dipole) antenna, was first shown by *Heinrich Rudolf Hertz* (1857–1894) in 1889 in his paper 'The forces of electric oscillations, treated according to Maxwell's theory'.

Hertz, at the university of Karlsruhe, Germany, was working on the experimental verification of Maxwell's theory. He started out working with two coaxially mounted flat induction coils,[1] very much like the set-up shown in figure 1.9 of the previous chapter [1] and shown here in figure 2.1.

When a Leyden jar (i.e. a charged capacitor) is connected to one of the coils, the second coil, by virtue of induction, produces a spark between its terminals. The Leyden jar connected to the coil forms a resonant electric circuit.

[1] Known at the time as *Knochenhauer spirals*.

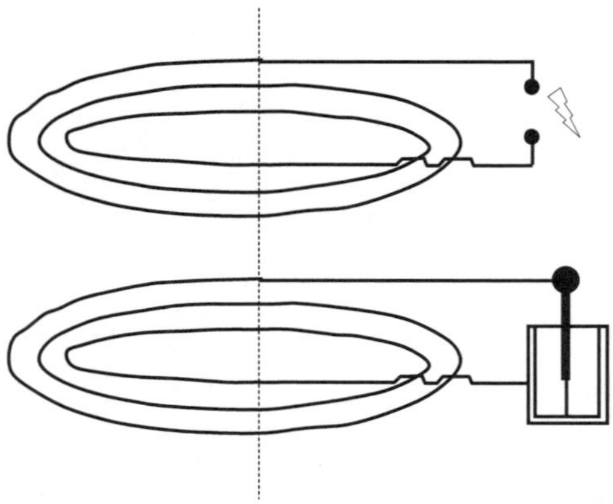

Fig. 2.1 Two coaxially mounted Knochenhauer spirals. When a Leyden jar is connected to the bottom one, a spark may be observed between the terminals of the top one.

2.1.1 Resonant Electric Circuit

In electric circuits, the discrete elements we most often encounter are the resistor (R), capacitor (C) and inductor (L).

A resistor directly resists the flow of electrons. The waveform of the voltage drop over the resistor is *in phase* with the waveform of the current through the resistor.

A capacitor does not directly resist the flow of electrons but opposes a *change* in voltage over the capacitor. The current 'through' the capacitor is a reaction against the *change* in voltage over the capacitor; the capacitor tries to retain its present voltage over the plates. If the voltage is at a maximum, the current will be zero, if the voltage is at maximum change, the current will be at a maximum. This is shown in figure 2.2 for sinusoidal waveforms.

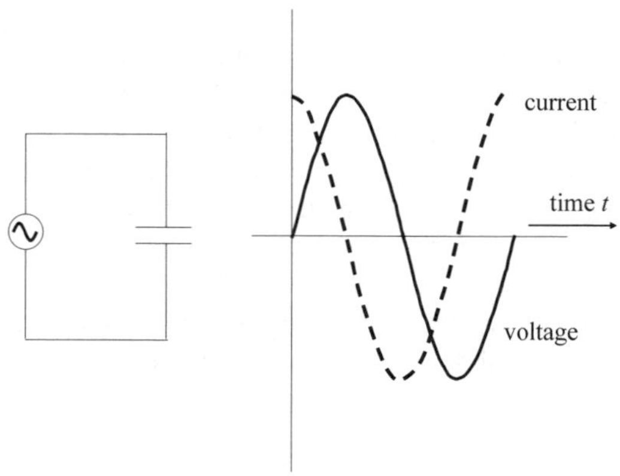

Fig. 2.2 Sinusoidal voltage and current waveforms for a capacitor.

For the sinusoidal waves shown, the voltage wave is −90° out of phase with the current wave.[2] The current leads the voltage and the voltage lags behind the current.

An inductor opposes the *change* in current through it. It does so by generating a voltage drop over the inductor. Since this voltage drop is a reaction against the *change* in current through the inductor, the polarity is such that it tries to maintain the current flow through the inductor. If the current is at a maximum, the voltage will be zero, if the current is at maximum change, the voltage will be at a maximum. This is shown in figure 2.3 for sinusoidal waveforms.

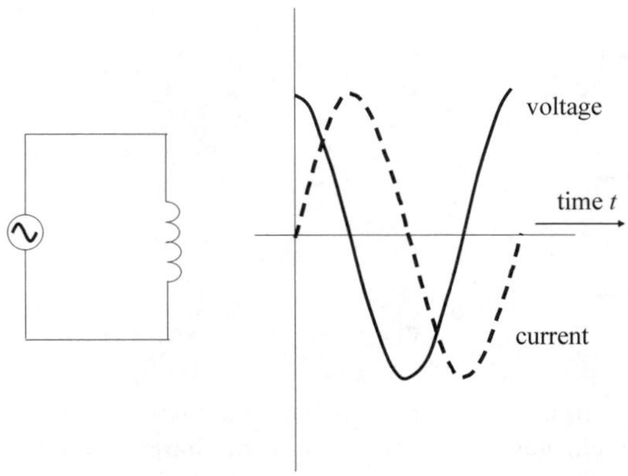

Fig. 2.3 Sinusoidal voltage and current waveforms for an inductor.

For the sinusoidal waves shown, the voltage wave is 90° out of phase with the current wave. The voltage leads the current and the current lags behind the voltage.

When a capacitor and an inductor are connected, for example in a parallel circuit, as shown in figure 2.4, after charging of the capacitor, energy will be exchanged back and forth between the capacitor and inductor.[3]

- When the capacitor is charged, see figure 2.4a, the voltage is at maximum and the current is zero.
- Then, see figure 2.4b, the capacitor starts to discharge; the voltage over the capacitor starts decreasing and the current 'through' it starts increasing.
- When the capacitor is fully discharged, see figure 2.4c, the voltage over it is zero and the current 'through' it is at maximum.

[2]Two signals are out of phase if they do not coincide as function of their argument. Our sinusoidal waves of figure 2.2 are of the form $\sin(At)$ and $\sin(Bt)$, where t is the time and A and B are constants. The argument of the sine function, i.e. At or Bt, is called the phase and has the unit of angles. In figure 2.2, the functions are plotted as function of t rather than as function of the whole argument.

[3]Our previous choice for sinusoidal voltage and current waveforms was not completely arbitrary. Since the current 'through' a capacitor is proportional to the rate of change of the voltage over it and since the voltage drop over an inductor is proportional to the rate of change of the current through it, it can be shown (mathematically) that the energy exchange between the connected capacitor and inductor will occur through sinusoidal voltage and current waves.

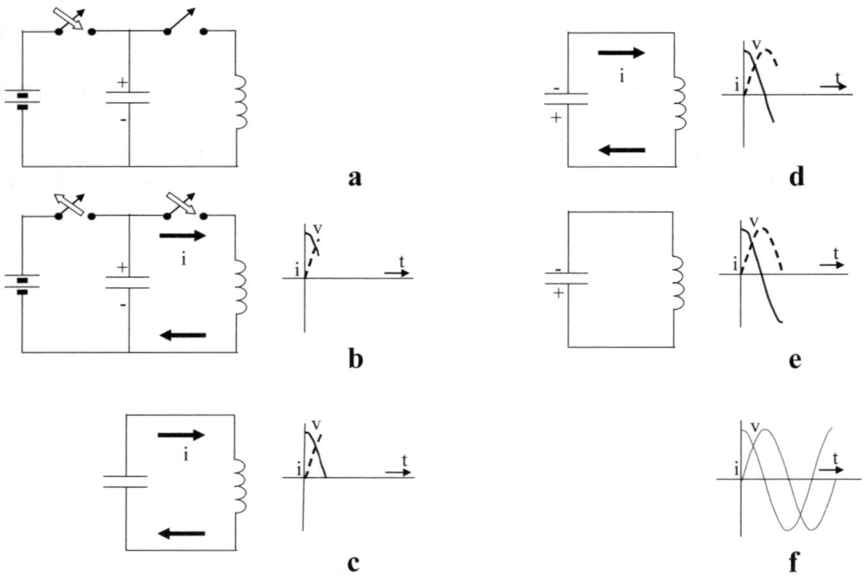

Fig. 2.4 Voltage and currents in a capacitor-inductor parallel circuit. a. Capacitor charged, voltage is at maximum, current is zero. b. Begin to discharge capacitor, voltage is decreasing, current is increasing. c. Capacitor fully discharged, voltage is zero, current is at maximum. d. The inductor is generating a voltage drop to keep the current flowing in the same direction, the polarity of the capacitor changes. e. Capacitor fully charged, voltage is at maximum, inductor fully discharged, current is zero. f. With the polarity of the capacitor changed, the cycle a–e repeats in the opposite direction.

- Then the inductor generates a voltage drop over the inductor having such a polarity that it keeps the current flowing in the same direction. The capacitor will change then from its neutral charge state into a polarity state opposite to the one it started with, see figure 2.4d. The voltage is increasing in opposite polarity and the current is decreasing.
- Then, see figure 2.4e, the capacitor gets fully charged (in the opposite polarity) and the inductor gets fully discharged. The voltage is at its negative maximum, the current is zero.
- Next, see figure 2.4f, the cycle as explained above repeats itself in the opposite direction.

This phenomenon, called *oscillation* or *resonance*, will continue with decreasing amplitude values, finally coming to a complete halt, the reason being the (distributed) resistance present in any real-world circuit. The resistance converts the electric energy into heat.

The period T and thus the *resonance frequency*[4] depends only on the dimensions of the capacitor and the inductor, not on the amplitude of the voltage and/or current.

[4]Frequency, f, is the reciprocal of the period: $f = \frac{1}{T}$.

It can be shown that at the resonance frequency, the complex admittance (one divided by the complex impedance) of the circuit shown in figure 2.4 becomes zero.

A similar *resonance* occurs in a series circuit consisting of a capacitor and an inductor.

2.1.2 Heinrich Hertz: The First Antenna and Radio System

To experimentally verify Maxwell's theory, Hertz transformed the *closed* resonance system of figure 2.1 into an *open* resonance system. He replaced the bottom coil with a pair of straight wires, at the centre connected to a spark gap. The spark gap was connected to the secondary windings of a conduction coil, see figure 2.5, and the straight wires were equipped with electrically conducting spheres that could move over the wire segments.

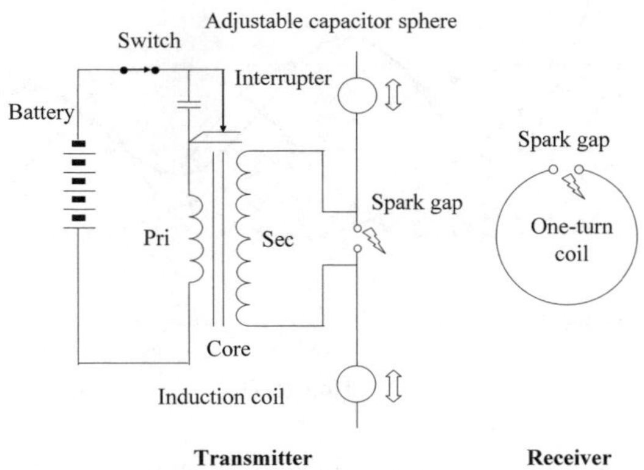

Fig. 2.5 Hertz's *open* resonance system. With the receiving one-turn loop, small sparks can be observed when the transmitter discharges.

By means of moving the spheres, the capacity of the circuit could be adjusted for resonance. The resonant circuit consisted of the conducting spheres and the straight wire segments. The former form a distributed capacitance, the latter a distributed inductance. Capacitor and inductor are connected in series. The induction coil with the mechanical interruptor transformed a low-frequency (pulsed) signal into a high-frequency (pulsed) signal.[5] By charging the two straight wire segments, eventually the breakdown voltage of air is reached and a spark is created over the small air-filled

[5] The operation of the interrupter shown in figure 2.5 is based on the electromagnet principle. When the switch is closed, a current is flowing through the primary windings of the induction coil and the core will be magnetised. Therefore, the flexible contact of the interrupter will be attracted to the core and the circuit will be opened. When the circuit is opened, the core will lose its magnetisation and the flexible contact will return to its original position, thereby closing the circuit again, after which the whole cycle starts again. By this mechanical interruptor a train of pulses is created in the primary windings of the induction coil and these pulses will be transformed to high-voltage pulses at the secondary windings that will discharge over the spark gap. Induction coils may be found nowadays in cars, where they raise the low voltage of the car battery to the high voltage necessary to produce sparks to ignite the fuel in the combustion engine.

spark gap. This spark makes a pulsed current flow through the two wire segments. Since a pulse signal actually consists of an infinite number of signals of different frequencies (or periods, $T = 1/f$), see figure 2.6, the resonant circuit picks out the current with the 'right' frequency and this current will oscillate in the circuit and create the radiation.

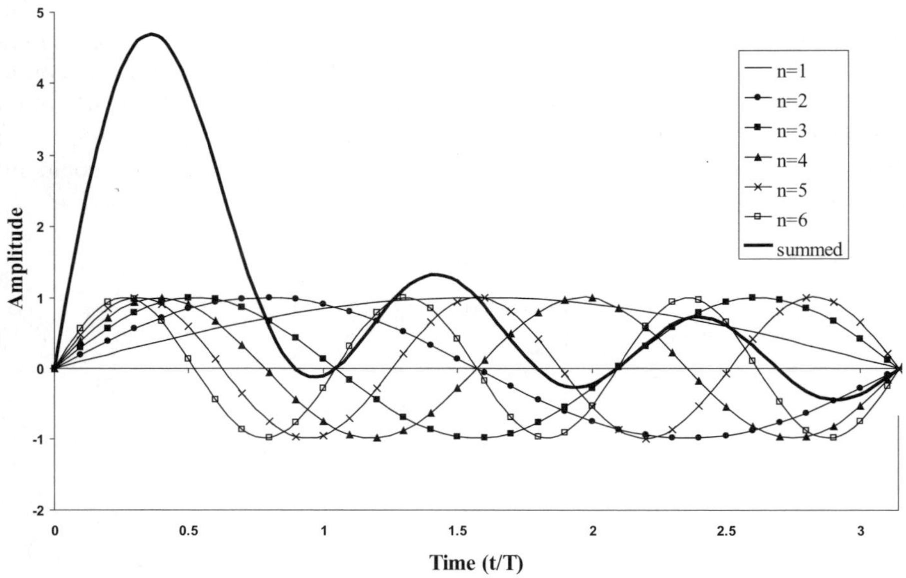

Fig. 2.6 A pulse consists of many signals of different frequency (period). Shown are signals of the form $\sin\left(n\frac{2\pi}{T}t\right)$ for $n = 1, \ldots, 6$, where T/n is the period and t is the time. We see that only six of these signals added, leads to a pulse-shaped signal.

The second coil of the set-up shown in figure 2.1, Hertz replaced with a single-turn square or circular coil with a small gap, see also figure 2.5.

By adjusting the perimeter of the one-turn receiving coil, Hertz was able to demonstrate resonance. With the receiver (this one-turn loop) placed several metres away from the transmitter (the pair of straight wires with the spark gap), small sparks could be observed in the gap of the receiver when the radiator discharged.

Hertz performed these experiments in 1886 and published the results in 1887. His equipment, shown in figure 2.5, may be regarded as the first radio (transmitter and receiver) system. His transmitter was equipped with the first dipole antenna, his receiver consisted of the first loop antenna.

By using a large, flat, electrically conducting plate, placed in front of his transmitting antenna, Hertz was able to create standing waves, thereby proving Maxwell right in his prediction of electromagnetic waves. The electromagnetic waves emitted by the transmitting antenna, interfered with the waves reflected from the flat plate, producing these standing waves. With his one-turn receiving coil, Hertz was able to detect the maxima of these standing waves. For this experiment he worked with a wavelength of 6m [1].[6] His transmitting antenna was approximately half a wavelength long. Half a wavelength is the shortest length for a dipole antenna to be *resonant.*

As we discussed before, Hertz's transmitting circuit consisted of a distributed capacitor and a distributed inductor, connected in series as shown in figure 2.7.

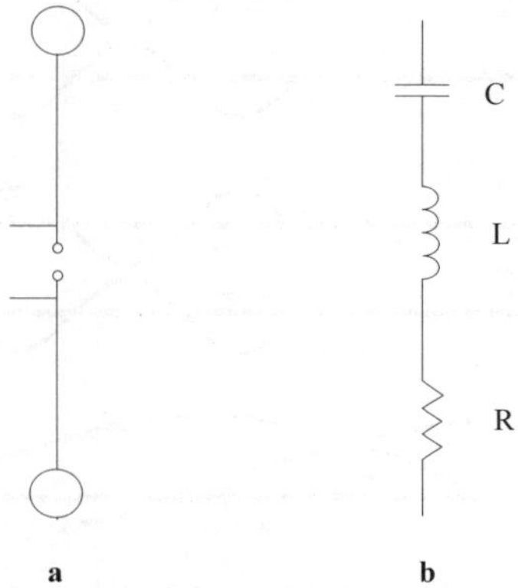

Fig. 2.7 Equivalent circuit of Hertz's transmitting circuit. a. Transmitting circuit. b. Equivalent series representation.

The spheres at the end of the straight wires make up the capacitor C, the straight wires themselves make up the inductor L and the resistor R stands for the radiation 'loss'. At a wire length of half a wavelength, the inductance equalises the capacitance, or in other words: 'Current is not leading the voltage, neither is voltage leading the current'; the antenna is resonant. To understand this, it is necessary to look at the current waves in a transmission line.

We will look at a two-wire transmission line, connected at one end to an AC current generator, open at the other end, see figure 2.8a.

The current has to be zero at the open end of the transmission line. At this open end a reflection of the current wave will occur and due to interference of the current wave going to the right and the reflected one going to the left a standing current wave will be generated.[7]

The radiated electromagnetic fields by the current through the upper wire of the two-wire transmission line of figure 2.8a are annihilated by those radiated by the current through the bottom wire, travelling in the opposite direction, due to the close proximity of the two wires. For the deformed part of the transmission line,

[6] One wavelength is the distance travelled by a wave-front in the time span of one period. This distance is equal to the speed of light (an electromagnetic wave) times this period.

[7] Initially, a sinusoidal current wave is travelling from the AC current source to the right, see figure 2.9. The figure shows the sinusoidal wave for three instances of time, $t_3 > t_2 > t_1$.

After complete reflection on the open end a similar wave is travelling to the left. The wave going to the right interferes with the reflected one going to the left, resulting in a standing wave, see figure 2.10.

Although the amplitude of a standing wave changes with time, the position of the maximum and minimum value and the zero crossings are invariant with distance along the transmission line.

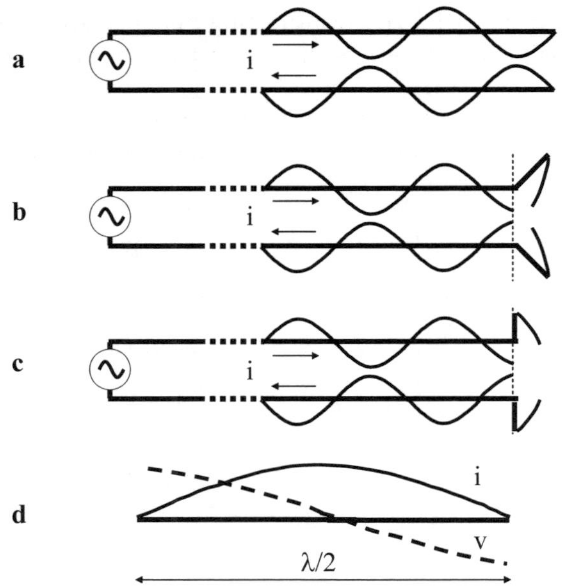

Fig. 2.8 Construction of a dipole from a two-wire transmission line. a. Current standing waves on an open-ended two-wire transmission line. b. Deformation of the transmission line. c. Half-wave dipole antenna. d. Current and voltage standing wave on a half-wave dipole antenna. The gap in the dipole is not drawn to scale, λ is the wavelength.

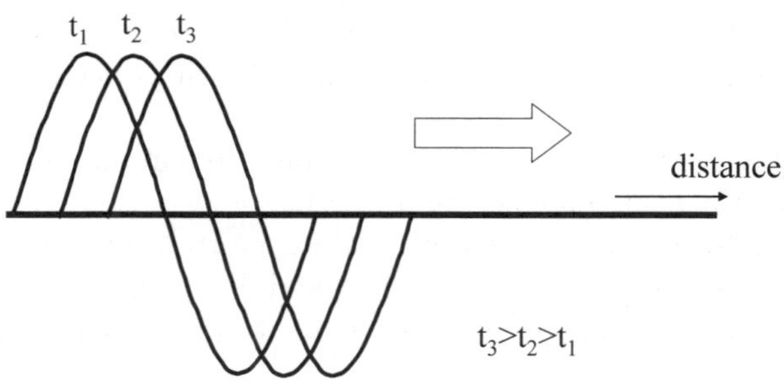

Fig. 2.9 Sinusoidal current wave travelling to the right from the source at the left.

figure 2.8b, this annihilation is not complete anymore and for the dipole, figure 2.8c, this annihilation does not occur anymore, and a net radiation stemming from the currents on the dipole remains. At the positions where the current is crossing the zero value (the end points of the dipole) and the positions where the voltage is crossing the zero value (feed point), see figure 2.8d, neither the voltage is leading the current nor the current is leading the voltage. These are the positions then from where radiation is stemming. For dipole lengths being an integer number of half wavelengths, such *resonances* occur. For lengths differing from these ones, the dipole antenna is not

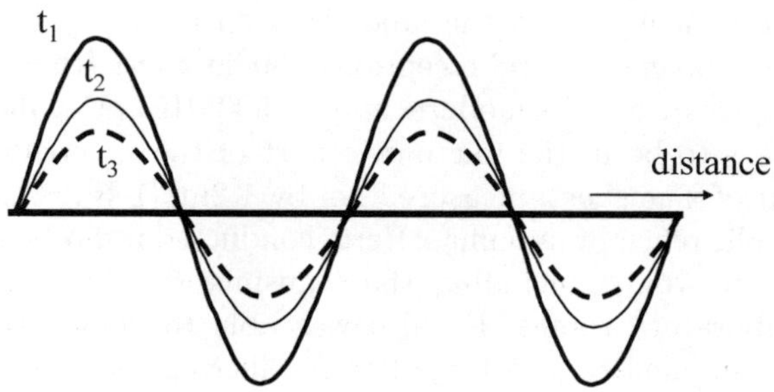

Fig. 2.10 Standing wave.

purely resistive, but also contains a capacitive or inductive component. The smallest possible length of a resonant dipole antenna therefore is one half of a wavelength.

James Clerk Maxwell already showed (theoretically) that light is an electromagnetic wave phenomenon. The radio waves demonstrated by Hertz are also electromagnetic waves, but operating at a (much) lower frequency. Pursuing the ray theory for light, Hertz realised that it would be possible to direct the electromagnetic energy radiated by his half-wave dipole, by placing this antenna in the focal point of a parabolic reflector, see figure 2.11.

The half-wave dipole antenna itself, we have seen, radiates in all directions with the exception of those along the dipole axis.

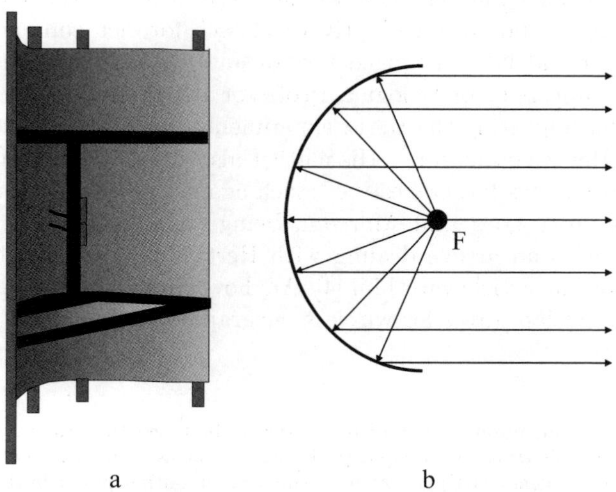

Fig. 2.11 To direct or 'beam' the energy of his transmitting antenna, Hertz placed the half-wave dipole in the focal point F of a parabolic reflector. a. Hertz's transmitting antenna. b. Ray-tracing principle of the transmitting antenna.

To make his equipment fit into the laboratory space allotted to him,[8] he scaled down his dipole for transmission and reception of 60cm waves (that means a frequency of 500Mc (Mega cycles), or - in modern units - 500MHz (Megahertz), a frequency considered nowadays to be in the microwave part of the spectrum!). The parabolic reflector, made out of zinc sheet, measured 2m by 1.2m, [1–3].

With two parabolic reflector antennas, Hertz conducted radio-beaming experiments over distances up to 16m,[9] including the transmission through a wooden door, connecting two laboratory rooms. He also was able to demonstrate with the two parabolic reflector antennas and a large flat conducting reflector, that the angle of reflection is identical to the angle of incidence, just as with light (which is also an electromagnetic wave phenomenon).

Heinrich Hertz - after having suffered from bad health for several years - died of blood poisoning in Bonn, Germany, in 1894, just before his 37th birthday. His name - *Hertz*, abbreviated as Hz - is given to the unit of frequency, replacing the *cycles per second* that was in use till the end of the 1960s. One cycle per second is equal to 1Hz.

The work of Heinrich Hertz was put into practice and commercial use by a young Italian, named *Guglielmo Marconi*.

2.1.3 Guglielmo Marconi, the Dawn of Wireless Communication

Guglielmo Marconi (1874–1937) spend his child- and young adulthood at his father's estate, Villa Griffone in Bologna, Italy. His father, once widowed, had remarried with Guglielmo's mother, an heiress of the Irish whisky distilling family, the Jameson's.

Although Guglielmo Marconi never went to university - he was denied permission to the Italian Navy Academy of Leghorn because he failed to finish secondary school, [4] - he had a big interest in electricity and magnetism and had read about the experiments of *Michael Faraday* and *Benjamin Franklin*.[10] His mother supported him in this interest and arranged private classes for her son on these subjects by *Vincenzo Rosa* and later arranged for consultations with *Augusto Righi*, [4, 5], professor at the university of Bologna. Professor Righi was at that time a leading person in physics. Repeating the Hertz experiments, he had succeeded in improving the efficiency of Hertz's oscillator.[11] His mother also supported Marconi in setting up his own laboratory at his father's estate, much against the will of his father.

Although the story goes that Marconi, being on a family holiday in 1894, by accident came across an article dealing with Hertz's experiments, it is more likely that Marconi took the article with him [4]. Anyhow, during this holiday he developed the idea to use Hertzian waves for wireless telegraphy.

[8] Hertz found that the dimensions of the mirror needed to be larger than the wavelength used.
[9] Hertz calculated that his transmitter *input* peak power went as high as 16kW. The spark being a very inefficient radio frequency (RF) generator, resulted in an output power in the order of one watt average, [1]!
[10] Benjamin Franklin (1706–1790), a president of the United States of America, had become famous for his 'thunderstorm experiment'. With a thunderstorm nearby, he flew a kite, picking up the static electricity, that he stored by means of a key attached to the rope into a Leyden jar or capacitor.
[11] Hertz called his transmitter *oscillator*.

The first experiments Marconi conducted dealt with the detection of lightning. He constructed an electrical circuit that contained an arrow-shaped zinc plate, a battery, a *coherer* and a bell [4]. The circuit might have looked like the one shown in figure 2.12a.[12]

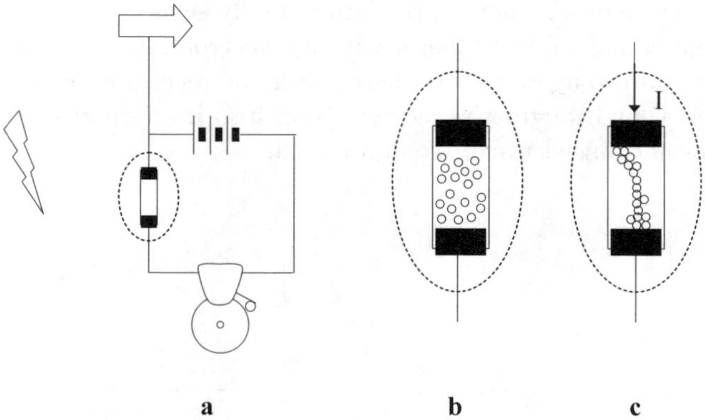

Fig. 2.12 Marconi's lightning detecting apparatus. a. Electric circuit including the coherer as a 'lightning switch'. b. Coherer in normal state, no current is flowing. c. Metal filings in the coherer subject to 'cohesion', a current is flowing.

A *coherer* is a device consisting of a glass tube filled with a mixture of different metal filings, between two metal probes, see figure 2.12b,c. In the normal state, figure 2.12b, the coherer has a very high impedance. Under the influence of electromagnetic waves, however (created, for example, by a spark in the neighbourhood of the coherer), the filings cohere, figure 2.12c, and the impedance drops to a low value. The coherer was developed by *Eugène Branly* (1844–1940) in 1890 and was perfected and named 'coherer' by *Oliver Lodge* (1851–1940) in 1894. The coherer put into the circuit of figure 2.12a acts as a switch. The electromagnetic waves created by lightning discharges in the neighbourhood would close the switch and make the bell ring. After the coherer is put into action, it needs to be tapped, to 'decohere' the filings and bring the coherer back into the normal state. For this purpose, Lodge added a timed hammer mechanism to tap the coherer back into normal state.

Although the electromagnetic waves caused by lightning are far more powerful than the Hertzian waves created by an induction coil, the fact that these waves were able to switch on the alarm bell over a distance of kilometres, convinced Marconi that communication by electromagnetic waves over many kilometres should also be possible. He set out in duplicating the Hertz experiments, eventually adapting the Righi oscillator for his experiments. In order to make his dream come true, he realised that he needed to work on increasing the power of the oscillator (transmitter) and increasing the sensitivity of the receiver.

[12]Marconi seems to have been inspired for these experiments by the Russian *Alexander Popov* (1859-1906), who elevated a wire connected to a coherer in an attempt to detect the energy in electric storms [6, 7].

He started out with the latter of these two tasks and found that the coherer could be greatly improved by using a mixture of nickel filings (95%) and zinc filings (5%), put into a vacuum tube having silver plugs [4]. The coherer, combined in a circuit with a battery and a bell, replaced the spark gap in the original Hertz receiver. The sparks from this spark gap receiver were hardly visible and usually reception of an electromagnetic wave was rather heard than actually seen.

Next, Marconi combined the coherer with an electromagnet that attracted a small hammer that would tap against the coherer after the reception of an electromagnetic pulsed wave and thus brought the receiver back into its normal state, ready for the reception of another pulsed wave, see figure 2.13.

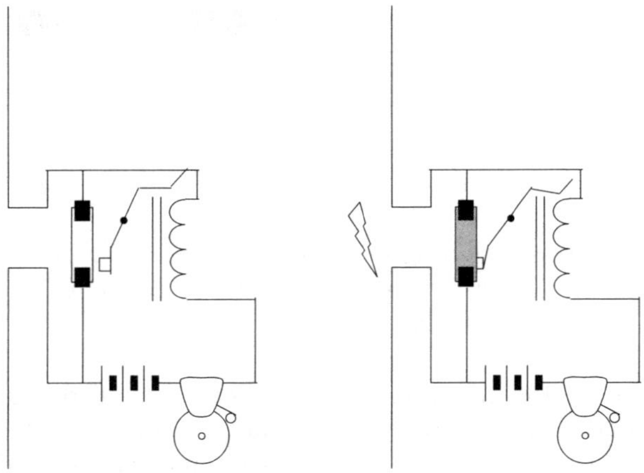

a b

Fig. 2.13 Receiver with the spark gap replaced by a coherer and an electromagnet-operated hammer for restoring the normal state of the coherer. a. Shortly before reception. b. Shortly after reception.

Since the current flowing through the coherer at reception was too weak to operate a Morse writer, Marconi next put a relay in the circuit to use the weak current to switch on a more powerful current needed to operate the Morse writer. A Morse writer consists of an electromagnet that attracts a lever, holding an inked pen. Underneath this pen, a clockwork motor transports a strip of paper. Every time the electromagnet is activated, the pen thus writes down 'dots' and 'dashes' - the code-elements of the Morse alphabet - on the paper, depending on the time (short for a 'dot', long for a 'dash') the electromagnet is activated. The Morse writer had been in use already for wire-operated electric telegraphs. Of course the pulse repetition frequency of the coherer restoring device (the hammer) dictated the speed of (Morse code) transmission.

Marconi started out his experiments, like Hertz before him, using cylindrical parabolic reflectors, fed in the focal point by half-wave dipole antennas at a frequency of 1.2GHz, [4, 6]. With his equipment, working at these microwave frequencies, he could transmit messages over the distance of the attic of his father's house; a distance far too low to his taste.

In 1895, however, he made an important change to his system that suddenly allowed him to transmit and receive messages over distances that progressively increased up to and beyond 1.5km, [4, 8]. At the reception of the Nobel prize in Stockholm, Sweden in 1909, Marconi said about this change [7, 8]:

> In August 1895 I hit upon a new arrangement which not only greatly increased the distance over which I could communicate but also seemed to make the transmission independent from the effects of intervening obstacles. This arrangement [figure 2.14a] consisted in connecting one terminal to the Hertzian oscillator or spark producer to earth and the other terminal to a wire or capacity area placed at a height above the ground and in also connecting at the receiving end [figure 2.14b] one terminal of the coherer to earth and the other to an elevated conductor.[13]

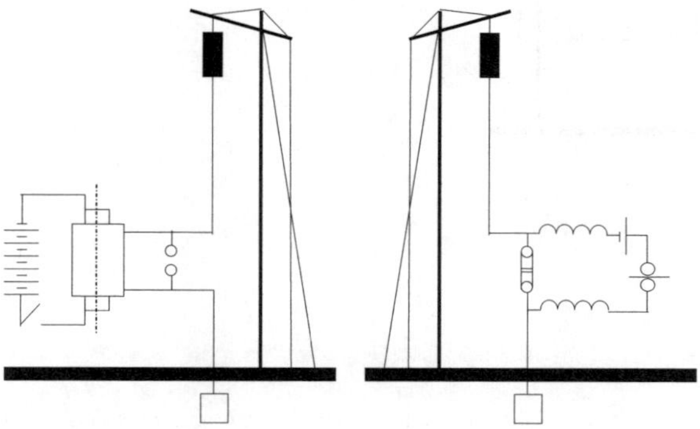

Fig. 2.14 Marconi antennas in 1895. a. Scheme of the transmitter used by Marconi at Villa Griffone. b. Scheme of the receiver used by Marconi at Villa Griffone.

By replacing the outside balls of the Righi oscillator by plates and subsequently, elevating one of the plates in the air and connecting the other one to earth, Marconi first of all had enlarged the antenna.[14] This monopole antenna, together with its image, see figure 2.15, formed a half-wave dipole antenna for a wavelength much larger than any that had been studied before.

[13]Although Marconi, in his speech in Stockholm, makes us believe that he came upon the antenna arrangement as shown in figure 2.14 by chance, this is hard to believe [4]. Considering his obsession in realising wireless communication and his reluctance in reporting failures, it seems more likely that his new arrangement was more the product of hard labour than mere luck. He may have been inspired by the work of *Franklin* and *Popov* and his own experiences in detecting lightning.

[14]The word *antenna* is actually introduced by Marconi [9]. Before he did so, the terms *areal* or *elevated wire* were used to describe the electromagnetic wave launching or receiving device. The word *antenna* stems from the art of navigation used in the Mediterranean sea and means 'the long wooden pole transversal to the mast which receives the upper end of the Latin sail'. Marconi introduced the term *antenna* in its present meaning in a lecture held on May 1909, before his Nobel speech in the same year, at the Koninklijk Instituut van Ingenieurs KIVI (Royal Netherlands Institute of Engineers) [9, 10].

We know now that electromagnetic waves propagating through the air (or even vacuum) experience a frequency-dependent attenuation. This attenuation increases with frequency. Thus by adapting his radio system for longer wavelengths (lower frequencies), Marconi made wireless communication possible over increased distances. The creation of the long wavelength electromagnetic waves was the key to his success.

Although Marconi succeeded in creating Hertzian waves at a wavelength larger than anyone before him, he did not derive his success from putting theory into practice. After all, Marconi was more of an inventor and engineer (and above all an entrepreneur) than an academic scientist. Up to the middle of the 1920s it was common practice to design antennas empirically (like Marconi did) and produce a theoretical explanation after the successful development of a working antenna [6].

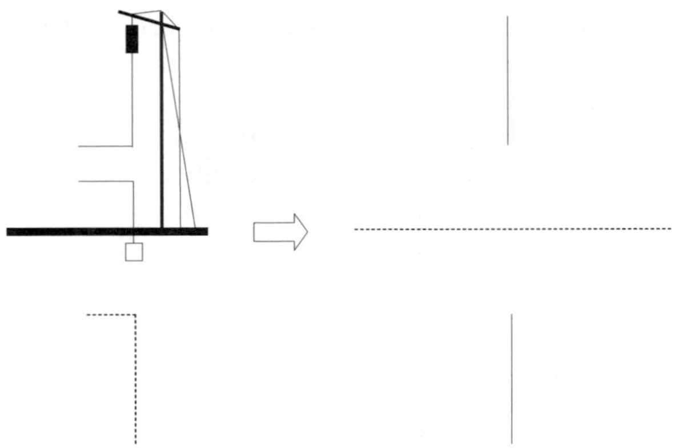

Fig. 2.15 The Marconi monopole, together with its image, created by the ground, forms a half-wave dipole and may be treated as such.

The above-mentioned explanation of Marconi's quarter-wave monopole antenna was not known to him when he made the antenna. The relation between antenna length and the operational wavelength of the radio system was later explained to him by his colleague, *Professor Ascoli*, who had calculated that the 'length of the wave radiated (was) four times the length of the vertical conductor' [11].

Since, at that time, the Italian government was not interested in Marconi's invention, at least not up to the point to financially support further research, in 1896 he moved - accompanied by his mother - to Great Britain. On 2 June of that same year his first patent was granted (No. 12,039, 'Improvements in Transmitting Electrical Impulses and Signals, and in Apparatus Therefor'). In Great Britain, the British Post Office was very interested in his invention and supported further research. Marconi was introduced to *William Preece*, chief engineer of the English Post and Telegraph Services through his cousin *Henry Jameson-Davis*, who had awaited the Marconi family upon arrival in London [4, 12].

Experiments, in cooperation with the British Post Office started on Salisbury Plane.[15] In May 1897 messages were exchanged over a distance of 14km (9mi) between

[15] Among the many people who witnessed the experiments and details of the equipment was Professor *Slaby* from Germany, who upon return to Germany, fulfilling the wish of the Kaiser to make the

Lavernock Point and Isle of Flatholm in the Bristol Channel and in 1899 a regular radio-telegraph service between England and France was put into service [8].

In 1897 the *Wireless Telegraph and Signal Co.* was founded with Marconi being one of the five directors and his cousin Henry Jameson-Davis being the first managing director. The company changed its name into *Marconi's Wireless Telegraph Co. Ltd.* in 1900. In 1899 the *Marconi Wireless Telegraph Company of America* was formed that was taken over in 1919 by the *Radio Corporation of America*, a company formed by *A.T. & T.*, *Westinghouse* and *General Electric* [4, 12, 17].

The next technical challenge Marconi was facing dealt with the limitation that only one transmitter could be used at a time, since all receivers in the 'influence sphere' of this transmitter would receive all signals transmitted. Up to 1898 the Marconi equipment was not able to deliver a satisfactory *syntony* or *tuning*. By initially inserting an adjustable resonance circuit - consisting of an inductor and a capacitor - into the receiver only, he was able to tune the receiver to the waves emitted by a particular transmitter with an antenna of a certain length [6, 8]. Later, the transmitter was also equipped with an adjustable resonance circuit, see figure 2.16, and the whole radio system became tunable. A patent was granted in 1900.[16]

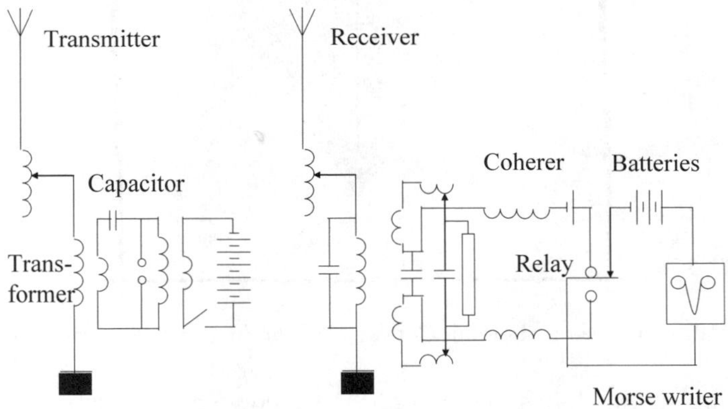

Fig. 2.16 The tunable radio system. By inserting in both transmitter and receiver an adjustable resonance circuit, the transmitter and receiver can be tuned into each other.

In his Nobel Prize speech, Marconi said about this [7]:

> In 1900 I constructed and patented transmitters which consisted of the usual kind of elevated capacity area and earth connection, but this was inductively coupled to an oscillator circuit containing a condenser, an inductance and spark gap, the conditions which I found essential for efficiency being that the periods of electrical oscillation of the elevated wire or conductor should be in tune or resonance with that of the condenser circuit.

The next and perhaps the most famous technical challenge Marconi encountered was that of establishing a transatlantic wireless telegraph connection. In the field trials

country independent from foreign communication systems, later, in 1903, participated in the forming of the *Telefunken* (Spark Telegraphy) Society [4, 12].

[16] The patent has become famous as the *four sevens patent*, named after the patent issue nr: 7777.

preceding the first transatlantic transmission experiment, Marconi had observed that the curvature of the earth would not obstruct long distance wireless transmission [6]. These observations were completely in contradiction with the beliefs of the scientific community at that time. Due to Marconi's stubbornness and perseverance he pursued his experiment and on 12 December 1901 the Morse code for the letter 'S' (...), sent from Poldhu in Cornwall, Great Britain was received at St. Johns in Newfoundland, Canada, having travelled a distance of 3684km.[17]

Although some confusion exists concerning the exact dimensions of the antennas and wavelength (intentionally or unintentionally) used [6, 8, 12–15], we may assume that the transmitting antenna at Poldhu consisted of 50 vertical wires in the form of a fan, connected to the ground through a spark transmitter. The wires were supported horizontally by a guyed wire between two 48m high wooden poles, 60m apart, see figure 2.17. The wavelength used was 366m, corresponding with a frequency of 820kHz.

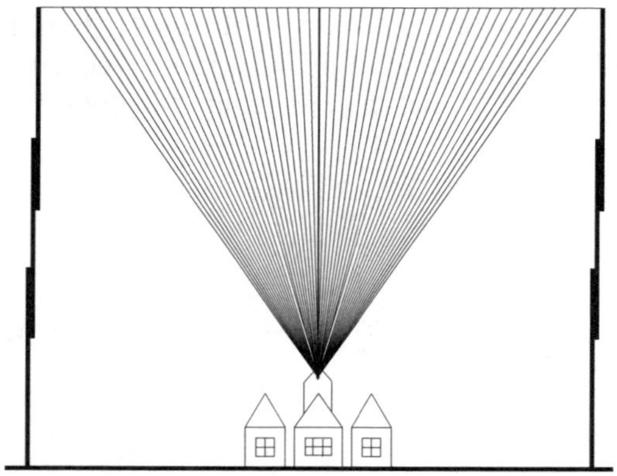

Fig. 2.17 The antenna used at Poldhu, Cornwall, in the first transatlantic transmission.

The antenna at St. Johns was a 122m long copper wire, supported by a kite. The strong wind prevented the use of balloons which was initially intended. Also initially intended was the use of a tuned or *syntonic* receiver. However, due to the same strong wind, the system would not stay in tune [12]. This is in agreement with a recent theoretical reconstruction of the transatlantic wireless experiment [16].

Not known at the time of the transatlantic transmission and reason for the scientific community to strongly disbelieve the possibility of transatlantic wireless telegraphy, was the refractive characteristic of the *atmosphere*. In the lower part of the atmosphere, the *troposphere* - which extends to a height of 10 to 15km [23] - the index of refraction (i.e. the ability of the medium to bend waves) decreases with increasing height above the earth (due to the decrease in permittivity [24]). This change in refraction index with altitude causes radio waves to bend toward the earth, as depicted in figure 2.18. In the same figure this refraction phenomenon is clarified

[17] Marconi had used the Morse code for the letter 'S' as a testing signal ever since the moment he had demonstrated the wireless transmission of this code at his father's estate, requested by his father to convincingly demonstrate the usefulness of wireless. This had been the turning point from where his father changed from a sceptic into an admirer of his son's work.

by visualising the process as occurring between discrete layers, although it should be understood that it is a continuous process. The variations of refraction index with height are strongly dependent on meteorological conditions, like temperature and humidity. This explains why, in the first years of transatlantic wireless communication, best results were obtained at night [4].[18]

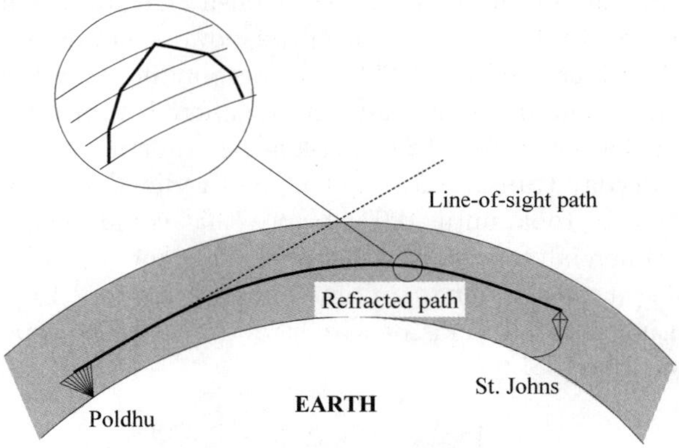

Fig. 2.18 By refraction in the atmosphere the wireless transmission between Poldhu and St. Johns became possible.

2.1.4 After the First Transatlantic Transmission

The first transatlantic transmission got an enormous media coverage that made Marconi a world hero and worked out well for the *Marconi's Wireless Telegraph Co. Ltd.* By 1903, six years after the beginning, the accomplishments were impressive. To name a few [12]:

- The Marconi system had been adopted by the British Navy, which employed 32 installations on her warships and by the Italian Navy which employed 20 installations on her warships.
- Lloyds had a contract to use only the Marconi system.
- England had 10 coastal stations, the USA also had 10 coastal stations and stations in Chicago, Cuba, Hawaii (4), Alaska and Milwaukee.

[18] In this chapter we have implicitly accredited Guglielmo Marconi as the inventor of the *wireless telegraph* or *radio*. It is well known, however, that (long) before Marconi, elements of his radio system had already been invented and used. Wire antennas, for example, were introduced by *Joseph Henry* in 1842. In 1885, Edison patented a communication system using vertical, grounded antennas [13]. Also *Popov* in Russia is known to have used elevated wire antennas before Marconi did. A tuning system had already been patented by Sir Oliver Lodge [17], the same person who, already in 1894, had demonstrated the wireless transmission of Morse code over a distance of 60m [18]. What makes Marconi unique, however, is that he took all the elements together and combined them into a working and commercially feasible system.

- The Dover-Calais and Ostend-Boulogne ships were equipped with Marconi systems.
- Thirty transatlantic liners were already equipped with Marconi systems.

Despite the first transmission in December 1901 and the following media coverage, it took Marconi till the autumn of 1907 to establish a regular, reliable transatlantic telegraphy service [18]. By that time, transmitter powers had increased to 300kW [4], and wavelengths had increased to 20,000m, corresponding to a frequency of 15kHz. The antennas had become huge and must have suffered from a low efficiency, since eye-witnesses from that era report to have seen the antennas glow red in the dark.

Although an everyday transatlantic exchange of radiotelegraphy was in existence from 1901 onwards, it took until 1907, before this could be put into a reliable and therefore commercially successful service. The key to this success was the introduction of the *suspended long-wire antenna*, see figure 2.19, a structure found - after experimenting with all kinds of wire structures - to be especially fit for long distance communication [6, 13].

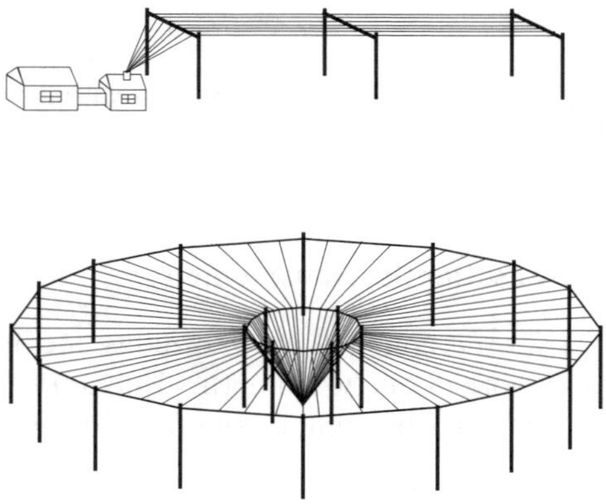

Fig. 2.19 Two versions of the suspended long-wire antenna, used for transatlantic communication.

The antenna had directive properties, meaning that is was more sensible for electromagnetic waves coming from certain directions than it was for electromagnetic waves coming from other directions. In using the antenna for transmission, this meant that it would radiate electromagnetic waves in certain directions more strongly than into other directions. This property was highly desirable for realising the very long distance to cover. The phenomenon of directivity will be covered in more detail, based on intuitive reasoning, in the next section. Besides the property of directivity, the antenna was also capable of sending and receiving over a wider range of frequencies than previous antennas.

A mathematical explanation of the antenna operation - based on the image theory, see figure 2.15 - was given in 1906 by *Ambrose Fleming* (1849–1945), professor at University College, London, who had been approved as scientific adviser to the Marconi Company in 1900 [4]. It is assumed that this must have been the first time that an antenna design has been accomplished both experimentally and theoretically [6].

Marconi made a commercial success of wireless telegraphy, cleverly using his technical insight, entrepreneurship and PR skills. In 1909 he received the Nobel Prize in physics. In 1912, the importance of wireless telegraphy at sea became painfully clear with the sinking of the - considered unsinkable - *SS Titanic*. The *Titanic* carried a Marconi wireless set aboard and had sent out CQD distress calls (CQD was the predecessor of SOS). In the accident, 1500 people died, 700 were saved. Tragically, the nearby *California*, also had Marconi equipment aboard, but only one operator. After a 16-hour shift, the operator had gone to sleep and therefore the distress calls of the *Titanic* were missed [4].

While Marconi prepared for the first transatlantic transmission and subsequently worked on establishing a regular, reliable transatlantic wireless telegraphy service, other researchers had accomplished important results. Certainly worthwhile mentioning is the work of Canada-born, USA Professor *Reginald Fessenden* (1866–1932), the inventor of wireless telephony.

When wireless telegraphy came into use, it was common practice for the operators to use headphones instead of the earlier mentioned Morse writer. Also Marconi, at St. John's, awaiting the first dots of the 'S' transmitted at Poldhu, used headphones since the human ear is much more sensitive than the Morse writer. When the Morse key of the transmitter was pressed, one could hear the crackle of the pulse train transmitted by the spark transmitter. Fessenden put a microphone in series with the antenna lead, see figure 2.20, and thus superimposed (low frequency) sound upon the (higher frequency) pulse train [18].

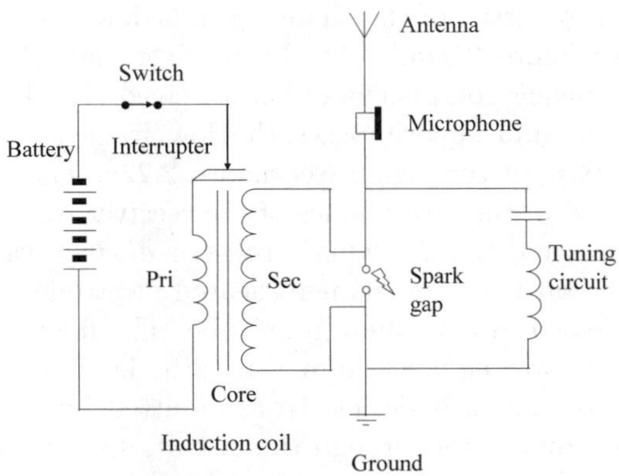

Fig. 2.20 An early version of Fessenden's spark-gap telephony transmitter.

In 1900, Fessenden was able to transmit, over 1600m, 'intelligible speech accompanied by an extremely loud, disagreeable noise due to the irregularity of the spark' [18]. Fessenden realised that what was needed for a proper audio transmission was *continuous waves* (CW) instead of a pulse train. He started out with the development of spark generators with a pulse repetition frequency (PRF) above audibility. Finally, in 1906, he succeeded in the creation of a true CW generator that could deliver enough power for transmission.[19] In December of that year the first radio broadcast took place from Brant Rock, Massachusetts and was received by operators onboard ships equipped with Fessenden receivers.

The Fessenden receiver was basically a Marconi receiver, but the coherer-detector was replaced with what Fessenden called a *barretter*. This barretter, or electrolytic detector, see figure 2.21 [14, 18] transforms an alternating current into a direct current. It may be considered as an early type diode.

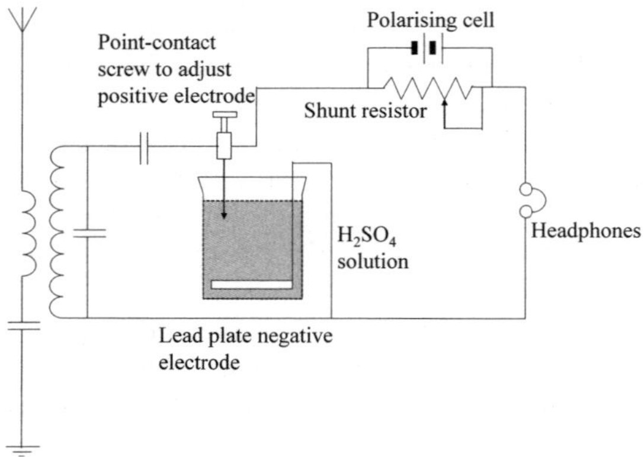

Fig. 2.21 An early version of a Fessenden receiver, employing a barretter or electrolytic rectifier [14, 18].

In the transmitter, a (relative) high frequency signal is *amplitude modulated* (AM) with the audio signal, by virtue of the changing impedance of the microphone when subject to sound, see figure 2.22a,b. At the receiver end, the rectifying detector converts the high-frequency component of the received signal, i.e. the CW carrier, into a constant current and superimposes the low frequency component, i.e. the information-carrying part of the signal, see figure 2.22c. This low-frequency signal is transferred into sound in the headphones of the receiver set.

In 1905 Fleming invented the *thermionic valve* or *diode*, a valve detector [4]. The diode, see figure 2.23 consists of a filament-heated electrode, called the cathode[20] and an anode, both sealed in a vacuum glass tube. The filament heats the cathode and therefore, electrons are emitted from the cathode. If the anode is positively charged with respect to the cathode, electrons, emitted by the cathode will travel to the anode and a current flows through the circuit, see figure 2.23a. If the anode is reversely charged (with respect to the cathode), electrons, ejected by the cathode will be decelerated and will not make it to the anode; a current will not flow in the circuit, see figure 2.23b.

The invention of the diode, a reliable, compact rectifying detector, paved the way for detection, reception and amplification of radio signals and thereby for radiotelegraphy and radio broadcasting. Amplification became available in 1907, when

[19]Marconi started experimenting with CW systems in 1912 [4].

[20]The names *cathode* and *anode* for, respectively, the negative and positive electrode, as well as the name *electrode* itself are given by Michael Faraday and originate from his (electro)chemical research.

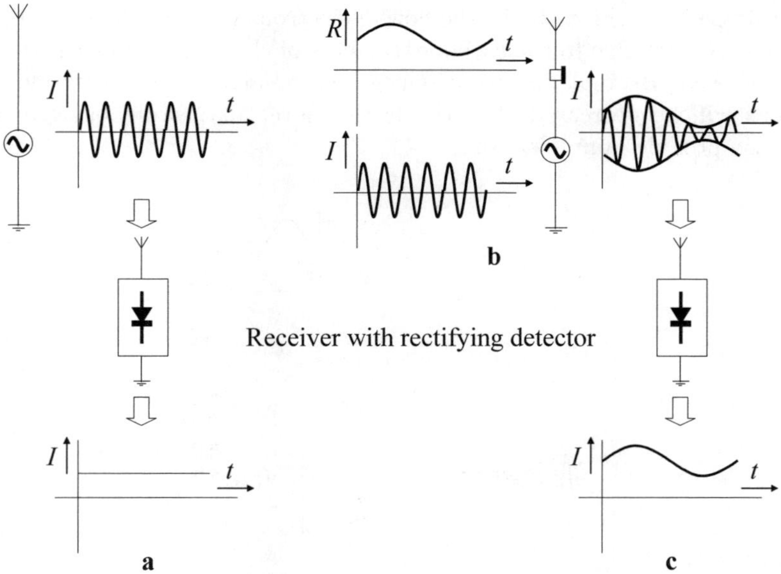

Fig. 2.22 Amplitude modulation. a. A (high) frequency CW signal is detected by a rectifying detector. b. A low-frequency signal is modulating a CW signal. c. A rectifying component filters the low-frequency signal from the modulated signal.

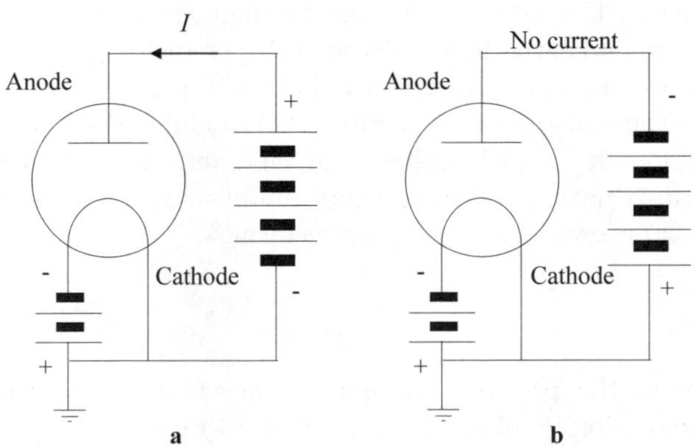

Fig. 2.23 Fleming's thermionic valve. a. Electrons emitted by the cathode (−) are travelling to the anode (+), a current flows in the circuit. b. Electrons emitted by the cathode (+) are repelled by the anode (−), no current flows in the circuit.

the American *Lee de Forest* (1873–1961) invented and used the *triode* or *audion* as a radio detector [17, 19, 20].

Lee de Forest added a third electrode to the diode. This electrode, placed between cathode and anode, had the form of a grid, see figure 2.24. The grid-shape allowed electrons to flow *through* the grid from cathode to anode. By changing the potential between this grid and the cathode, the flow of electrons will be influenced. By making the grid positive, relative to the cathode, the flow of electrons from cathode to anode will be increased. By making the grid negative, relative to the cathode, the flow of electrons will be decreased. The triode therefore, apart from being a rectifying element, also provides amplification.

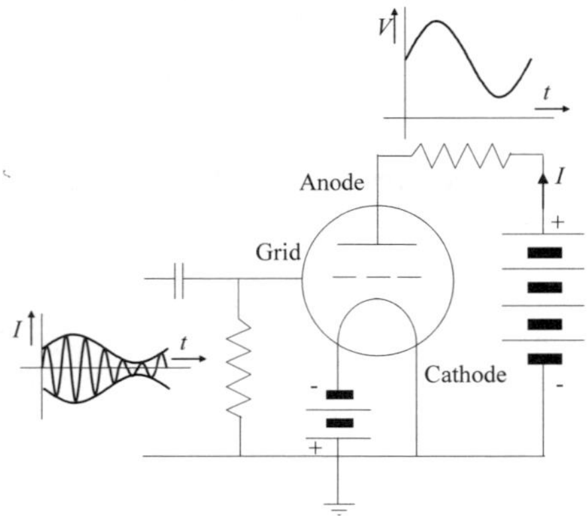

Fig. 2.24 Triode or audion. Besides being a radio detector, the triode is also an amplifier.

In figure 2.24, on the left, we show the small AM modulated current picked up by the receiving antenna. This current is flowing through the resistor providing a varying voltage between grid and cathode. By virtue of this changing grid voltage, the current amplitude in the circuit on the right, containing a resistor, a battery, the cathode and the anode, follows the envelope of the rectified AM modulated signal. Since the current amplitude in this circuit is much higher than the one picked up by the antenna, the triode has performed both rectification and amplification. The amplified current is led through a resistor, creating a voltage waveform.

2.1.5 Directivity

As we have seen in the previous chapter, an accelerated charged particle, emits electromagnetic radiation in all directions, except those coinciding with the direction of the particle movement. A charged particle, accelerated back and forth over an infinitesimal distance, emits radiation continuously. This configuration is known as the elementary or Hertzian dipole. On a macroscopic level, an alternating current (i.e. a net replacement of charged particles), over a half wavelength wire, fed in the centre, also behaves as a dipole, the charge accelerations/decelerations originating from the tips of the half wavelength wire. This is shown again, for two-dimensional cuts, in figure 2.25. Of course the radiation phenomenon is three dimensional, but since the radiation is symmetrical around the dipole axis, two-dimensional cuts in planes containing this dipole axis will contain all the information needed. The Marconi antenna, being a quarter of a wavelength long acts as a half wavelength antenna through the addition of the image in the ground of the quarter wavelength monopole.

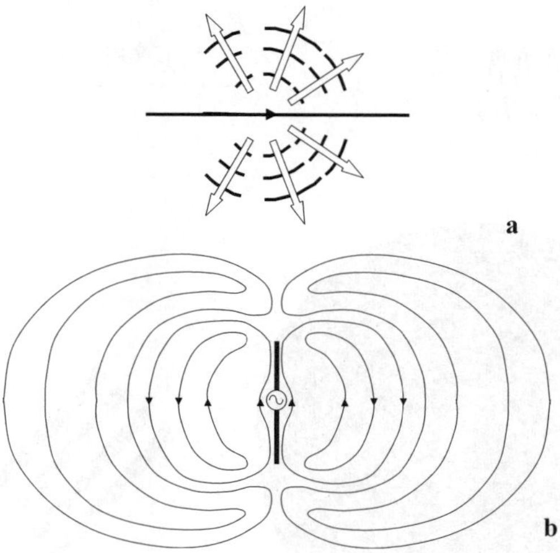

Fig. 2.25 Electromagnetic radiation. a. Radiation emitted by a charged particle. b. Radiation emitted by an elementary or a half-wave dipole antenna.

Based on the radiation properties as observed in the above figure, we may regard the radiation stemming from two so-called *isotropic radiators* positioned at the tips of the dipole. An isotropic radiator is a radiator that emits electromagnetic radiation equally in all directions.

Let's now consider a sinusoidal continuous wave (electric field **E** in figure 2.26). For educational purposes only, we convert this sinusoidal wave into a rectangular pulse wave as shown in figure 2.26. Still assuming that we deal with an isotropic radiator, we may - in a two-dimensional plane - represent the emitted wave as shown in the figure by the concentric white and black rings. The wavelength is indicated in the figure.

A half wave dipole antenna exhibits maximum charge accelerations (i.e. changes in current) at the two tips. In the top part of figure 2.27 this half wave dipole antenna with the current distribution over the antenna is shown (enlarged). Taking isotropic radiation from the two antenna tips we observe a resulting radiation from the half wave dipole antenna that exhibits *directivity*: a preference for certain directions over other directions, see figure 2.27, bottom part. This directivity is the result of constructive and destructive interference from the isotropic radiators, spaced half a wavelength apart.

We see that this radiation agrees with what we have seen for an accelerated charged particle, see figure 2.25a. Waves are emitted in directions perpendicular to the antenna axis and no waves are emitted in directions along the antenna axis.

From this isotropic radiators exercise we may conclude that, since radiation is inherently associated with charge displacement, isotropic radiators cannot exist in real life. Isotropic radiators are physical abstractions that help us in the understanding of antenna functioning.

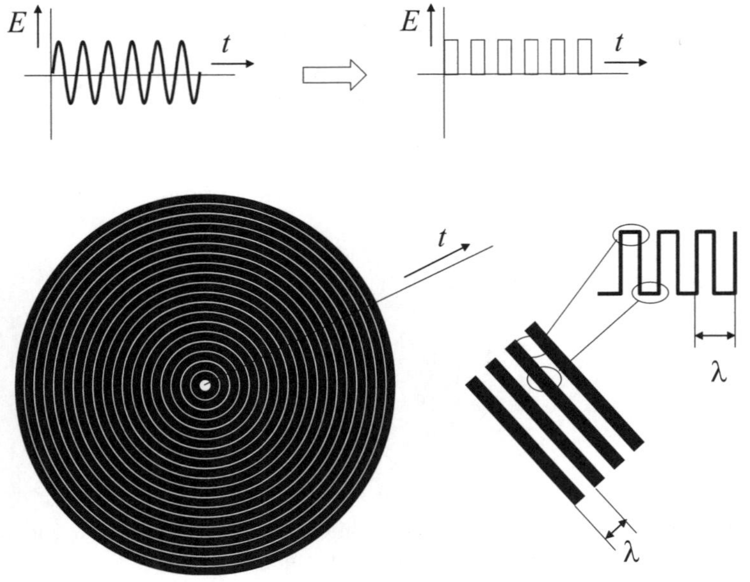

Fig. 2.26 Two-dimensional representation of isotropic radiation. A sinusoidal varying electric field is - for educational purposes - transformed into a block wave.

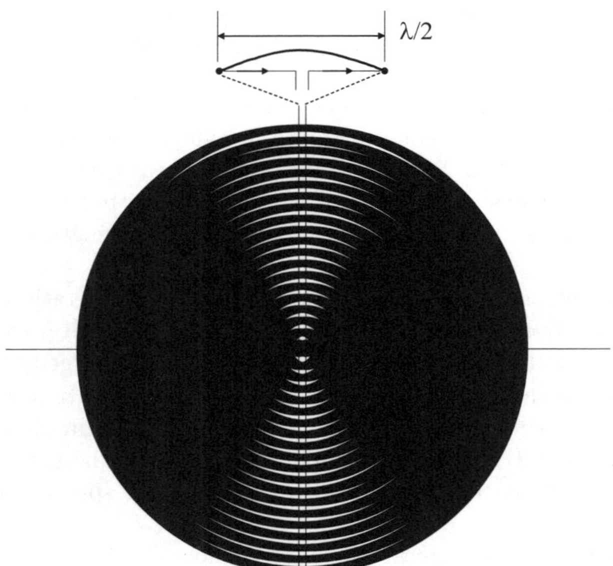

Fig. 2.27 Radiation from a half wave dipole antenna represented as interference from two radiating isotropic radiators.

So, the half wave dipole antenna, or actually the quarter wave monopole antenna, employed by Marconi, already had some directivity. This directivity was such that waves were transmitted equally around the antenna, having a maximum in directions perpendicular to the antenna axis, loosing amplitude in directions going from perpendicular to parallel to the antenna axis. Equally, the antennas were most sensitive for waves coming in perpendicular to the antenna axis, losing sensitivity for angles deviating from this direction, finally being insensitive for waves coming in parallel to the antenna axis.

By enlarging the antenna to a full wavelength, we can distinguish three points of maximum current change on the antenna and associated with these maximum current change positions are three isotropic radiators, spaced half a wavelength apart. The net radiation from these three (non-existing) isotropic radiators is shown in figure 2.28.

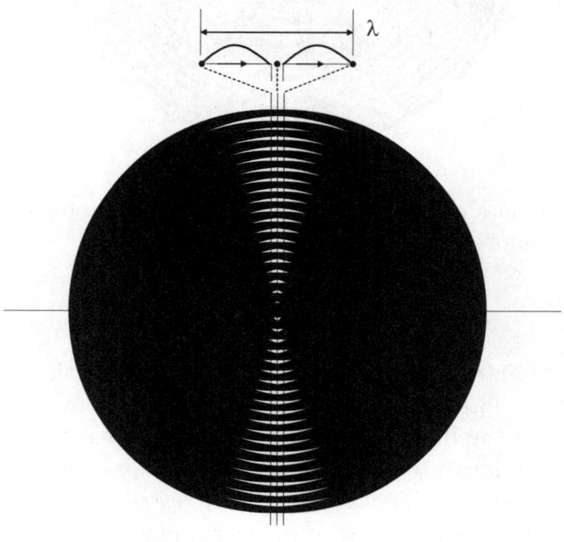

Fig. 2.28 Radiation from a full wave dipole antenna represented as interference from three isotropic radiators.

We see that through constructive and destructive interference of the three isotropic radiators a stronger directivity is obtained. The preferred radiation directions have become more restricted; the radiation beams have become smaller. Thus, by *enlarging* the antenna, the directivity has increased.

If we increase the antenna further to one and a half wavelength, see figure 2.29, we can distinguish four distinct points of maximum current change. However, at every point, the current undergoes a direction reversal (180° phase shift). Therefore, the two inner points have a relative phase that is the reverse of the phase of the two outer points. The isotropic radiators at the two inner points should therefore be represented with concentric rings of black and white with the colours reversed. For clarity reasons - the graphical possibilities in print being restricted - we have left out the contributions of the two inner points since their contribution is highly outweighed by those of the two outer points. The resulting radiation pattern is shown in figure 2.29.

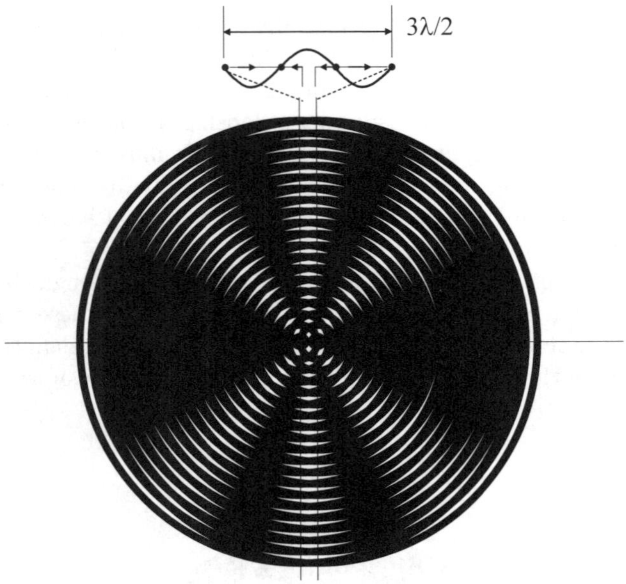

Fig. 2.29 Radiation from a one and a half wave dipole antenna represented as interference from four isotropic radiators.

We see that for an antenna enlarged beyond one wavelength, additional beams are formed. These results, obtained by physical reasoning and graphical representations, are in agreement with the dipole antenna's transmitted electric fields, calculated as function of the angle relative to the antenna axis, see for example [21].

In the above explanation of directivity, we have restricted ourselves to single dipole antennas only. Of course, for other wire (or metal plate) structures or combinations of wire structures (arrays!) the same physical reasoning applies. The general idea is to combine isotropic (or real) radiators in such a way that by the mechanism of constructive and destructive interference of these individual radiators a net radiation results that has the desired direction preference, i.e. *directivity*.

Directivity allows for an efficient use of available power in long distance communication. Having reached the limit in generating more power, it finally was the characteristic of directivity that allowed Marconi in 1907 to set up a reliable transatlantic wireless communication service.

2.2 ANTENNA DEVELOPMENTS DURING THE FIRST WORLD WAR

The First World War was not the first armed conflict in which wireless communication was used. That dubious honour is reserved for the Boer War in South Africa (1899–1902). The conflict (over independence) was between the Boers of Transvaal and Orange Free State and the British.

Five mobile wireless installations were constructed by the Marconi Company and, with six engineers from the company, delivered to the British Army. Although the wireless sets seem to have been intended for ship-to-shore communications - to assist disembarkment of reinforcement troops in 1899 - they were sent to the front. The employment was far from a success for a multitude of reasons. First of all, the appropriate poles to support the wire antennas were not available and makeshift poles were made from bamboo, or kites were used to elevate the wire antennas.[21] The wireless sets used were of the 'untuned' type, meaning that communication depended heavily upon the correct lengths of transmit and receive antenna. Furthermore, lightning activity - not uncommon in South Africa - set off the coherers, making wireless communication very difficult if not impossible. Finally, the - as we now know - low conductivity of the South African soil was the reason for a bad earth connection, further degrading wireless communication. In 1900, the sets were transferred from the Army to the Navy - the masts of the cruisers were extended to accommodate the long wire antennas - and performed very satisfactory [11].

The First World War started with the assassination of the Archduke Franz Ferdinand, heir to the Emperor of Austria-Hungary, in Sarajeva, Bosnia-Herzegovina, on 28 June, 1914. The reason for the world conflict was a complicated mixture of economic interests and old sentiments. What was intended to be a quick, mobile, military conflict ended up in a four-year trench-war on the European mainland, costing the lives of millions of people, military and civilian.

At the outbreak of the war, wireless telegraphy and telephony was still in its infancy, especially considering compactness and mobility. The 1918 British 'Trench Set' still required 12 men to carry its components. The bulk of the components were the heavy, wet-cell batteries [22]. Marconi's focus on long-wave, high-power transmission turned out to be an obstacle in creating compact wireless sets for use on land. For long-distance communication, however, the long-wave sets performed well. At the beginning of the war, the British Navy had dug up the German overseas telegraphic cables and cut them and the Germans, in return, had dug up and cut Britain's telegraphy cables to India in Turkey. Connections were restored by means of wireless telegraphy.

At sea, the British Navy already had experience in using wireless. In 1914, the wireless telegraph as means of long-range communication was used with success in the campaign against the German commerce raiders. For shorter ranges, as in fleet actions, the wireless was less successful. The long-wave networks overcrowded and the Navy was forced to fall back to the old-fashioned flag-signalling with all its drawbacks [22].

Guglielmo Marconi had volunteered in 1915 for the Italian Army and was appointed reserve lieutenant in the airship engineers with the task of directing telecommunications [4]. Based upon the then current (long-wave) wireless equipment, he built (one of) the first apparatus for aviation, see figure 2.30.

The equipment was for transmission (of Morse code) only and weighed about 150kg. Pilots described it as 'the very fat third passenger'. (The second passenger was the operator of the wireless transmitter.) It was obvious that transmitters and receivers operating with shorter wavelengths (i.e. at higher frequencies) were

[21] The kites were manufactured with the assistance of *Baden Powell*, the founder of the boyscouts [4].

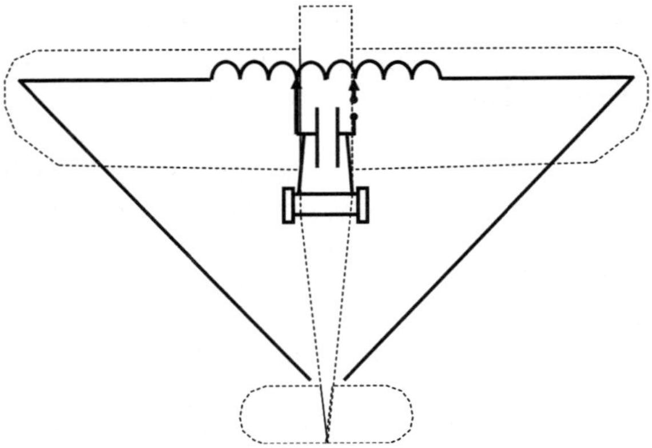

Fig. 2.30 Aircraft wireless telegraph transmitter.

desperately needed. The introduction of short wavelengths was withheld however - after the initial experiments by Hertz, Marconi and others - by the availability of high-frequency, high-power sources. Now, initiated by the war, research interest again turned to short wavelengths, not only because it would allow the realisation of compact wireless equipment, but also because it would serve the need for security by the ability of making very directive antennas.

In 1916, Marconi constructed, from tuned wires, a cylindrical parabolic reflector. A monopole aerial was positioned in the focal line of this reflector and the reflector of the transmitter was made revolvable, see figure 2.31 [13]. Thus, Marconi realised a (secure) point-to-point wireless connection, using wavelengths of two and three metres.

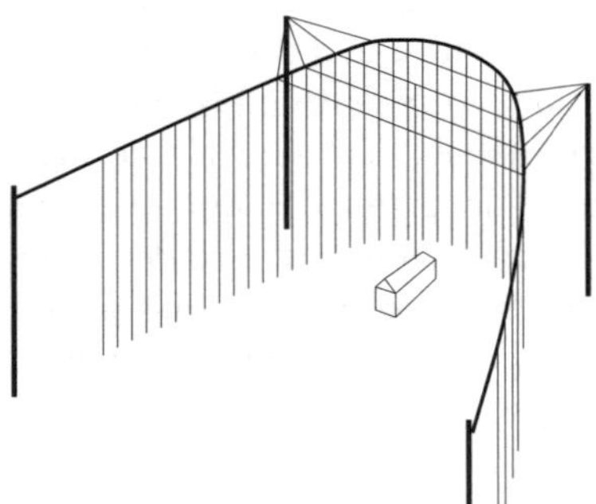

Fig. 2.31 Marconi's 1916 parabolic reflector.

The other way of securing wireless communication was using ciphers. At the outbreak of the war, direction finding, message interception techniques, cryptoanalysis and jamming techniques were all at their infancy but they developed rapidly throughout the war. The *Tannenberg* campaign in 1914 at the Eastern Front became a German success because of the lack of Russian radio security. The difficulty in distributing code books along the Russian lines, forced the Russians to transmit messages unciphered, or *en clair*, permitting the Germans to listen in to the tactical plans [22]. The *Zimmermann telegram* accident [22] is another example of the importance of using secure communications. The then German foreign minister *Zimmermann*, sent a telegram - in diplomatic code - to the German ambassador in Washington. The message was intended for the Mexican government, trying to get them to enter the war against the USA with Texas, New Mexico and Arizona promised as reward. The message was cracked by the British and the contents were - very secretly - passed to the US government. This so-called *Zimmermann telegram*, made public on 1 March, 1917, helped bring the USA into the war.

2.3 ANTENNA DEVELOPMENTS IN BETWEEN THE WARS

In 1910, a carbon-arc transmitter of Lee de Forest was used by the *De Forest Radio Telephone Company* for broadcasting music - including a performance of Caruso from the Metropolitan Opera House - in the New York area [13, 17]. De Forest did not realise that his *audion* could be used for voice transmission and in 1913 he sold the rights to a lawyer, who turned out to work for the American Telephone and Telegraph Company (AT&T) [17].

2.3.1 Broadcasting

In 1916 music was being broadcast from Monday till Saturday at New Rochelle, NY. In 1917, broadcasts were carried out in Europe and in 1919 the Marconi Company conducted broadcast tests, even though Marconi himself considered broadcasting a misuse of his invention [13]. In 1920, broadcasts started on a regular basis and in 1922 the British Broadcasting Corporation (BBC) was formed [5]. In 1923, AT&T in cooperation with RCA succeeded in accomplishing the first transatlantic wireless telephony between Rocky Point, NJ, USA and New Southgate in England. A frequency of 60KHz and a transmitter power of 60kW were used [25].

The early antennas in the broadcasting business were makeshift antennas, derived from the designs used in point-to-point communication. Later, T-configured antennas were used for the transmitters [26] and eventually the vertical radiators became standard, due to their circular symmetrical coverage (directivity) characteristic [13, 26]. The receiver antennas, used by the public, were backyard L-structures, see figure 2.19, and T-structures, see figure 2.32.

Even nowadays, L- and T-configured antennas are employed for transmission and reception aboard ships [26].

In 1919, Marconi had conducted experiments at 20MHz (15m wavelength), using a vacuum tube generator. He found that even at this frequency, communication beyond the horizon was possible [4, 13]. This led to a regained interest in microwave research after the initial experiments by Hertz and Marconi in the previous century.

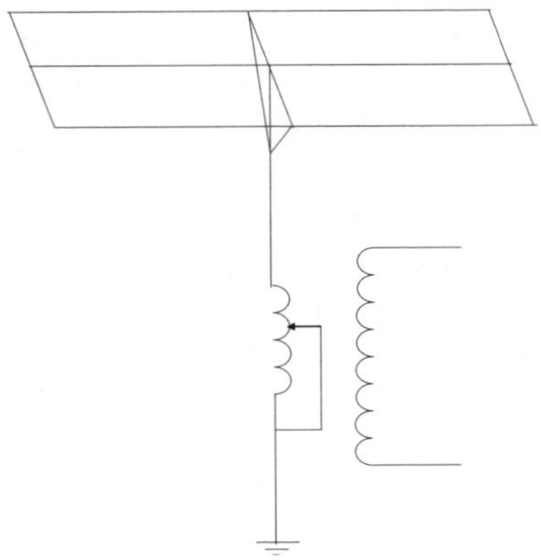

Fig. 2.32 T-configured antenna.

2.3.2 Microwaves

In the 1930s a return to the higher end of the radio spectrum took place. Apart from the antennas already described, like Hertz' dipole and loop antenna, also the electromagnetic horn antenna, invented by *Chunder Bose* before 1900 [6, 13] and the open-ended waveguide, used by *Oliver Lodge* in the same period, were already used for wavelengths ranging from 5mm to 20cm, or for frequencies ranging from 1.5GHz to 60GHz. Figure 2.33 shows a rectangular open-ended waveguide radiator and a pyramidal horn radiator, which is identical to the open-ended waveguide, except for the flared transition to free space. These antennas are both member of the class of *aperture antennas.*

In both radiators, a guided wave between the conducting plates of the rectangular waveguide is launched by means of the coaxial probe at the left. This guided wave travels to the right, the launched wave travelling to the left is reflected at the short circuit plate and added in phase to the wave going to the right. At the end of the waveguide or horn, i.e. at the *aperture*, the guided wave transfers into a radiated or unguided wave. In contrast to the wire antennas we have been discussing so far, the radiation from the aperture antennas shown in figure 2.33 is not symmetrical; energy is mainly radiated into the directions to the right. This is indicated in figure 2.33d, in a way similar to that we have been using to show the directive properties of a group of isotropic radiators.

With the return to the short wavelengths, also the parabolic reflector antenna - first used by Heinrich Hertz - made its reappearance. It may be said that many of the antennas that (re)appeared in the 1930s were classical, i.e. still being seen today. That even goes for the *array antenna*, already proposed in 1899 and implemented in 1906 [13].

An array antenna is a system of individual antennas, that, together, form another antenna. To avoid confusion, from now on we will call the individual antennas *radiators* and will only refer to the whole system or array as *antenna* or *array antenna.* The purpose of arraying radiators is to shape (increase) the directivity.

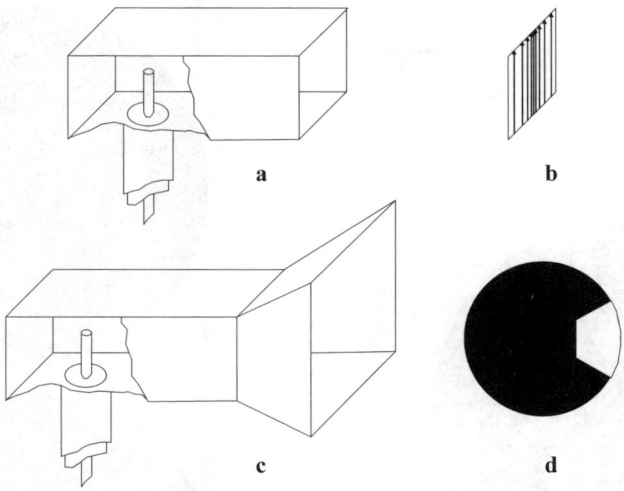

Fig. 2.33 Aperture antennas. a. Open-ended rectangular waveguide. b. Electric field lines in the aperture of the waveguide. c. Pyramidal rectangular horn antenna. d. Indication of the sensitivity of a horn antenna.

We may demonstrate this in a manner similar to the one we have been using for demonstrating directivity before. What applied to two or more isotropic radiators, also applies to a system of clusters of isotropic radiators, or in other words: 'The principal of superposition of isotropic radiation patterns also applies to non-isotropic or directive patterns'. This is demonstrated for an array of half-wave dipole radiators in figure 2.34.

We clearly see that by adding radiators, we can increase the directivity, i.e. confine the radiation to a smaller angular region.

We have to bear in mind though that the reason for Marconi and others to concentrate on long wavelengths thus far was due to the fact that long wavelength electromagnetic waves undergo less attenuation than short wavelength waves do upon travelling through the atmosphere. The difficulty in the creation of high-power, high-frequency generators had turned the attention away from short to long wavelengths. In the 1930s, however, technology had evolved to a level that made a regained interest in short wavelengths feasible.

To name a few of the accomplishments: in 1931 a 17cm microwave link, using reflector antennas, was established between England and France. In 1932, Marconi discovered tropospheric scatter communication at 50cm. In 1936 TV broadcast service started in England at 45MHz and in 1937 a radiotelescope reflector was built by *G. Reber*, using a horn feed at 160MHz [13].

In 1933 Marconi spoke to the Royal Institution of Great Britain [27]. His presentation, reprinted by the IRE in 1957, deals largely with the design of reliable, high power transmitters for short wavelength communication. High power and directivity were obtained by arraying dipole radiators, each one connected to a transmitter and placed in the focal point of a wire-structure parabolic reflector [27].

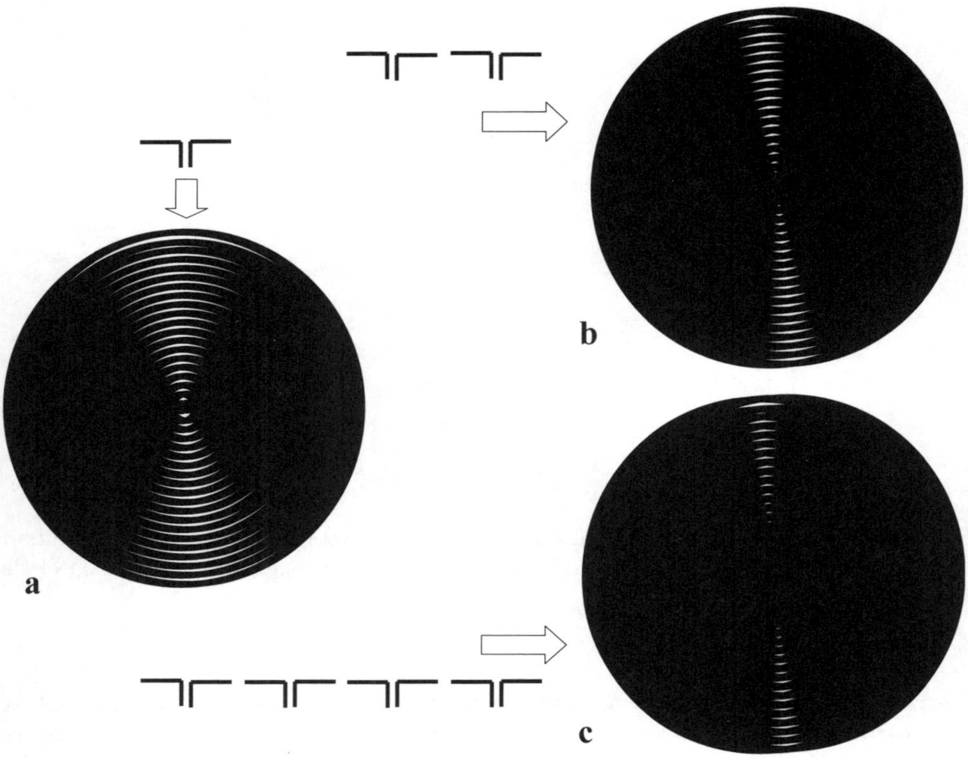

Fig. 2.34 Directivity of a dipole array antenna. a. Directivity of a single half-wave dipole radiator. b. Directivity of a two-element half-wave dipole array antenna, radiators positioned one half of a wavelength apart. c. Directivity of a four-element half-wave dipole array antenna, radiators positioned one half of a wavelength apart.

In 1932, a below 1m wavelength communication system was installed at the Vatican and at Castel Gandolfo, the summer residence of the Pope, at a distance of 20km from Vatican City.

2.4 ANTENNA DEVELOPMENTS DURING THE SECOND WORLD WAR

The outbreak of World War II, in 1940, intensified the need for reliable, high-power, high-frequency sources. These were needed for compact communication equipment as well as for compact (airborne) and high-resolution radar. *Radar* is an acronym for *ra*dio *d*etection *a*nd *r*anging.

2.4.1 Radar

The use of radio waves reflected from conducting objects, was already demonstrated for the detection of ships in 1903 and patented the next year by the German *Christian*

Hülsmeyer (1881–1957). The German admiralty, however, was not interested due to the fact that technology in those days limited the usable range to about 1.5km, well within the range for visible detection [28].

In 1922, Marconi, in a speech to the (American) Institute of Electrical Engineers (IRE), mentioned the use of radio waves for the detection of metallic objects [28, 29]:

> As was first shown by Hertz, electromagnetic waves can be completely reflected by conducting bodies. In some of my tests I have noticed the effects of reflection of these waves by metallic objects miles away. It seems to me that it should be possible to design apparatus by means of which a ship could radiate or project a divergent beam of these rays in any desired direction, which rays, if coming across a metallic object, such as another steamer or ship, would be reflected back to a receiver screened from the local transmitter on the sending ship, and thereby immediately reveal the presence and bearing of the other ship in fog or thick weather.

In that same year, *A.H. Taylor* and *L.C. Young* of the Naval Research Laboratory (NRL) made a CW interference radar, operating a wavelength of 5m. Transmitter and receiver were physically separated (just like Marconi proposed in his 1922 speech). With their CW interference radar, Taylor and Young demonstrated the detection of a wooden ship. A proposal for further research was not accepted however [28].

The CW interference radar worked by receiving both the transmitted signal and the Doppler-frequency-shifted reflected signal from the object. The reflected signal interfered with the signal received directly from the transmitter and this interference was detected as a 'flickering' or 'flutter' in the received signal. This 'flutter' therefore announced the presence of an object in the transmitted beam.

In 1930, *L.A. Hyland* of NRL while experimenting with a 33MHz direction-finding apparatus, noticed that aircraft could be detected by the principle of CW interference and in 1934, after slow progress due to a lack of funds, a 60MHz CW interference radar was demonstrated by NRL. The mentioned lack of funds was mainly due to the fact that the CW interference radar could be used to detect the *presence* of an object (airplane), but extraction of *position information* was extremely difficult. For that purpose, pulse radar was needed. In 1936, NRL demonstrated a pulse radar with a range of about 40km [28].

Radar developments in Britain started later, but with a war drawing near and being in a very vulnerable position, much effort was being put into radar development. Developments started in 1935 when the British Government asked *Sir Robert Watson-Watt* (1892–1973) of the National Physical Laboratory to investigate the feasibility of the so-called *death-ray* [28, 29].

The idea was born in the pre-war period - with all the progress being made in the development of high-power radio transmitters - that high-power radio beams could be used to sabotage the electrical ignition systems of aircraft engines. It turned out that the amounts of power needed for creating a 'death-ray' made the concept unfeasible. Together with this result of his study, Watson-Watt proposed the investigation of *radio detection* as opposed to *radio destruction*, since *detection* seemed to be technically feasible.

This proposal led to an experiment in 1935, involving the BBC's radio transmitter at Daventry. A van was equipped with a radio receiver, which was connected to an oscilloscope and stationed near the Daventry transmitter. When a bomber flew by, oscillations caused by the BBC signal reflected from the bomber and interfering with the signal directly received from the transmitter, were visible on the oscilloscope [29]. This CW-interference experiment used a 6MHz signal and demonstrated a 13km range [28].

In the same year, pulse technique was demonstrated at 12MHz and in 1936, the range had been extended to 135km, using a frequency of 25MHz [28]. In 1937, a number of radar stations - operating at frequencies in the range 20MHz to 30MHz were erected along the British east coast and in the same year these stations were put into 24-hour duty. The 24-hour duty of the so-called Chain Home (CH) radar lasted till the end of the war. A CH radar station is depicted in figure 2.35 [29, 30].[22]

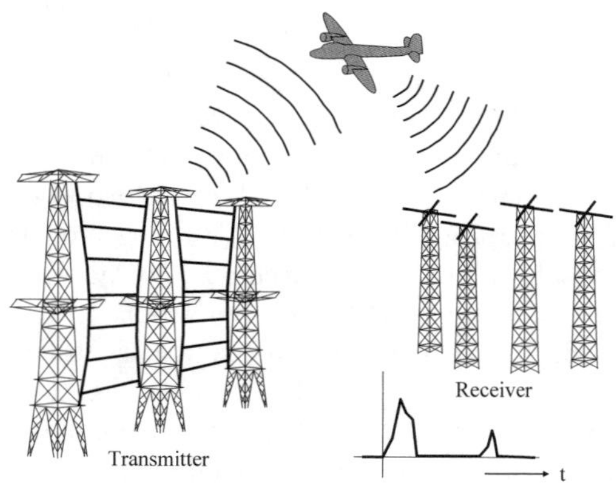

Fig. 2.35 A Chain Home radar station with separate transmitting and receiving antenna. The time elapsed between transmitted pulse and received echo is a measure for the distance of the plane.

The transmitter towers were 105m high lattice steel masts, with the transmitting antennae in between the masts. The receiving towers were 24m high wooden lattice masts, carrying crossed-dipole antennas.[23] The distance of the aircraft was obtained by measuring (on an oscilloscope, see figure 2.35) the time elapsed between a pulse transmitted and a pulse received after reflection. The height was obtained separately, by an operator using a directional coil antenna after a target had been detected on the oscilloscope screen [29]. The azimuth direction was obtained in the *filter room*, i.e. the office were all CH radar information was collected and processed before it

[22]The line drawing is an interpretation of the author, based on photographs and descriptions in literature. The problem with old or reproduced photographs is that they suffer in resolution, therefore making wire structures nearly invisible. Most non-technical literature - if showing a CH radar station at all - focus on the masts, disregarding the fact that the actual antenna is hung in between the masts. Similar problems occur with photographs of Marconi's 'transatlantic' antennas, [16]. Photographs, being patched up by artists not understanding the picture are still in circulation, showing antennas that have probably never been in existence.

[23]The lattice structure of transmitting and receiving masts and antennas proved to be a valuable characteristic in the Battle of Britain, making it a very hard target to bomb since pressure waves caused by non-direct hits would hardly damage the structure.

was relayed to the headquarters of *fighter command*.[24] In the filter room, the distance information of two nearby CH radar stations was combined to get the azimuthal angle information of the target, see figure 2.36 [31].

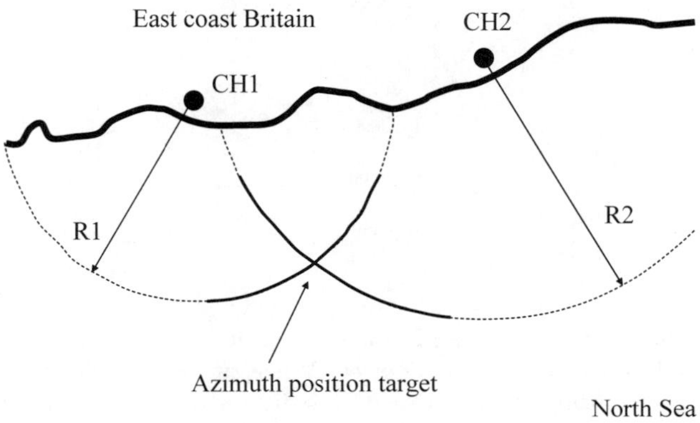

Fig. 2.36 The azimuth position of a target is found at the intersection of the circle with centre CH radar station 1 (*CH*1) and radius the reported distance of CH station 1 (*R*1) and the circle with centre *CH*2 and radius *R*2.

Because of this processing and to hide the real function of the CH stations, the system was first called *radio direction finding* (RDF). This name later changed into *radio location* (RL) and finally it became *radio detection and ranging* (radar), [31].

The CH wavelength (about 10m) and the position of the antennas (horizontal) were chosen by Watson-Watt to coincide with span and position of the wings of a bomber [31]. Next to the CH stations a chain of CH-Low (CHL) stations was also erected, working with rotating antennas at a wavelength of 1.5m for the detection of low flying targets [31]. The range of the CHL radar was much less than that of the CH radar.

Later radar systems - operating at higher frequencies - were of the duplexing type, sharing a common, highly directive antenna for transmission as well as for reception.

2.4.1.1 Radar Principle The radar principle is best explained using such a pulsed, duplexing system, see figure 2.37.

A highly directive antenna is mounted such that it can be moved around (in azimuth) and up and down (in elevation). Immediately after a pulse is transmitted, the antenna is switched to the receiver. The time between a transmitted and a received, reflected pulse is a measure of the distance of the target. The distance is *half* that time (the pulse travels from the antenna to the target, the reflection travels from the target to the antenna) multiplied by the speed of light. Since the antenna is highly directive, i.e. transmitting the electromagnetic waves in a narrow beam, the angular position of the antenna indicates the angular position of the target. When the antenna is rotated and a fluorescent cathode ray tube (CRT) is configured such that

[24]At the headquarters of fighter command, finally, decisions were made concerning the employment of (the scarce) fighter squadrons against the incoming bombers.

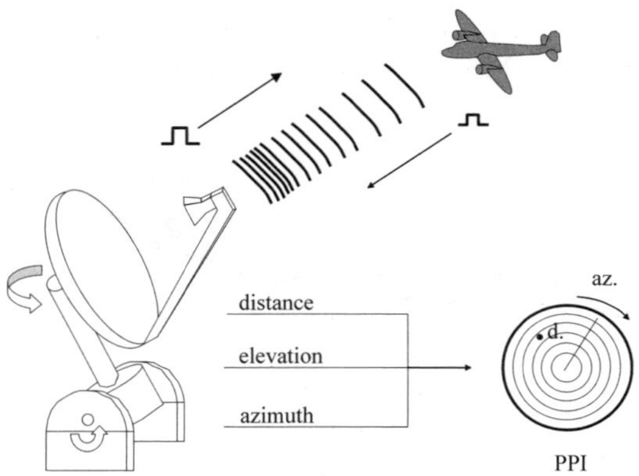

Fig. 2.37 The radar principle. A pulse is transmitted and the antenna is disconnected from the transmitter and connected to the receiver. When an echo (reflection) is received, the time elapsed between transmission and reception indicates the distance, the angular positions (azimuth and elevation) are those of the highly directive antenna.

the deflected electron beam follows this rotation, see figure 2.37, the so-called plan position indicator (PPI) is created that was introduced later on in the second world war. Now, the angular position on the PPI screen indicates the azimuth position of the target and the distance can be read from the concentric circles on the screen. (The centre of the screen corresponds with the transmitted pulse, the radius of the screen corresponds to the time interval in between transmitted pulses. The radial position of the echo on the screen is thus linearly related to the distance that can be read as a fraction of the PPI radius.)

2.4.1.2 Further Pre-war Developments In the USA, the mentioned duplexing system, wherein a common antenna is used for both transmission and reception was demonstrated at NRL in 1936. This system operated at a frequency of 200MHz, but was limited in range to about 20km. By 1938, this range was extended to 80km. In that same year, the US Army also got involved in radar. The US Army Signal Corps employed its first operational radar for antiaircraft fire control, the *SCR-268*; a pulsed radar system. This SCR-268 was used in conjunction with search lights because of the poor angular accuracy. The SCR-268 became the US Army Signal Corps standard till January 1944 when it was replaced by the *SCR-854* microwave radar. The SCR-854 had a better angular accuracy and could be employed without the necessity of employing search lights at the same time [28].

In 1939, the long-range *SCR-270* radar, see figure 2.38, was brought into service for early warning purposes. Six of these radars were positioned on Hawaii (as well as 16 SCR-268 radars). One of the six SCR-270 radars detected the attack on Pearl Harbor in December 1941, but the significance of this detection was not realised until it was too late [28].

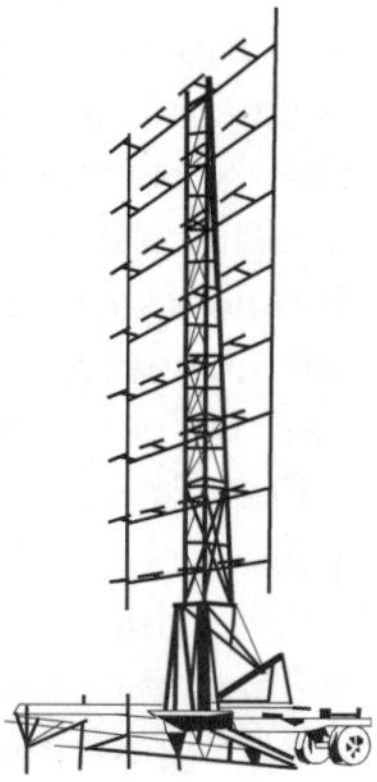

Fig. 2.38 Antenna of an SCR-270 radar system. Note that the antenna is an array antenna consisting of 8 × 4 horizontally positioned dipole radiators. The SCR-270 radar worked at a frequency of about 100MHz.

The reason for using an array antenna is, as said before, to increase directivity. In general, directivity increases with the size of the antenna. Thus, the larger the antenna, the smaller the beam of electromagnetic waves.

In Germany, serious radar experiments commenced in 1934, even before Watson-Watt was asked to investigate the 'death ray'. In that year, *Künhold* of the German navy demonstrated the detection of ships on 12km distance [31]. In 1937, the pocket battleship *Graf Spee* was equipped with the *Seetakt*, a fire control radar. In 1938, the Germans employed the long-range surveillance radar *Freia* and in 1939 the *Würzburg-A*, an antiaircraft fire control radar [31]. The Freya radar operated at 126MHz, the Würzburg radar, see figure 2.39, operated at a frequency of 570MHz [29].

a b

Fig. 2.39 Antenna of a mobile Würzburg radar installation, photographed in the Imperial War Museum, London. Note the dipole in the focus of the parabolic dish. a. Front view. b. Side view.

In the years preceding the Second World War, radar developments were conducted simultaneously in a lot of countries, including the USA, Great Britain, Germany, Japan, Russia, France, Italy and The Netherlands. Therefore it is not possible to accredit the invention of radar to a single country or individual.

All of the mentioned radar developments did not exceed frequencies of 600MHz. The invention of the microwave radar, however, may be accredited to Great Britain. Before we discuss the microwave radar, we conclude the prewar developments with a remarkable radar development from The Netherlands.

As a result of rumours in the press concerning mysterious accidents involving incapacitated car engines, thought to be due to a secret, German 'death ray', the then Netherlands Minister of War was forced to take action. He established a research institute that got the specific task of improving military equipment by applying physical principles. In 1927 the 'measurement building' was erected near The Hague.[25] This establishment has become the later TNO Physics and Electronics Laboratory.[26] Professor *J.L.W.C. von Weiler* working at this institute, discovered that planes could be detected by means of reflection of electromagnetic waves. His discovery led to the realisation, in 1939, of the 'electrical listening device', see figure 2.40.[27]

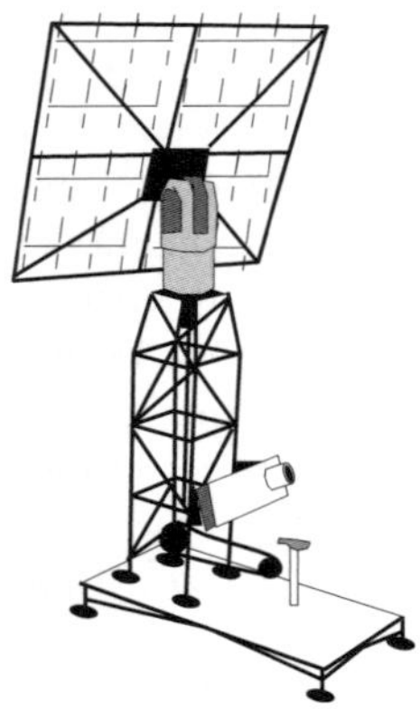

Fig. 2.40 The 1939 'electrical listening device' of von Weiler.

In this early radar system we again see the employment of an array antenna! The array antenna consists of 32 (4×8) dipole radiators, positioned in front of a planar wire mesh reflector. The radar system operated a frequency of 423MHz.[28]

[25]In the years of disarmament and pacifism between the two World Wars, the name 'laboratory' was - for reasons of public relations - carefully avoided.

[26]TNO (instituut voor Toegepast Natuurwetenschappelijk Onderzoek) is the Netherlands Institute for Applied Scientific Research.

[27]The name *electrical* was used to distinguish this device from the then existing *audio* devices. These sound locators, consisting of artificially enlarged ears, could obtain a range - under favourable conditions - of about 30km and were not fit for distinguishing course information.

[28]The antenna looks very similar to the antenna of the SCR-270 radar, see figure 2.38, having roughly a quarter of its size which corresponds to the frequency being four times as high as that of the SCR-270 radar. It may be assumed though, that developments in the USA and in The Netherlands took place completely independent of one each other.

The antenna motion is delivered by the radar operator, note the bicycle saddle and pedals in figure 2.40. Radar information is presented on a CRT in front of the cyclist operator.

In May 1940, when Germany invaded The Netherlands, a number of these radar systems were transferred to Great Britain and at least one system is known to have been operational during the war on board one of the Royal Netherlands Navy's warships that had taken refuge in Great Britain.

2.4.1.3 The Cavity Magnetron Although the Chain Home radar performed well in daytime under good weather conditions, its accuracy was too low to guide fighter aircraft completely to interception at night or in bad weather. The CH-Low was more accurate, but lacked range. In order to overcome these problems, in 1939, an aircraft interception (AI) radar (AI Mark II), operating a wavelength of 1.5m was installed in an aircraft [28, 30]. The development of the AI Mark II had resulted in the observation that an airborne radar could be used for the detection of ships from the air. After the Battle of Britain when the air war subsided (around spring 1941), radar development was directed to the detection of submarines, the next threat to Britain [30].

Since the need for mobile, airborne and high accuracy radar was getting more and more urgent, the British Government in 1939 brought together scientists from various universities to work on the creation of high-frequency sources [29]. The production of enough RF power at high frequencies was the limiting factor in the development of these desired radar systems. The frequencies aimed for ran up to *X-band.*

The frequency spectrum had been divided into frequency bands. The lower bands followed the natural historic exploration of continuously higher frequencies. The frequency bands opened up during the Second World War are designated randomly for reasons of military secrecy. The frequency bands are listed in table 2.1 [28]. In this table also the new band designations - that are more systematic - are listed [32]. Although the new band designations are the recommended ones, the old ones are still used by microwave engineers and it will probably take at least a whole generation before the old habit dies out.

Table 2.1 New and old frequency band designations.

Old band designation	Frequency range	New band designation	Frequency range
HF	3-30MHz	C	500–1000MHz
VHF	30-300MHz	D	1–2GHz
UHF	300-1000MHz	E	2–3GHz
L	1–2GHz	F	3–4GHz
S	2–4GHz	G	4–6GHz
C	4–8GHz	H	6–8GHz
X	8–12GHz	I	8–10GHz
K_u	12–18GHz	J	10–20GHz
K	18–27GHz	K	20–40GHz
K_a	27–40GHz		
mm	40–300GHz		

John Randall (1905–1984) and *Henry Boot* (1917–1983) were two of the scientists in the British 'think tank' at the physics department of Birmingham University. Since they had been assigned to work on receiver valves, they started with the creation of a microwave source for testing the receivers. Based on a low power device from 1921, called the *magnetron*, they got the brain-wave to create a resonant structure: the *cavity magnetron*. The first model was tested in early 1940 and immediately performed very well. It was possible now to create 500kW at 3GHz or 100kW at 10GHz with an extremely compact device [28, 29]. The early type cavity magnetron is shown in figure 2.41 [28, 32].

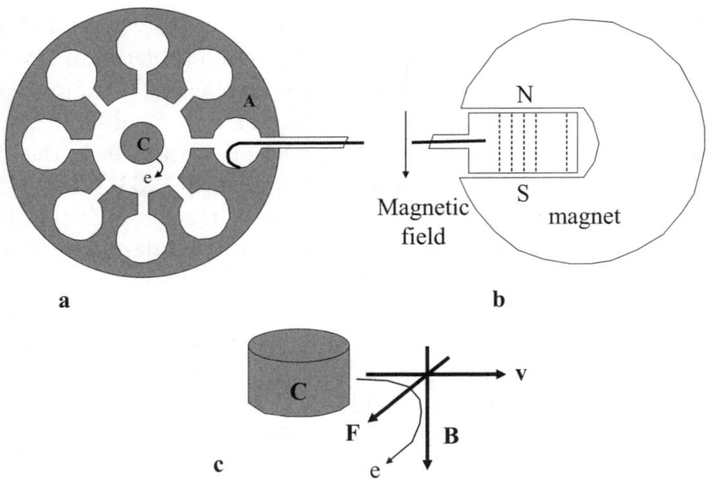

Fig. 2.41 Cavity magnetron. a. Top view. b. Side view. c. Electron movement.

The cavity magnetron consists of a cathode, coaxially placed inside an anode block. Inside this anode block, holes and slots are cut. The slots connect the holes to the interaction space, i.e. the space wherein the cathode is centrally located, see figure 2.41a. The holes form resonant electromagnetic cavities. The cavity magnetron is placed in a permanent magnetic field, see figure 2.41b. When electrons are emitted by the cathode and moving towards the anode, the magnetic field - being perpendicular to this electron motion - causes a force to act upon the electrons in a direction perpendicular to both the electron velocity and the magnetic field as shown in figure 2.41c. Therefore, the electrons, upon leaving the cathode, will bunch and move along in a circular motion. The presence of the holes or resonant cavities allows for the existence of an electromagnetic field moving in a circular motion at the same speed as the electrons. Therefore the electrons can deliver radio frequency (RF) power to the electromagnetic wave.[29] This power may be extracted by a coaxial line loop as shown in figure 2.41a,b or by means of a waveguide [28, 32].

The theory explaining the operation of the cavity magnetron became available after the functionality had been demonstrated. Randall has said about this:

[29]This is often compared with blowing over the opening of a bottle. The cavity (the bottle), when brought into resonance then, creates a whistling sound.

> It was fortunate that we didn't have time to study all the learned papers available on the magnetron, otherwise we would never have started.

The invention of the cavity magnetron made a large improvement in the AI radar possible. The AI Mark IV became a success. Later versions of the Mark IV could provide 10kW of power and provided a detection range of up to 6km [30]. From the AI radar came the *air to surface vessel* (ASV) radar. The ASV Mark II (the equivalent of the AI Mark VIII), that went into service in 1943, could detect submarines on the surface at distances up to 10km [30].

The magnetron and ASV design had been brought to the USA in the so-called *Tizard Mission*[30] in 1940 - a mission to exchange war-time secrets for production capacity - and led to the introduction of microwave radar in the USA.

Tests with the high-frequency, cavity magnetron powered ASV radar had revealed that landscape features like coast lines, rail lines, roads and rivers could be easily identified (even through clouds) on a PPI. This led to the development of a navigational radar for bombers, the *H2S*, formerly known as *blind navigation* (BN) radar. Several explanations for the name *H2S* exist. The name is said to be an acronym for *height to slope* or *home sweet home*. It is also said that the name stands for the chemical symbol for hydrogen sulphide, referring to the fact that the device 'stinks' and another explanation is that the 'S' is referring to the frequency, 'S-band'. Most likely though the name does not refer to anything for military security reasons.

The H2S radar was compact enough to be fitted in a blister on the belly of a bomber. The antenna, a compact parabolic reflector antenna, would rotate, see figure 2.37, and thus scan the terrain underneath the bomber and display the terrain features on the PPI CRT inside the bomber. At the start, the British were very reluctant to bring the cavity magnetron aboard operational aircraft, for fear of the enemy obtaining this device from a shotdown aircraft. In February 1943, a Pathfinder[31] was shot down near Rotterdam, The Netherlands, and parts of the H2S radar, including the cavity magnetron were captured (and copied) by the Germans. After that accident the H2S was used on a large scale, although it did not increase bombing accuracy up to the level expected [30].

Another negative aspect of H2S was that since it was continuously transmitting pulses, enemy night fighters could and did home in on these signals (using the *NaxosZ* unit, see figure 2.42) and created havoc on the bomber formations [30].

2.4.1.4 German Airborne Radar Although, Germany did not have a cavity magnetron, the Germans had developed an airborne 50cm radar, known as the *Lichtenstein* radar. Since countermeasures were introduced to 'blind' this radar, larger wavelength Lichtenstein radars were introduced as counter-counter measure, see figure 2.43.

The mentioned countermeasures consisted of ejecting clouds of metal foil strips having each a length equal to the wavelength of the radar. These clouds of strips, ejected from a high flying plane, when floating down to earth, would therefore generate

[30] Named after the mission leader *Sir Henry Tizard*.

[31] Pathfinders were bombers fitted with the cavity magnetron radar that had the task of flying ahead of the bomber waves, identify the target and drop down incendiaries or flares on this target to mark it for the bomber formations ahead.

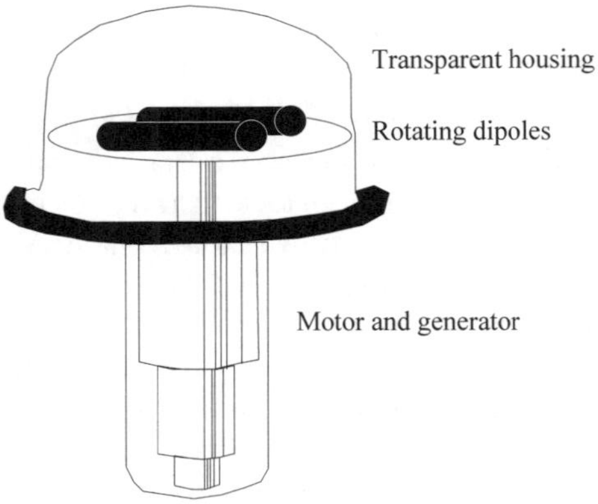

Fig. 2.42 The NaxosZ unit was small enough to be fitted in Junkers Ju 88G or Messerschmitt 109G airplanes. The Naxos unit was also used aboard submarines to locate merchant and other vessels [33].

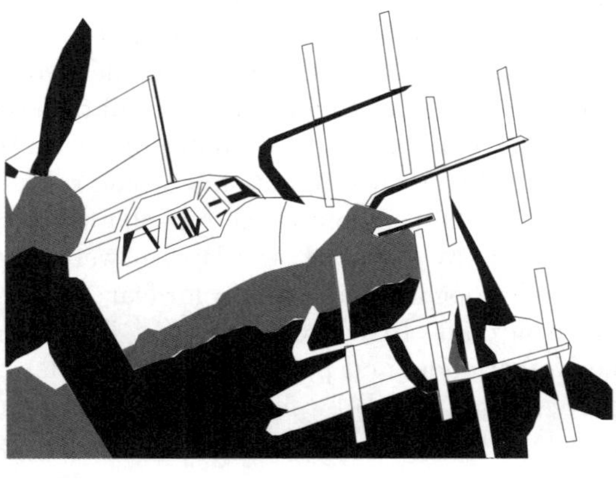

Fig. 2.43 The double-dipole antennas of a two metre wavelength Lichtenstein radar mounted on the nose of a Junkers Ju 88G night fighter. The four antennas are sequentially switched to the radar transmitter/receiver and thus provide a conical scan of the area in front of the aircraft [30].

a massive reflection, cluttering the PPI, making it possible to fly behind the cloud, undetected by the radar. The British called this system *window*, the Germans called it *Düppel* and the Americans called it *chaff*, the name that is still in use today [29].

2.4.2 Other Antenna Developments

Apart from radar, electromagnetic waves - and thus antennas - were employed for communication and other war needs, such as *identification friend or foe* (IFF), navigation and direction finding. In the subsequent paragraphs we will briefly outline these applications and the antennas used.

2.4.2.1 IFF When radar became operational, the immediate need evolved to distinguish between received echoes from friendly and enemy vehicles. At the outbreak of the war, the need was most prominent for the identification of aircraft. The first identification friend or foe (IFF) systems aboard aircraft consisted of dipole antennas, see figure 2.44, tuned to the ('friendly') radar frequency and alternately short- and open-circuited. This switching caused a fluctuating radar echo to be received at the radar station and served the purpose of identifying the aircraft as being 'friendly' [34].

Fig. 2.44 IFF antennas on a Supermarine Mark I Spitfire.

This system was not very reliable. Another system to identify friendly aircraft was known as the *pipsqueek* system [31, 34]. For the pipsqueek system the aircraft was equipped with a mechanical timer connected to the already present wireless communication system. Fourteen seconds of every minute, the switch would turn on the transmitter and transmit an unmodulated signal. The combined output of ground-based direction-finding stations then could determine the aircraft position, see figure 2.36. Also this system was far from perfect. It would, for example, prohibit the pilot from using the transmitter or receiver for 14 seconds every minute.

The next development was the *IFF Mark I* which consisted of a complete transmitter/receiver. The unit was normally in receive mode. Upon receiving a (friendly) radar signal, it went into transmitter mode and returned a signal to the radar that interfered with the radar echo and so identified the carrier of the IFF Mark I as being 'friendly'. The frequency of the set was swept over all radar frequencies being used so that it could operate in cooperation with all current radar systems. Its successor, the *IFF Mark II* was updated for the new frequency bands being put into use and its return was a coded response, [34].

The *IFF Mark III* became the standard being used by the Allies. At the ground radar station an *interrogator* transmitted in between the radar pulses. The unit aboard the interrogated vehicle, i.e. the *transponder*, responded with a coded signal. This signal was received by the *responder* at the ground station, processed and displayed on the radar screen [34].

The *IFF Mark III* was also used aboard ships. The interrogator antenna most commonly consisted of system of four folded dipole radiators, see figure 2.45.

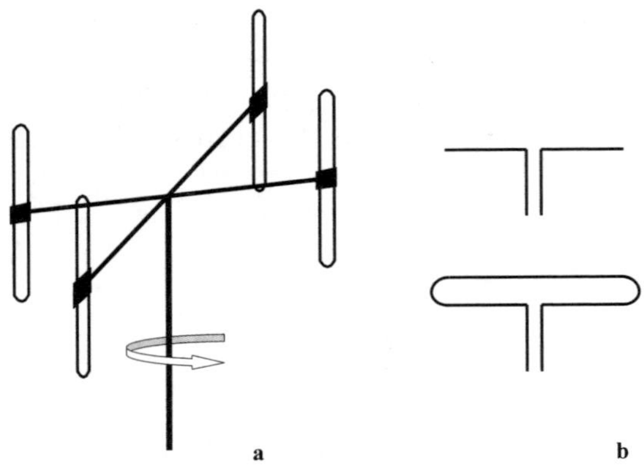

Fig. 2.45 Shipboard IFF antenna. a. Rotatable array of four *folded dipole* radiators. b. A folded dipole antenna evolved from a 'standard' dipole antenna. A folded dipole antenna exhibits an input impedance about four times that of a dipole antenna, making it fit to be connected directly - i.e. without impedance transformer - to a two-wire transmission line.

2.4.2.2 Navigation The start of the Second World War was also the beginning of the air war. The way of determination of position and course of aircraft (bombers) thus far had just been dead reckoning. So an immediate need for more accurate methods existed.

Knickebein In Germany, the Lorenz company had devised a navigation system (also called *Lorenz*) for Lufthansa in 1934. At the outbreak of the war this system had been upgraded into the *Knickebein* (crooked leg) system, named after the shape of the antenna, see figure 2.46.

Again we see an array of dipole radiators. As we have seen in figure 2.34, the array configuration is used to increase the directivity or create a narrow electromagnetic beam. This particular antenna got its crooked shape for a particular reason.

The antenna, which could be rotated around its axis, would alternately transmit *two* beams, see figure 2.46b, one carrying an audio tone representing Morse 'dots', the other one an audio tone representing Morse 'dashes'. An aircraft receiving the Knickebein signal could decide on the basis of the received signal if and how it had to correct its course [35]. When on course, the dots would add up to the dashes, producing a continuous tone. Since the Knickebein antenna could be rotated, the centreline between the two beams could be directed to a desired course. When drifting to the right, see figure 2.46b, dashes would be heard aboard the aircraft and a correction to the left should be made. When drifting to the left, dots would be heard and a correction to the right should be made.

A second transmitter, on a separate location, was used in the Knickebein system. A bomber was keeping its course by listening to the dots and dashes of the first

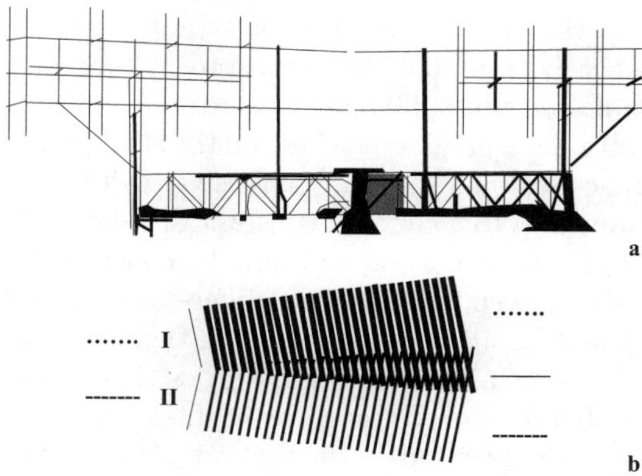

Fig. 2.46 The Knickebein system. a. Knickebein antenna. b. One of the beams is transmitting Morse 'dots', the other one Morse 'dashes'. At the centreline in between the lobes a continuous tone will be received.

transmitter. Upon receiving the signal from the second transmitter, that was also aimed at the target, the target was reached and the bombs would be dropped.

Gee Early in the war, the British had developed the *Gee*[32] navigation system [35]. Gee is a so-called *hyperbolic* navigation system. It operates on the basis of timing the delays between sets of received signals. When a master ground station M transmits a pulse and a slave station A transmits a pulse after a prescribed time interval, a receiver can determine the time difference between the pulse received from M and the one received from A, see figure 2.47.

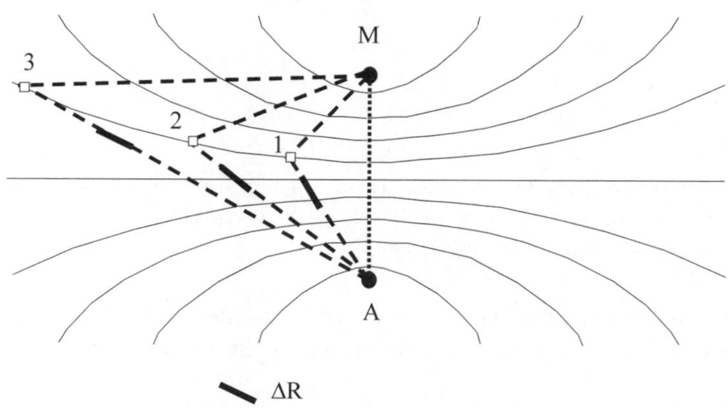

Fig. 2.47 The Gee system. Curves of constant time delay translate linearly to curves of constant distance difference between two points. These curves form hyperbola.

[32]Or *AMES Type 7000*. Gee was short for 'Grid'.

Since the velocity of the electromagnetic waves is finite and constant, the time difference translates linearly to a difference in distance relative to the master and to the slave. If, for example the time difference is zero, the receiver will have the same distance to M as to A. It then will be somewhere on the straight line in between M and A. For any time difference unequal to zero, the receiver will be on a curve of constant time difference that translates to a curve of constant distance difference relative to M and A. These curves form hyperbola. As an example, consider the positions *1*, *2* and *3* in figure 2.47. As can be seen, the distance differences relative to M and A (the thick line pieces) are equal for all three positions on this hyperbola.

To straighten out the remaining uncertainty in position, a second slave is needed on a separate position. If the master transmits a pulse and after a prescribed time, slave A transmits a pulse, the position on the first set of parabola may be determined. If then, after a prescribed time, the master transmits a double pulse and a prescribed time after this double pulse, a slave station B transmits a pulse, the receiver position may be determined on a second set of parabola. The exact position of the receiver is then found where the two parabola intersect, see figure 2.48.

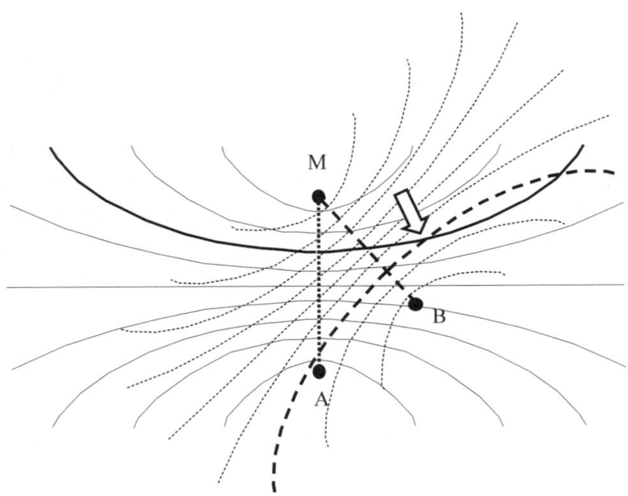

Fig. 2.48 The Gee system. The position relative to M and A is on the fat hyperbola. The position relative to M and B is on the fat dashed hyperbola. The exact position is at the interception point of these two hyperbola.

The *double* pulse of the master station is needed to distinguish between determining the position relative to M and A and determining the position relative to M and B.

X-Gerät The *X-Gerät* was the German derivative of Knickebein [35]. It operated at a higher frequency - 60MHz instead of 30MHz - and could therefore employ a less clumsy, more compact antenna. This antenna did not generate a single small beam but 14 individual, equal strength, small beams. These beams were used for staying on course and, also just as in the Knickebein system, a second beam was used to determine the drop point. Because of the multiple beams of the X-Gerät, the bomber had to pass multiple beam interception points or checkpoints before reaching the correct dropping point.

By 1941, the X-Gerät, like its predecessor, Knickebein, had become useless because of the jamming activities in Great Britain.[33]

Y-Gerät The *Y-Gerät* was the successor of the X-Gerät. In the Y-Gerät the need for a second transmitter, generating a beam to intersect with the 'course beams' to determine the dropping point, was eliminated. The course beams were still present, but now, from the same transmitter location a second signal was transmitted that was reradiated by the equipment aboard the bomber. The phase difference between the signal transmitted by the ground station and the signal reradiated by the bomber increased with the bomber getting further away from the ground station. This phase difference was monitored at the ground station and when the bomber reached its target, the bomber crew was informed so by the ground station over radio. Also early in 1941 this system had become useless due to the British jamming activities.[34]

Oboe *Oboe* or *AMES Type 9000* was developed by the British to improve targeting accuracy. Gee was good enough for getting an estimate of one's position, but lacked the accuracy needed for targeting. The Oboe system was developed on the basis of Chain Home technology, using *transponders* [35]. Two stations in Britain, at separated locations, would transmit a signal to a light bomber (a Mosquito pathfinder). The transponder aboard the bomber would reradiate or reflect the signal and from this reflected signal, picked up again by the stations in Britain, the distance to the bomber would be determined. The bomber then would fly along the circumference of the circle of constant distance defined by one of the stations in Britain, called the *Cat* until it reached the intersection with the circle of constant radius defined by the other station, called the *Mouse*, see figure 2.49, were it was told to drop the flares for marking the target so that heavy bomber formations could sight on these flares.

For the obvious reason of the predictable fly path, the Germans called the system *Boomerang*.

A drawback of the system was that one pair of *Cat* and *Mouse* stations could only direct one plane. This drawback was annihilated with a new scheme wherein the plane carried the transmitter and the ground stations were equipped with transponders [35]. This updated Oboe system was called *Gee-H*.

Decca Another hyperbolic navigation system, meant to replace Gee, was the *QM* or *Decca* navigator. Decca was developed, implemented and maintained by Decca Radio Ltd, a commercial company. In contrast to Gee, Decca used continuous waves rather than pulses, on a lower frequency - 70kHz to 130kHz as opposed to 20MHz to 85MHz - and was less accurate but could be used over longer distances. Like the Y-Gerät, it used phase to determine distance. To be more precise, it used phase differences between signals received from different stations to determine possible positions on hyperbola

[33]The signals were injected with Morse code patterns to degrade them beyond use.

[34]The jamming of the Y-Gerät consisted of receiving the signal and radiating it back at high power. For this purpose, amongst others, the large BBC transmitting station in London, set up for national television broadcasting just before the war, was being used.

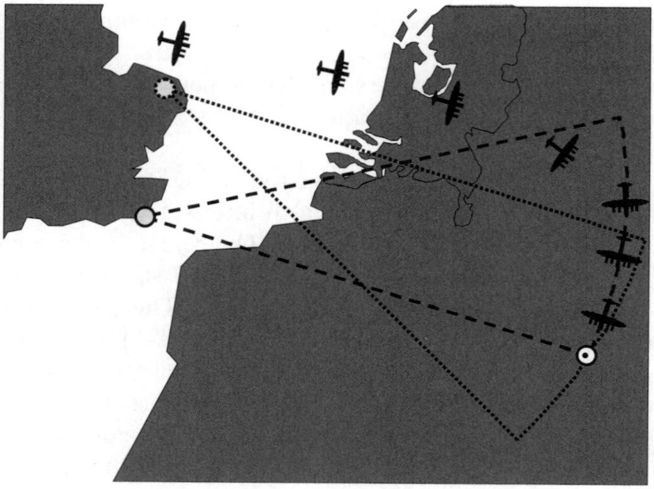

Fig. 2.49 The Oboe system. The *Cat* directs the bomber to the target, the *Mouse* indicates the dropping point.

laid out over a map. Since CW signals are periodic, the same phase relationships repeat with growing distance. Therefore, the Decca receiver needed to keep track of the number of passings through the same phases. The Decca system, which was developed for the British Admiralty was not used until D-day. It was used however till long after the war. It remained in service until the year 2000.

LORAN The LOng RAnge Navigation (*LORAN*) system was the United States of America version of the British Gee system [35]. The development work on Gee had been reported in the USA by the Tizard mission, and LORAN was the result of the work performed on the idea at the Rad Lab.[35] LORAN used longer wavelengths than Gee and therefore had a greater range, at the expense of less accuracy. LORAN was first put in service in 1943 and at the end of the war a navigational coverage of 30% of the earth's surface had been reached. LORAN is still in existence today.

2.4.2.3 Huff Duff One lesson learned from the First World War was the danger that submarines form for merchant vessels. As a result of this threat, the convoy system had been invented that made it very difficult and dangerous for a single submarine to attack a merchant vessel. Therefore, the German Admiral *Dönitz*, in building the new U-boat fleet came up with the *wolf-pack* strategy. In the wolf-pack strategy individual

[35] After the secret Tizard mission in 1940 had brought the cavity magnetron to the USA, the Massachusetts Institute of Technology (MIT) was chosen in that same year to house an independent laboratory staffed by scientists from every discipline to aid in the war effort. The newly formed Radiation Laboratory (Rad Lab) investigated aspects of microwave electronics. Large-scale research was devoted to the development of microwave radar. The Rad Lab closed on 31 December, 1945, but many of the staff members remained for another six months or more to work on the publication of the results of five years of microwave research and development. This resulted in the famous 28 volumes of the Rad Lab series, many of which are still in reprint today.

U-boats were sent to sea in a search pattern to find a convoy. Once a convoy was found, the U-boat crew would transmit a coded message to shore informing about position and course of the convoy and this message would be broadcasted from the shore to the other U-boats. Then the U-boats would gather and make a coordinated attack [36, 37].

The only weak spot in this strategy was the necessary radio communication, which had been chosen to use high frequency (using ionospheric reflection) to minimise the risk of detection. This high frequency (HF) signal was exploited by the allies to detect the U-boats with the *high frequency/direction finding* (HF/DF) equipment. The HF/DF soon got nicknamed *Huff Duff*.

An excellent antenna for direction finding is the loop antenna. The loop antenna is the dual of the dipole antenna. The loop antenna receives a maximum signal when the plane containing the loop is parallel to the direction of wave propagation and it receives a minimum signal (a 'null') when this plane is perpendicular to the direction of wave propagation, see figure 2.50.

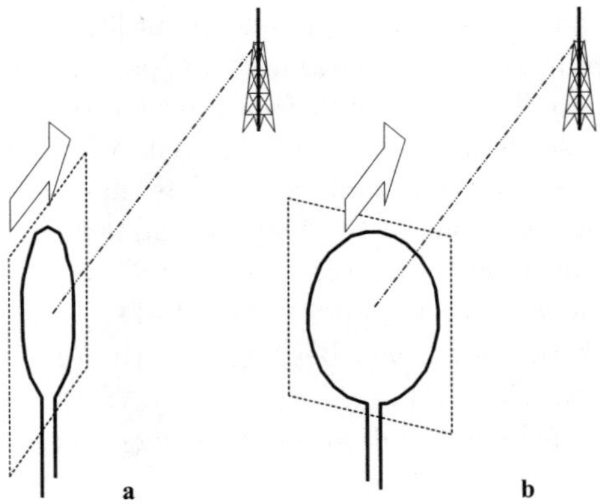

Fig. 2.50 By detecting the maximum (a) or a null (b) in the received signal, while rotating a loop, the direction of a transmitter may be found.

However, rotating the antenna takes time and requires a high mechanical precision for reading the antenna position. This may be overcome by using two antennas, positioned perpendicular to one each other, connecting the terminals to coils which are also placed perpendicular to one another and then rotating a third coil in the magnetic fields of these two coils to obtain a direction by either looking for a null or a maximum.[36] This configuration is know as a *radio goniometer*, [36], see figure 2.51.

The magnetic fields of the coils are linearly related to the received electromagnetic fields by the loops. Thus by comparing the amplitudes and knowing that the fields are perpendicular, the bearing may be found by vector addition, see also figure 2.51. Of course this setup only gives the line bearing, the actual transmitter can be on either direction along this line bearing.

This ambiguity may be solved by adding a omnidirectional reference antenna with unambiguous phase relationship in between the two antennas. In figure 2.52 this

[36]In general a null is easier to detect than a maximum.

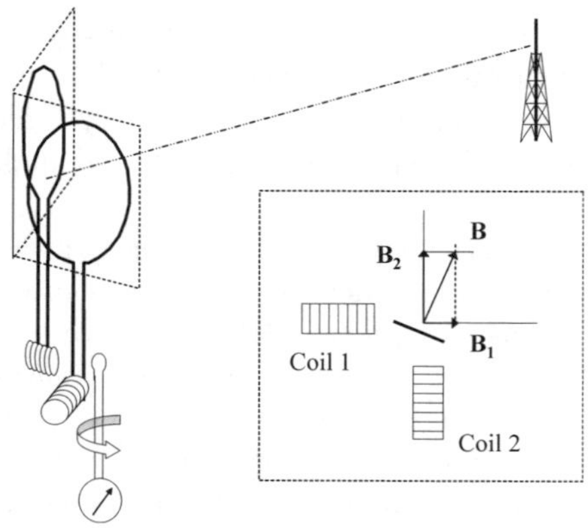

Fig. 2.51 Determination of the line bearing using a goniometer.

setup is in what is called a Watson-Watt arrangement [38]. The sensitivity of the loop antenna is at maximum in the plane containing the loop and at minimum in the plane perpendicular to the loop. The sensitivity as a function of azimuth angle is therefore in the form of an 'eight', see figure 2.52. The sensitivity of the omnidirectional antenna as a function of the azimuth angle is constant and is therefore represented by a circle around its axis. The output of one of the loop antennas is now - in the Watson-Watt arrangement - fed to the x-deflection plates of a CRT and the output of the other loop antenna is fed to the y-deflection plates of the CRT. The omnidirectional antenna output is used to blank the reciprocal bearing and the direction of the transmitter may be read directly from the CRT.

Other configurations of the huff duff antenna, using different radiators exist as well.

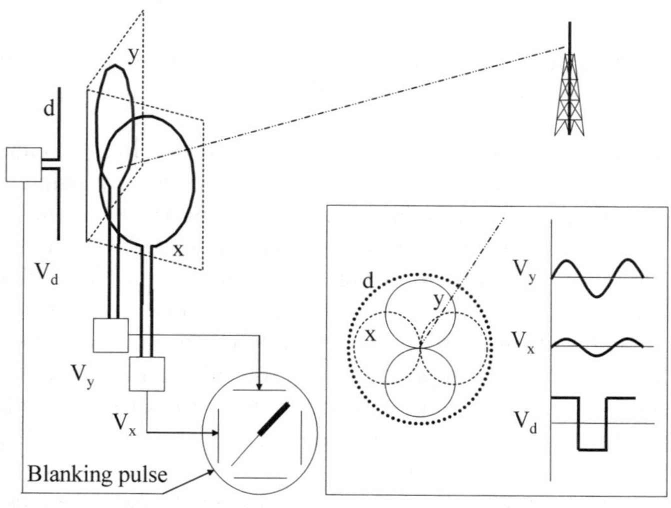

Fig. 2.52 Huff duff antenna arrangement with Watson-Watt instantaneous CRT display.

2.4.2.4 Examples A nice opportunity to see an important range of antenna developments during the Second World War and shortly thereafter is offered by a visit to the London warship HMS Belfast. She may be found in the Thames in front of the Tower Bridge, see figure 2.53.

Fig. 2.53 HMS Belfast.

The cruiser *HMS Belfast* was launched in March 1938 and served throughout the Second World War, playing a leading part in the destruction of the German battle cruiser *Scharnhorst* at the Battle of North Cape and in the Normandy Landings. After the war, she supported United Nations forces in Korea and remained in service with the Royal Navy until 1965. In 1971 she was saved as a unique and historic reminder of Britain's naval heritage in the first half of the twentieth century.

To see these antenna developments at a single glance, we take a look at the foremast of the ship, see figure 2.54.

We have numbered the different antennas we find in the foremast:

1 *VHF Communication Aerial.* This antenna takes the form of a half wave dipole.

2 *Target Indicating Radar Antenna.* This antenna is a so-called *pillbox* or *cheese* antenna. A pillbox antenna belongs to the class of parabolic cylindrical reflectors [13, 38]. The feed of this antenna may be a dipole antenna or a horn antenna. The focal point of the reflector is usually found in the centre of the aperture.

We have seen in discussing the arraying of dipole radiators that adding more dipoles to an array led to a narrower antenna beam. The relationship is in fact broader than that: the larger the antenna aperture, the smaller the antenna beam will be and even more specifically: the larger an aperture dimension, the

Fig. 2.54 The foremast of HMS Belfast.

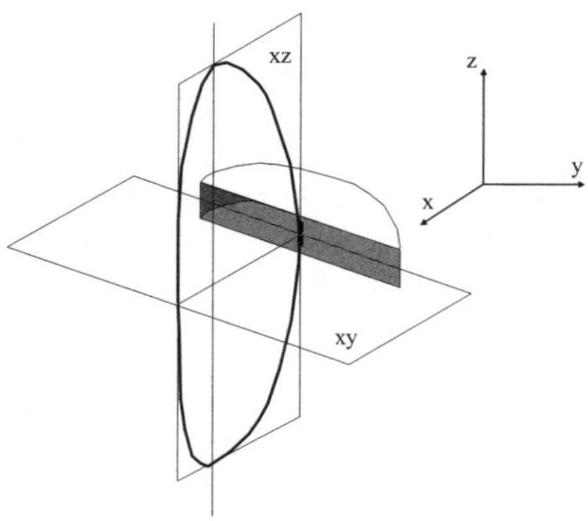

Fig. 2.55 The beam of a pillbox antenna is narrow in the plane along the long dimension of the aperture and broad in the plane along the small dimension of the aperture.

smaller the antenna beam will be in the plane along this dimension. This is shown in figure 2.55. We call this type of antenna beam for obvious reasons a *fan beam*.

When the pillbox antenna is rotated along the z-axis, see figure 2.55, a very accurate azimuth position of a target may be obtained. An accurate elevation position needs to be determined by other means.

3 *Navigational Radar Antenna.* This antenna is also of the pillbox type and will be rotated in azimuth. For navigational purposes, only azimuth and distance information is important. Therefore a wide beam in elevation, combined with the small beam in azimuth will be useful. This antenna needs only be rotated in azimuth (and not in elevation) to obtain all relevant information.

4 *Height Finding and Search Radar Antenna.* This is a doubly curved, paraboloidal type of reflector antenna. In general, the feed antenna (dipole or horn) is placed in the focal point, see also figure 2.37. The antenna beam is rotationally symmetrical and provided the aperture is large enough, this beam may be very small, see figure 2.56. We call this type of antenna beam, again for obvious reasons, a *pencil beam.*

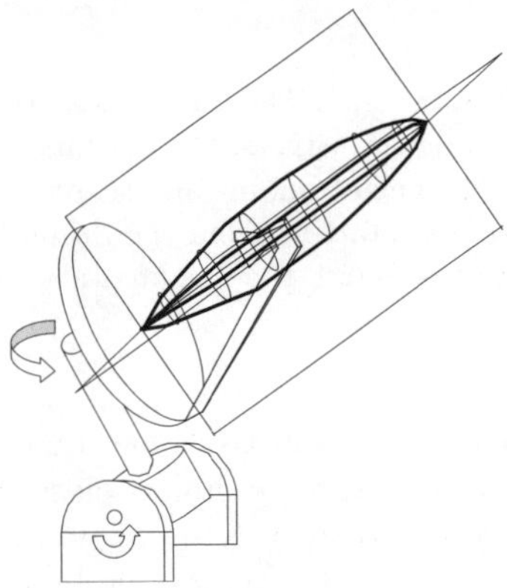

Fig. 2.56 The beam of a paraboloid reflector antenna is narrow in the planes along every aperture dimension.

In figure 2.56, we have shaded the feed and feed support, for reasons of simplifying the figure. In practise, feed and support blocking may seriously degrade the antenna functioning and therefore often an offset feed is being used.

Due to the narrow beam, this type of antenna may be used in a radar system for obtaining both accurate azimuth and elevation information of a target. Searching the whole hemisphere by rotating the antenna in azimuth and elevation will take too much time. Therefore, a fan beam antenna will be employed for target indication (see 2) and when a target position is found in azimuth, a height finding and search radar, using a pencil beam antenna, will be employed to get accurate angular position information and follow or track the target, see figure 2.57.

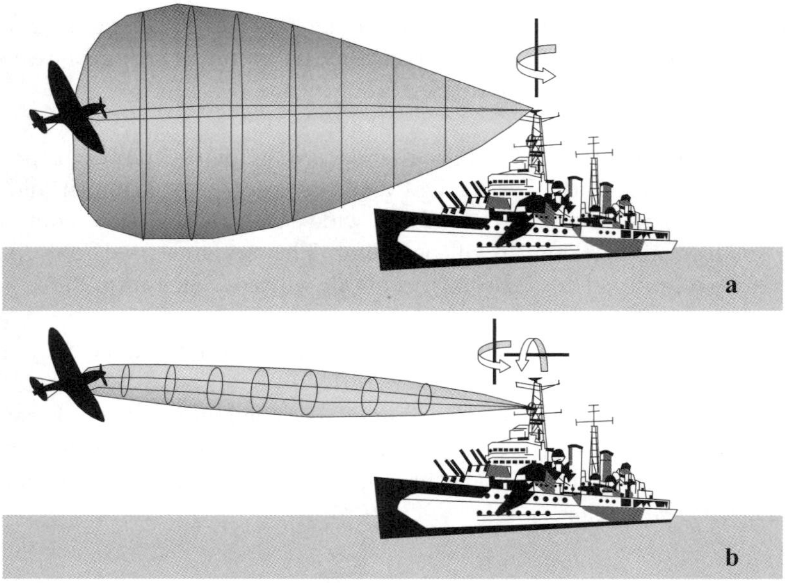

Fig. 2.57 Target searching and tracking. a. The target indication radar, employing a fan beam antenna, is rotated in azimuth to detect targets. Once a target is detected, the height finding and search radar takes over. b. The height finding and search radar, employing a pencil beam, is employed for searching the target, starting from the azimuth position found by the target indication radar. When the target is found, the height is also determined and the target may now be tracked.

5 *UHF Communication Aerial.* Low-gain UHF antennas may be realised in discone-monopole form. The length of the antenna is close to a quarter of a wavelength, the infinite ground is replaced by a finite size disk and the monopole (half a dipole) is cone-shaped to widen the monopole bandwidth.

Other antennas developed during the second world war worth mentioning are slot antennas and arrays of slots and metallic and dielectric lens antennas. Examples of these antennas are shown in figure 2.58.

The functioning of lens antennas may be explained by using optics, just as we used optics to explain the functioning of (parabolic) reflector antennas. By virtue of *Babinet's principle* [40], slots may be seen as the dual of metal strip antennas, just like the loop antenna is the dual of the dipole antenna.

2.5 POST-WAR ANTENNA DEVELOPMENTS

Although it is not the purpose of this book to give an exhaustive overview of all possible kinds of antennas - there are much better books dealing with that subject, for example [38] - it is the author's belief that at least commonly encountered antenna types should be briefly introduced in this introductory chapter.

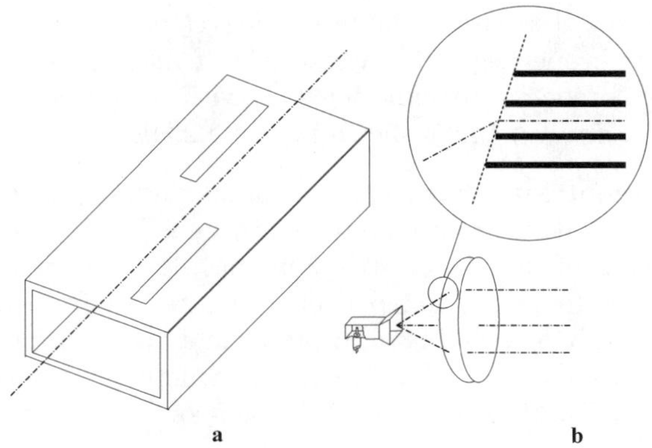

Fig. 2.58 Slot and lens antennas. a. Two slots cut in the broad wall of a rectangular waveguide. Also slots in the small side are possible. b. Lens antenna, fed by a pyramidal horn antenna. The lens may be dielectric or made out of parallel metal plates as shown in the inset.

Most antenna types in use today have seen their origin in the late nineteenth century or were developed during the Second World War. The monopole antenna we use for our car stereo has been in use since the early experiments of Marconi, the parabolic dish reflector antenna we use for the reception of satellite television signals has seen its origin in Hertz's experiments and has been re-introduced in radar systems during the Second World War. The antennas in our cellular phones may be traced back to the monopole of Marconi and the antennas that may be on top of our roofs for the reception of TV broadcast signals are in some form or other consisting of a feed antenna and a reflector, like the many reflector antenna configurations developed during the war.

After the war, a number of completely new antenna configurations have been developed that are not directly derived from war or pre-war antennas. In the following, we briefly describe a choice of these antennas.

2.5.1 Frequency Independent Antennas

At the end of the 1950s, three principal ideas concerning the creation of *frequency independent* antennas emerged. With the term *frequency independent* is meant that the antenna pattern and impedance remain constant over a relatively wide (10:1) frequency bandwidth [38, 39]. These three principal ideas are:

- **Angular condition** When an antenna is satisfying the *angular condition*, its form is completely determined by angles and not by any dimension. It means that a basic antenna structure may be transformed into an identical structure, except for a possible rotation. Examples of such an antenna are the *conical antenna*, see figure 2.59a and the *planar equiangular antenna*, see figure 2.59b. These two antennas may be thought to consist of an infinite number of resonant antennas, each resonant at a single frequency. Antennas possessing the angular condition are also referred to as *continuously scaled structures* [38].

- **Log-periodicity** If a basic structure is *discretely* scaled by a factor $\frac{1}{\tau}$, it will show the same properties at the frequencies f and τf. Therefore the antenna characteristics become a periodic function with period $|\log(\tau)|$. An example of such a *log-periodic antenna* is shown in figure 2.59c.
- **Self-complementarity** When a planar conducting strip antenna is joined by its complementary or dual structure (created by interchanging conducting and non-conducting parts of the planar strip antenna), the impedance of the complete structure will be frequency independent. This results from Babinet's principle [40], from which it follows that the product of the impedance of the original structure and that of the complementary structure is constant, independent from frequency [38, 39]. An example of a self-complementary antenna is shown in figure 2.59b.

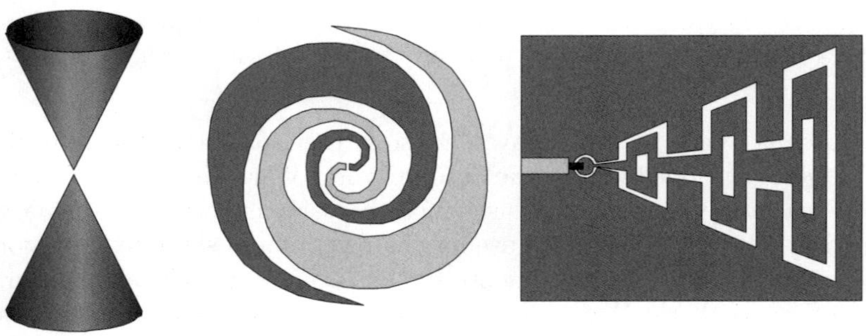

a b c

Fig. 2.59 Frequency independent antennas. a. Biconical dipole antenna. b. Equiangular complementary (strip-slot) spiral antenna. c. Slot log-periodic antenna.

2.5.2 Helical Antenna

The helical antenna or *helix* emerged from the helical wave-guiding structure used in *travelling wave tubes* (high-frequency (radar) sources). In 1946, *John Kraus* (1910–2004) made the first helical antenna in his cellar [40] and showed a high gain (narrow), circularly polarised beam.[37]

The helical antenna, see figure 2.60, may be regarded as the general case of the linear antenna and the loop antenna, both linear antenna and loop being special cases of the helical antenna [40]. The antenna may be operated in three modes, called respectively *normal mode*, *axial mode* and *conical mode*, referring to the shape of the radiation pattern, see figure 2.60.

[37]Polarisation describes the position of the electric field vector in time. An electric field vector changing with time along a line is called a *linearly polarised* electric field. An electric field vector describing a circle in time is called a *circularly polarised* electric field. Polarisation will be treated in more detail in the next chapter.

- **Normal mode** In normal mode, the radiation reaches a maximum in directions *normal* to the helix axis, see figure 2.60a. This mode occurs for helix diameters being small with respect to the wavelength [38–40].

- **Axial mode** In axial mode, see figure 2.60b, the radiation is at maximum in the *axial* direction of the helix. This mode occurs when the helix circumference is of the order of one wavelength [38–40]. The axial mode is the most commonly used mode.

- **Conical mode** The conical mode, see figure 2.60, will be obtained for helix dimensions exceeding those needed for axial mode operation [38, 40]. These dimensions will result in higher-order radiation modes that are characterised by multilobed or conical radiation patterns.

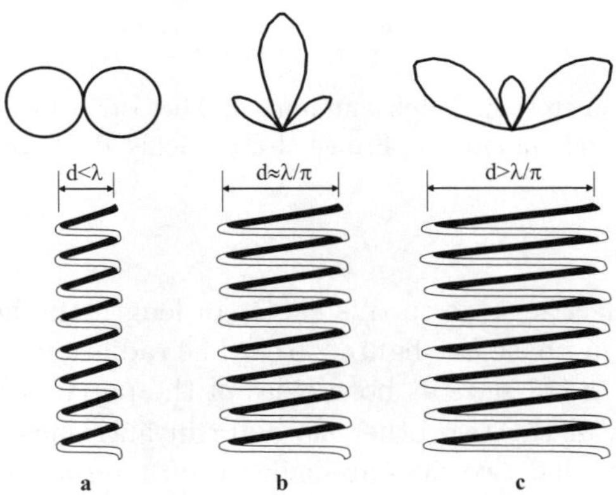

Fig. 2.60 Helical antenna. a. Normal mode. b. Axial mode. c. Conical mode.

2.5.3 Microstrip Patch Antenna

Microstrip patch antennas have evolved from microstrip transmission line technology in the 1950s. In the 1970s, the need for low profile, conformal (i.e. curved) antennas led to extensive research in the field of these printed antennas [41].

Microstrip patch antennas are constructed using printed circuit board (PCB) technology. The basic rectangular microstrip patch antenna consists of a thin rectangular electric conductor of length L and width W on top of a grounded dielectric slab of height h and relative permittivity (dielectric constant) ε_r, see figure 2.61a,b.

The microstrip patch may be considered as a cavity with electrical conducting top and bottom and magnetically conducting sidewalls. The electrical conducting top and bottom force the electric field lines to be perpendicular to them, the magnetically conducting side walls force the magnetic field lines to be perpendicular to them.

Since the walls are not *perfectly* conducting, the fields will fringe at the edges of the patch and these so-called *fringe fields*, see figure 2.61b, are responsible for the radiation. The rectangular microstrip antenna length L is chosen to be a half

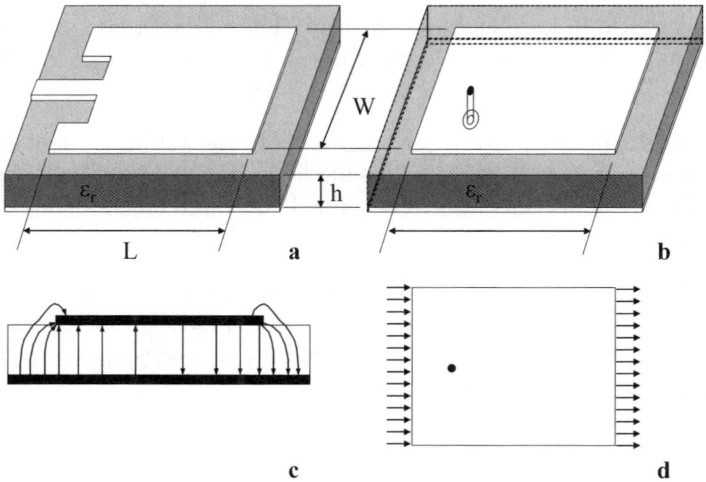

Fig. 2.61 Rectangular microstrip patch antenna. a. Microstrip line excited patch antenna. b. Probe (via) excited patch antenna. c. Fringe electric fields. d. Radiating slots.

wavelength in the dielectric. Due to this choice of length the horizontal components of the fringe fields are in phase, see figure 2.61d. The radiation may be thought of then to originate from two small slots at both sides of the patch, separated a distance L. The field components of the two other slots are in anti-phase and thus cancel one another. Therefore, the first two slots are called the *radiating slots* and the latter two are called the *non-radiating slots.*

The fields inside the 'cavity' are excited by means of a microstrip transmission line, see figure 2.61a, by means of a coaxial probe or via, see figure 2.61b, or by means of a slot in the ground plane that in turn is excited by a microstrip transmission line. The input impedance of the microstrip patch antenna may be tuned by varying the feed position. The width W should be smaller than one and a half wavelength in the dielectric material to avoid the excitation of higher order modes [38]. Besides the rectangular microstrip patch antenna, many other microstrip antennas exist, see for example [42].

2.5.4 Phased Array Antenna

Although not really an antenna type in the sense of frequency independent antennas, helix antennas and microstrip patch antennas but rather being an antenna system, it is appropriate to briefly discuss the *phased array antenna* at this point.

We have already seen that a half-wave dipole antenna may be represented by two isotropic radiators, spaced apart by half a wavelength. An isotropic radiator radiates equally in all directions. By virtue of the non-zero spacing of the isotropic radiators, the radiated fields of the isotropic radiators will possess a phase difference, dependent on the direction of radiation. This results in a direction-dependent constructive or

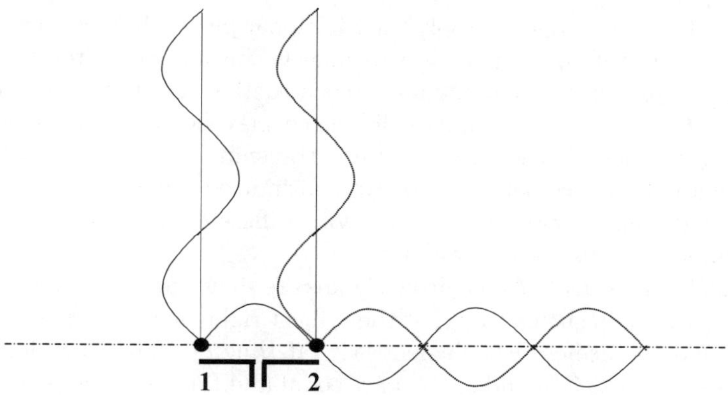

Fig. 2.62 Half-wave dipole antenna radiation represented as the radiation of two isotropic radiators spaced apart by a half wavelength.

destructive interference. This phenomenon is shown in figure 2.62 for the direction along the dipole axis and the direction perpendicular to the dipole axis.

We see that in the direction along the dipole axis, the radiated fields are 180 degrees out of phase and therefore cancel one another. A half-wave dipole antenna does *not* radiate in the directions along the dipole axis. The phase difference of 180 degrees is the result of the half-wavelength spacing of the two isotropic radiators.

In the direction perpendicular to the dipole axis, the fields radiated by isotropic radiators one and two are in phase and therefore add. This means that the half-wave dipole radiation will be at maximum for directions perpendicular to the dipole axis. The zero degrees phase difference is the result of the fact that in the direction perpendicular to the dipole axis, the two isotropic radiators do not possess any spatial difference.

For all other directions the radiated field amplitudes will be in between zero and maximum.

The physical reasoning we applied in the above for two isotropic radiators applies equally well for more than two isotropic radiators. It also is not restricted to isotropic radiators, but may be equally well applied to clusters of isotropic radiators, meaning non-isotropic radiators. We have already shown this in figure 2.34 for a linear array consisting of four half-wave dipole radiators.

By grouping radiators into an array we were able to increase the directivity, i.e. narrow the angular region wherein energy is radiated. If we want to point the thus formed antenna beam, we need to rotate the whole array around one of its axes, just as we need to rotate the parabolic reflector antenna of a search and track radar system.

However, we have seen that the principle of an array antenna is based on the fact that the spacing of the elements causes phase differences between the waves transmitted by the individual radiators. Due to these phase differences, in some directions constructive interference occurs and in other directions destructive interference occurs.

Therefore, if we were able to apply an additional phase shift between radiators - added to the one caused by the non-zero distance between the radiators - we should be able to influence the positions of maximum constructive and destructive interference. From figure 2.62 we see that a phase difference may be regarded as a time delay between the radiators. If one radiator starts transmitting a while after a previous radiator, a phase difference between the two transmitted signals results. A time delay between two radiators may be created by having different lengths of transmission line from a common source to various radiators.

In figure 2.63 the separate transmitted signals are shown for four isotropic radiators. The radiators are fed in such a way that the utmost right element starts radiating first (at time t_1) and the utmost left element last (at time t_4). The time delays between the radiators ($t_2 - t_1$, $t_3 - t_2$ and $t_4 - t_3$) are equal and therefore the phase differences between the subsequent radiators are equal too.[38]

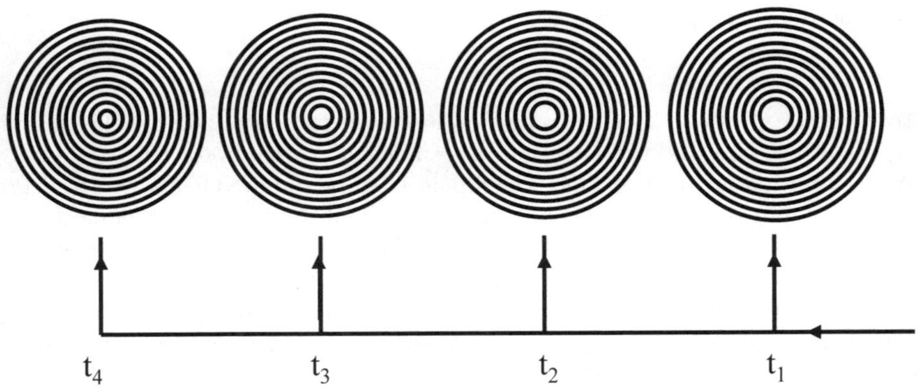

Fig. 2.63 Phase differences created between four isotropic radiators.

By adding this *phase taper* over the elements of the linear array, we now have created a *phased array antenna* consisting of four isotropic radiators. The combined radiation of the four isotropic radiators with phase/time differences applied is shown in figure 2.64.

We see that the chosen phase/time delay taper has led to a radiation that has moved from a direction perpendicular to the line connecting the four radiators, to a direction that is tilted with respect to this former direction. The 'tilt' is dependent on the applied phase taper (differences) over the array elements. By changing the phase taper we can change the beam position, without moving the array antenna itself!

The grouping of radiators into a (fixed beam) array to increase the directivity had already been used in the First World War and in between the two World Wars. We have seen that a lot of the World War II radar systems employed array antennas.

Phased array antennas, i.e. steered beam array antennas, have been employed in the Second World War by the British, Americans and Germans [28]. The phase/time

[38]By looking at the first 'black ring' from the centre, for every element, we see that indeed every element has started to radiate before any of the elements to its left.

Fig. 2.64 Radiation of the four isotropic radiators, time-delay fed as shown in figure 2.63.

delay in these early phased array antennas was accomplished by (electro)mechanical means.[39]

By changing the direction of radiation without moving the (often voluminous and heavy) antenna, a greater flexibility in employing radar has been accomplished. Beams can be steered from one position in space to another much faster than by mechanical means and therefore it has become possible not only to track targets easier but also the detection and tracking of multiple targets has become a possibility.

In the 1950s the mechanically operated phase shifters were replaced by electronic phase shifters. This major advance in phased array antenna technology increased switching and steering speed and thereby flexibility. In the 1960s, digitally switched phase shifters were introduced, increasing the flexibility even further, making it possible to electronically steer the antenna beam, controlled by a digital computer. This step forward made it relatively easy to steer the array antenna beam into *two* orthogonal angular coordinates [28].

The most recent step forward in phased array antenna technology is the introduction of the *active* phased array antenna in the 1980s. The basic idea behind the active phased array antenna is to get rid of the often complicated, bulky and lossy feed network behind the radiators. This feed network, see figure 2.63, is needed to deliver the energy from a common source in a prescribed manner to all radiators. In an active phased array antenna, each radiator is equipped with its own transmitter and only digital steering signals and power supply lines are distributed to all these

[39]One can think about changing the length of a transmission line by mechanically moving a telescopic section, often in the form of a 'U', just like the movable part of a slide trombone [28].

transmitters. An RF feed network that was subject to severe attenuation has become obsolete by the introduction of the active phased array antenna.

With the advance of analog to digital (A/D) converter technology it is not even necessary anymore to determine and fix phased array antenna beam shape during the array antenna development. For an active receive phased array antenna (an array antenna where every element is equipped with its own receiver) we can connect every receiver to an A/D converter. The digital signals may be connected to a digital bus and then be processed by a digital computer to form any beam or multiple beams with prescribed shape(s) and direction(s) [43].

REFERENCES

1. John D. Kraus, Heinrich Hertz - Theorist and Experimenter, *IEEE Transactions on Microwave Theory and Techniques*, Vol. 36, No. 5, May 1988, pp. 824–829.

2. D.J. Cichon and W. Wiesbeck, The Heinrich Hertz Wireless Experiments at Karlsruhe in the View of Modern Communication, *Proceedings IEE International Conference on 100 Years of Radio*, 5–7 September 1995, pp. 1–6.

3. Helmut V. Friedburg, Heinrich Hertz at Work in Karlsruhe, *Proceedings IEEE-MTT-S International Microwave Symposium*, 25–27 May 1988, pp. 267–270.

4. Giancarlo Masini, *Marconi*, Marsilio Publishers, New York, 1995.

5. John Hamilton (ed.), *They Made Our World, Five Centuries of Great Scientists and Inventors*, Broadside Books, London, 1990, pp. 125–132.

6. A.D. Olver, Trends in Antenna Design over 100 Years, *100 Years of Radio*, 5–7 September 1995, Conference Publication 411, IEE, 1995, pp. 83–88.

7. Guglielmo Marconi, Wireless Telegraphic Communications, *Nobel Lectures in Physics*, 1901–21, Elsevier, 1967.

8. Gian Carlo Corazza, Marconi's History, *Proceedings of the IEEE*, Vol. 86, No. 7, July 1998, pp. 1307–1311.

9. Giuseppe Pelosi, Stefano Selleri and Barbara Valotti, Antennae, *IEEE Antennas and Propagation Magazine*, Vol. 42, No. 1, February 2000, pp. 61–63.

10. Guglielmo Marconi, The Most Recent Developments of Wireless Telegraphy, *De Ingenieur*, Nr. 22, May 1909, pp. 431–439.

11. B.A. Austin, Wireless in the Boer War, *100 Years of Radio*, 5–7 September 1995, Conference Publication 411, IEE, 1995, pp. 44–50.

12. R.W. Simons, Guglielmo Marconi and Early Systems of Wireless Communication, *GEC Review*, Vol. 11, No. 1, 1996, pp. 37–55.

13. Jack Ramsay, Highlights of Antenna History, *IEEE Communications Magazine*, September 1981, pp. 4–16.

14. John S. Belrose, Fessenden and Marconi: Their Differing Technologies and Transatlantic Experiments During the First Decade of this Century, *International Conference on 100 Years of Radio*, 5–7 September 1995.

15. Constantine A. Balanis, Antenna Theory: A Review, *Proceedings of the IEEE*, Vol. 80, No. 1, January 1992, pp. 7–23.

16. A.G.P. Boswell, Guglielmo Marconi's Antenna Design, 10th *International Conference on Antennas and Propagation ICAP*, 14–17 April 1997, pp. 546–549.

17. Ralph Stein, *The Great Inventions*, The Ridge Press, 1976, pp. 116–130.

18. John S. Belrose, Reginald Aubrey Fessenden and the Birth of Wireless Telephony, *IEEE Antennas and Propagation Magazine*, Vol. 44, No. 2, April 2002, pp. 38–47.

19. Albert Kloss, *Von der Electricität zur Elektrizität* , Birkhäuser Verlag, Basel, 1987 (in German).

20. Harold Sobol, Microwave Communications - An Historical Perspective, *IEEE Transactions on Microwave Theory and Techniques*', Vol. MTT-32, No. 9, September 1984, pp. 1170–1181.

21. John D. Kraus, *Antennas*, McGraw-Hill, New York, 1950.

22. John Keegan, *The First World War, An Illustrated History*, Pimlico, London, 2002.

23. Edward A. Wilson, *Electronic Communications Technology*, Prentice Hall, Englewood Cliffs, New Jersey, 1989.

24. David M. Pozar, *Microwave Engineering*, John Wiley & Sons, New York, 1998.

25. William E. Gordon, A Hundred Years of Radio Propagation, *IEEE Transactions on Antennas and Propagation*, Vol. AP-33, No. 2, February 1985, pp. 126–130.

26. William F. Crosswell, Some Aspects of the Genesis of Radio Engineering, *IEEE Antennas and Propagation Magazine*, Vol. 35, No. 6, December 1993, pp. 29–33.

27. Guglielmo Marconi, Radio Communication by Means of Very Short Electric Waves, *IRE Transactions on Antennas and Propagation*, January 1957, pp. 90–99. Reprinted from *Proceedings of the Royal Institution of Great Britain*, Vol. 22, 1933, pp. 509–544.

28. Merill I. Skolnik, *Introduction to Radar Systems*, McGraw-Hill, Auckland, 1981.

29. R.J. James, A History of Radar, *IEE Review*, October 1989, pp. 343–349.

30. R. Schroer, Airborne Radar, *IEEE AES Systems Magazine*, Vol. 18, No. 7, July 2003, pp. 43–48.

31. Len Deighton, *Fighter, The True Story of the Battle of Britain*, Dutch translation, Elsevier, Amsterdam, 1983.

32. Robert E. Collin, *Foundations for Microwave Engineering*, McGraw-Hill, New York, 1992.

33. Navy Department, Office of the Chief of Naval Operations, Washington, Report on the Interrogation of Survivors from U-172 Sunk 13 December 1943, *Final Report* - G/Serial 29.

34. The Story of IFF (Identification Friend or Foe), *IEE Proceedings Pt.A*, Vol. 132, No. 6, October 1986.

35. R. Schroer, Navigation & Landing, *IEEE AES Systems Magazine*, Vol. 18, No. 7, July 2003, pp. 27–36.

36. P.G. Redgement, Chasing Wolves, An Engineer's View of the Battle of the Atlantic, *IEE Review*, January 1995, pp. 35–37.

37. R.W. Burns, Impact of Technology on the Defeat of the U-Boat, September 1939 - May 1943, *IEE Proceedings Sci. Meas. Technol.*, Vol. 141, No. 5, September 1994, pp. 343–355.

38. R.C. Johnson (ed.) *Antenna Engineering Handbook, third edition*, McGraw-Hill, New York, 1993.

39. Warren L. Stutzman and Gary A. Thiele, *Antenna Theory and Design*, John Wiley & Sons, New York, 1998.

40. John D. Kraus, Antennas Since Hertz and Marconi, *IEEE Transactions on Antennas and Propagation*, Vol. AP-33, No. 2, February 1985, pp. 131–137.

41. R.E. Munson, Conformal Microstrip Antennas and Microstrip Phased Arrays, *IEEE Transactions on Antennas and Propagation*, Vol. AP-22, No. 1, January 1974, pp. 74–78.

42. J.R. James (ed.) and P.S. Hall (ed.), *Handbook of Microstrip Antennas*, IEE Electromagnetic Waves Series, 28, 1998.

43. R.C. Hansen, *Phased Array Antennas*, John Wiley & Sons, New York, 1998.

3

Antenna Parameters

Now that we have developed a basic understanding of the operation of antennas, based on physical reasoning, it is time to quantify and interrelate the different antenna parameters that describe the antenna functioning. Therefore, in this chapter we will introduce mathematical expressions that describe and interrelate these parameters. Instead of mathematically deriving these expressions, which is beyond the scope of this book, we will limit ourselves to simply posing these relations and explaining the use of them. The mathematical background needed for understanding and working with the expressions posed in this chapter is on a level as educated in secondary school. This chapter will finish the development of the basic antenna knowledge needed for understanding the basics of array and phased array antennas.

3.1 RADIATION PATTERN

In the previous chapters we have seen that the most basic antenna, i.e. a short dipole antenna, may be considered as a radiator consisting of two isotropic radiators.

An isotropic radiator is a (hypothetical) radiator that radiates equally in all directions.

Due to the fact that even the most basic antenna consists of more than one isotropic radiator, placed apart by a non-zero distance, the radiation will no longer be equally distributed over all directions. The non-zero distance between the isotropic radiators - that may be thought of as making up the antenna - cause phase differences in the fields radiated into the different directions. These phase differences will cause constructive interference in some directions and destructive interference in other directions.

Array and Phased Array Antenna Basics Hubregt J. Visser

3.1.1 Field Regions

When talking about radiated fields, we have to take into account the distance relative to the antenna where these fields are evaluated. We have seen in the previous chapters, that close to the antenna, a region exists where energy is stored and returned to the antenna. This region is called the *reactive near-field region* of the antenna [1, 2], see figure 3.1. The reactive near-field region is indicated in this figure by *I*.

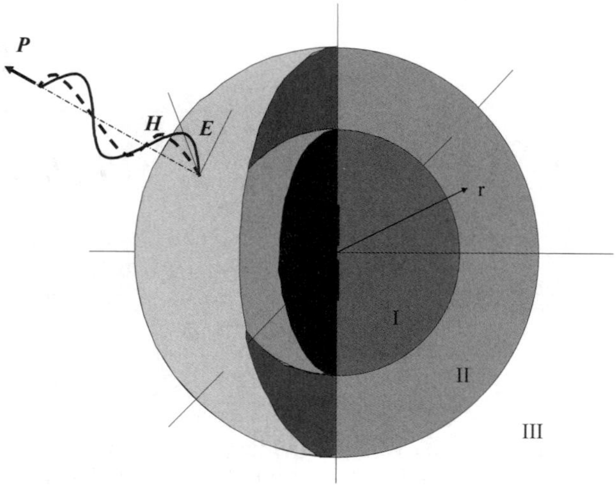

Fig. 3.1 Field regions. I. Reactive near-field region, II. Radiating near-field region. III. Far-field region.

Moving away from the antenna, through the reactive near-field region, the next region encountered is called the *radiating near-field region* or *Fresnel region* [1, 2]. The radiating near-field region is characterised by the fact that the radiation fields dominate the reactive fields and that the angular distribution of this radiated field is dependent on the distance from the antenna. This distance is $|\mathbf{r}|$, the length of vector $\mathbf{r}$, see figure 3.1.

The angular distribution may be described by the spherical coordinate angles ϑ and φ, see figure 3.2.

The angle ϑ is the angle between the positive z-axis and the observation direction. The angle φ is the angle between the positive x-axis and the perpendicular projection of the observation direction on the x,y-plane. The interrelation between the spherical coordinates r, ϑ, φ and the rectangular coordinates x, y, z, are given by, see also figure 3.2

$$\begin{aligned} x &= r\sin(\vartheta)\cos(\varphi), \\ y &= r\sin(\vartheta)\sin(\varphi), \\ z &= r\cos(\vartheta). \end{aligned} \tag{3.1}$$

The radiating near-field region is indicated in figure 3.1 by *II*.

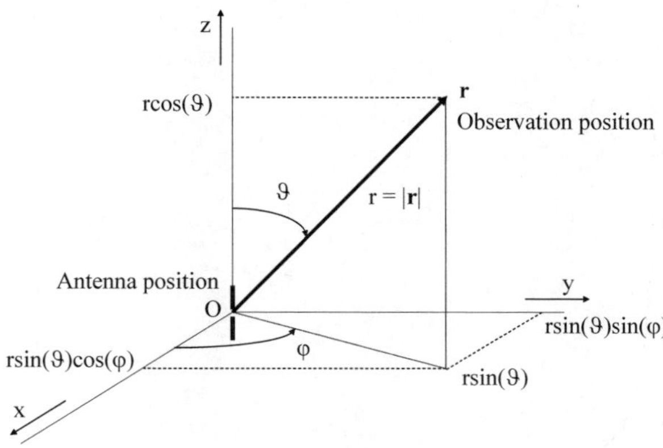

Fig. 3.2 An antenna in a rectangular coordinate system and the relation to spherical coordinates.

The *far-field region*, indicated by *III* in figure 3.1, is that region where not only the radiating fields predominate, but where also the angular field distribution has become independent from the distance to the antenna [1, 2]. In the far-field region, the electric field vector **E** and the magnetic field vector **H** are perpendicular to the observation direction **r** and to each other, see figure 3.1. Power flow in the far-field region is therefore only in the direction of **r**, in contrast to the situation in both near-field regions.

The radiated fields of an antenna will be evaluated in the far-field region. In real-life situations, the separation between transmitting antenna and receiving antenna will (almost) always be such that the antennas are in each other's far-field regions.

Although we have indicated sharp boundaries between the reactive-near field region, the radiating near-field region (together called the *near-field region*) and the far-field region in figure 3.1, in practice these boundaries are not that sharp and precisely determined.

The reactive near-field components attenuate, going away from the antenna, with a distance-dependency of $\frac{1}{r^3}$. The radiating near-field components attenuate, going away from the antenna, with a distance-dependency of $\frac{1}{r^2}$ and, finally, the far-field components attenuate with a distance-dependency of $\frac{1}{r}$. If we plot the functions $\frac{1}{r^3}$, $\frac{1}{r^2}$ and $\frac{1}{r}$ as function of r, we see that we can distinguish the areas where, respectively, the near-field components and the far-field components predominate, see figure 3.3 and figure 3.4. The figures also show that the boundary between the near- and far-field regions are not well-defined. Strictly speaking, the far-field only exists at distances r approaching infinity.

For most antennas, the boundaries are empirically established as a function of the antenna dimensions and the used wavelength. The reactive near-field (rnf) region is commonly taken to be [1]

$$r_{rnf} < 0.62\sqrt{\frac{D^3}{\lambda}}, \tag{3.2}$$

where D is the largest dimension of the antenna and λ is the used wavelength.

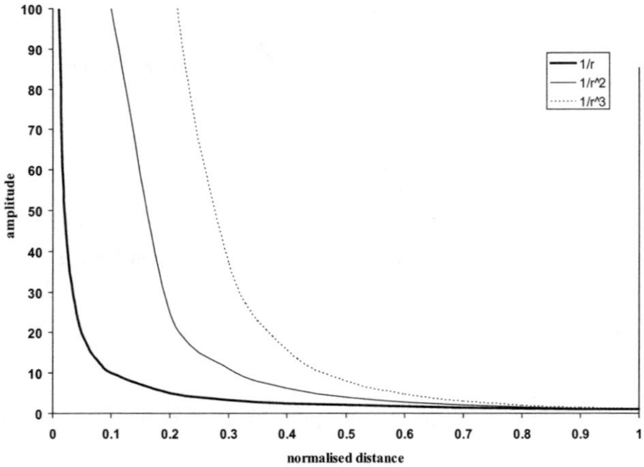

Fig. 3.3 Functions $\frac{1}{r^3}$, $\frac{1}{r^2}$ and $\frac{1}{r}$ as function of r, for $0.0 \leq r \leq 1.0$.

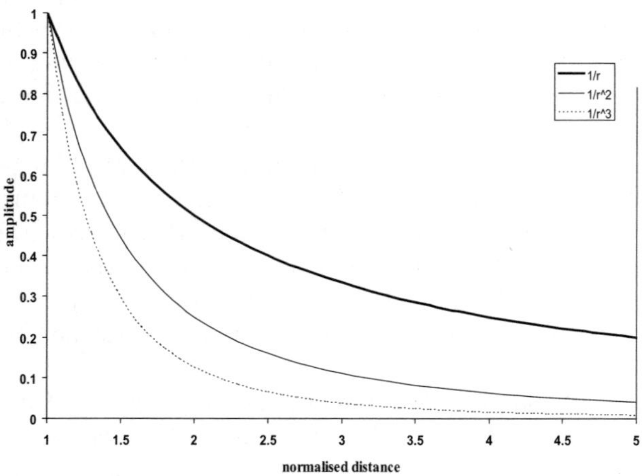

Fig. 3.4 Functions $\frac{1}{r^3}$, $\frac{1}{r^2}$ and $\frac{1}{r}$ as function of r, for $1.0 \leq r \leq 5.0$.

The radiating near-field or Fresnel (Frs) region is taken for distances [1],

$$0.62\sqrt{\frac{D^3}{\lambda}} < r_{Frs} < \frac{2D^2}{\lambda}, \tag{3.3}$$

and the far-field (ff) region is taken to exist for distances

$$r_{ff} > \frac{2D^2}{\lambda}. \tag{3.4}$$

The far-field condition yields a maximum phase error of 22.5° [1]. When phase accuracy is important, the far-field criterion $r_{ff} > \frac{nD^2}{\lambda}$, with n an integer ranging from 3 to 10, is employed.[1]

We return now to the phase differences that will cause constructive interference in some directions and destructive interference in other directions, knowing that we have to evaluate these effects in the far-field region.

3.1.2 Three-Dimensional Radiation Pattern

For the half-wave dipole antenna we have seen for instance that no radiation occurs in directions along the dipole axis. To explain this in more detail, we consider a horizontal half-wave dipole that we represent by two isotropic radiators, 1 and 2 in figure 3.5, spaced apart by half a wavelength. We will look at the electric field in the far-field region.[2]

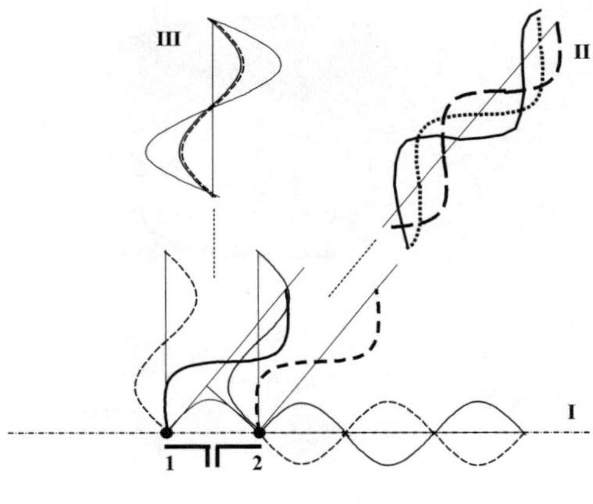

Fig. 3.5 A half-wave dipole is represented by two isotropic radiators, spaced apart by half a wavelength. Along the dipole axis, the waves emitted by radiators 1 and 2 are 180 degrees out of phase. Perpendicular to the dipole axis, the waves emitted by the isotropic radiators are in phase.

For situation *I*, that is in the direction along the dipole axis, the waves emitted by isotropic radiators 1 and 2 are 180 degrees out of phase and therefore cancel each other.

For situation *III*, that is in the direction perpendicular to the dipole axis, the waves emitted by isotropic radiators 1 and 2 are in phase and add, giving the maximum possible amplitude. When we evaluate these signals at infinity - what we, strictly speaking, should do to have a far-field region evaluation - the two distinct directions stemming from isotropic radiators 1 and 2 become one. This is indicated in figure 3.5.

[1]For very small antennas, the far-field criterion $r_{ff} > \frac{2D^2}{\lambda}$ would lead to erroneously short distances. For small antennas therefore the far-field criterion $r_{ff} > \frac{\lambda}{2\pi}$ is employed. [3].

[2]We could also have chosen to look at the magnetic field. The magnetic field is perpendicular to and in phase with the electric field.

For situation *II*, that is for a direction in between parallel and perpendicular to the dipole axis, the two waves add, but with a phase difference due to the fact that one wave reaches a far-field evaluation point before the other one does. Therefore, the amplitude of the combined waves will be less than maximum.

In figure 3.5, we have restricted ourselves to the evaluation in the plane parallel to the dipole axis and containing the dipole. If we evaluate the far electric field amplitude for all possible angular positions ϑ, φ and normalise the amplitude to the maximum value, the three-dimensional *radiation pattern* or *antenna pattern* of figure 3.6 results. Note that this pattern is for a half-wave dipole antenna directed along the Cartesian z-axis, also indicated in the figure. Contour plots of the radiation pattern (equi-amplitude lines) are shown, projected on the x,y-plane.

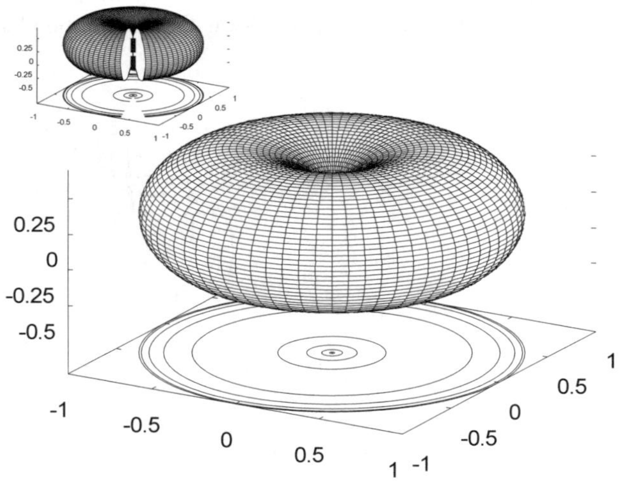

Fig. 3.6 Three-dimensional electric field radiation pattern of a z-directed half-wave dipole antenna.

We see that the radiation pattern is symmetrical around the dipole axis, at maximum in the directions perpendicular to the dipole axis and zero in the directions along the dipole axis.

Although we are dealing here with a basic antenna having a relatively simple radiation pattern, we already encounter some difficulty in reading and interpreting this radiation pattern. This will become worse for more complicated antennas. To illustrate this we created the artificial (normalised) radiation pattern shown in figure 3.7. This could be the pattern of a pyramidal horn antenna. The position and orientation of this horn is indicated in the figure.

Although we encounter even more difficulties in the interpretation of this pattern, we can still distinguish some salient features of this pattern. First of all we observe angular regions of strong radiation, surrounded by regions of weak radiation. We call these regions of strong radiation *lobes*, see 1, 2 and 3 in figure 3.7. The biggest lobe, number 1 in figure 3.7, is called the *main lobe* or *main beam*. The main lobe contains the direction of maximum radiation. Here, that direction is given (in spherical

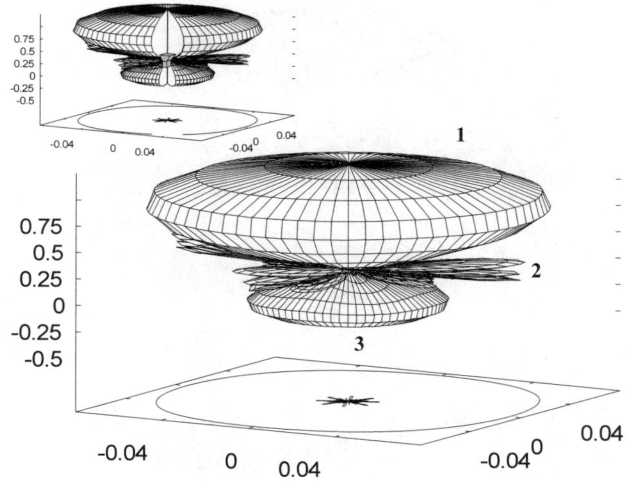

Fig. 3.7 Three-dimensional electric field radiation pattern of a hypothetical pyramidal horn antenna.

coordinates) by $\vartheta = 0$. All lobes other than the main lobe are called *minor lobes* or *side lobes*. Side lobes, numbers 2 and 3 in figure 3.7, are radiation lobes pointing in other directions than the main lobe. The side lobe number 3 in the figure is also called a *back lobe*, due to its position relative to the antenna: 180° rotated with respect to the main lobe.

To allow the comparison of radiation intensities in different directions relative to each other for this and other antennas, it is better to represent the three-dimensional information in another format. In figure 3.8 the amplitude of the radiated electric field is shown along the z-direction, while the ϑ- and φ-coordinates are shown in a polar format in the x,y-plane. The coordinate ϑ varies between 0° and 180° along the radius of the circle, the coordinate φ is the angle between the x-axis and the ϑ-axis. The **E**-field amplitude is plotted along the z-axis, as shown in figure 3.8.

More detailed information can be seen if we plot only part of the angular domain. In figure 3.9, ϑ varies between 0° and 90° along the radius of the circle.

We see that for $\vartheta = 0°$, the electric field is zero and that for every angle φ, the electric field amplitude is at maximum for $\vartheta = 90°$. This is in agreement with figure 3.6. Note that more detail has become visible at the expense of not having all the information in one plot.

The radiation pattern of figure 3.7 transforms, in this first alternative representation, into figure 3.10.

To obtain more information from this graph, we again narrow down the ϑ-domain down to 90°. The result is shown in figure 3.11.

It is clear that figure 3.11 reveals more information than the original three-dimensional pattern, shown in figure 3.7. However, all information in the back lobe region is not plotted in this figure. Furthermore, additional mental processing is required to interpret the results shown. The three-dimensional plot serves best for

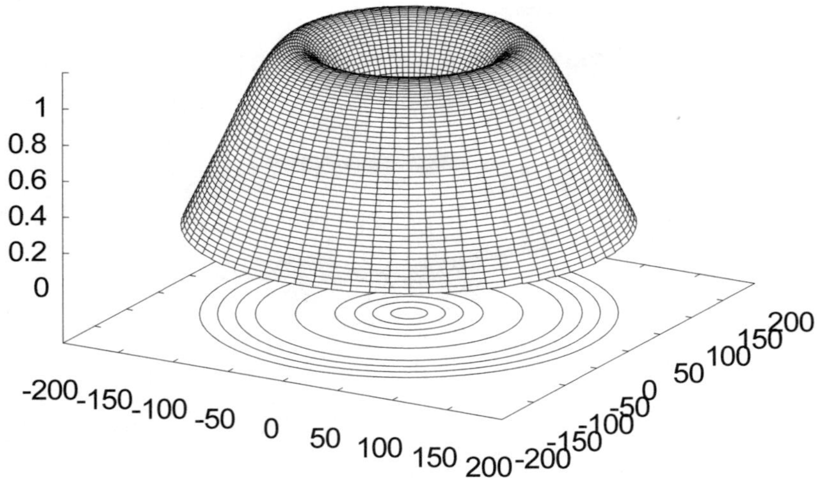

Fig. 3.8 Alternative three-dimensional electric field radiation pattern of a half-wave dipole antenna.

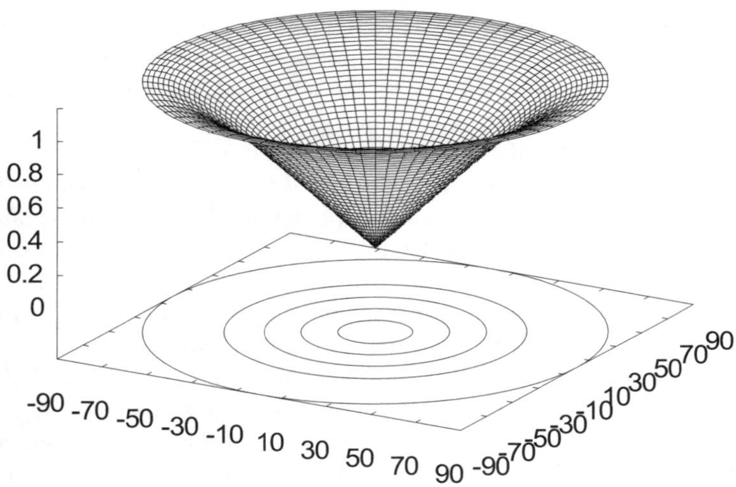

Fig. 3.9 Alternative three-dimensional electric field radiation pattern of a half-wave dipole antenna, ϑ varies between 0 and 90° along the radius of the circle.

inspection purposes, e.g. the identification of side lobes and their angular distribution. It is not easy to obtain qualitative information from this representation. For that last purpose it is customary to use planar cuts of the three-dimensional radiation patterns.

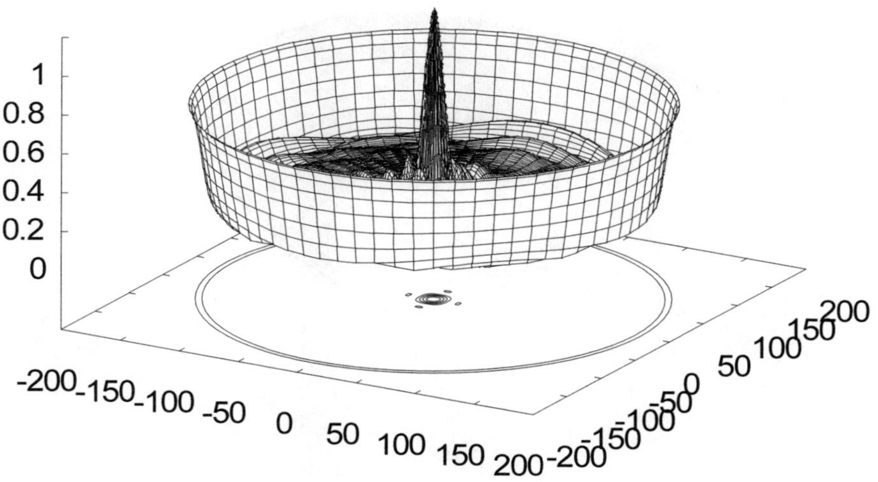

Fig. 3.10 Alternative three-dimensional electric field radiation pattern of a hypothetical pyramidal horn antenna.

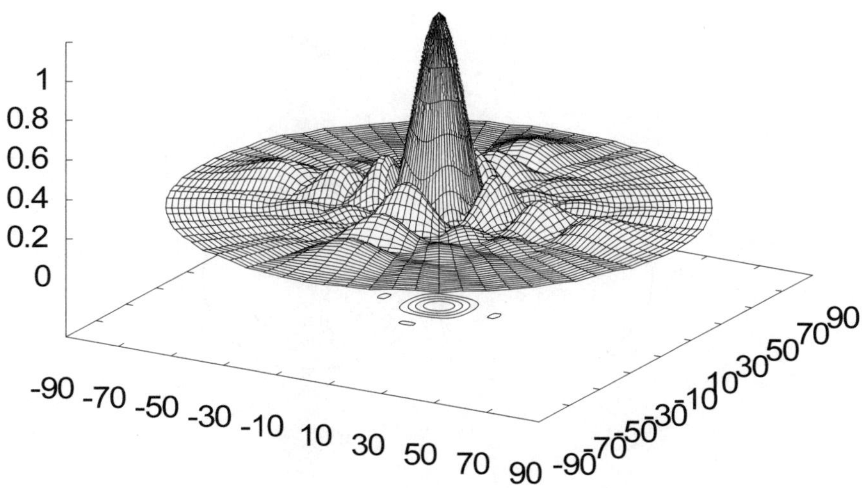

Fig. 3.11 Alternative three-dimensional electric field radiation pattern of a hypothetical pyramidal horn antenna, ϑ varies between 0° and 90° along the radius of the circle.

3.1.3 Planar Cuts

In figure 3.12 a planar cut of the three-dimensional radiation pattern of a half-wave dipole antenna is shown. This figure should be compared with figure 3.6.

This planar cut, transformed to the two-dimensional domain, is shown in figure 3.13.

The elevation angle ϑ increases, going clockwise around the circle. The amplitude of the electric field is plotted along the radius of the circle.

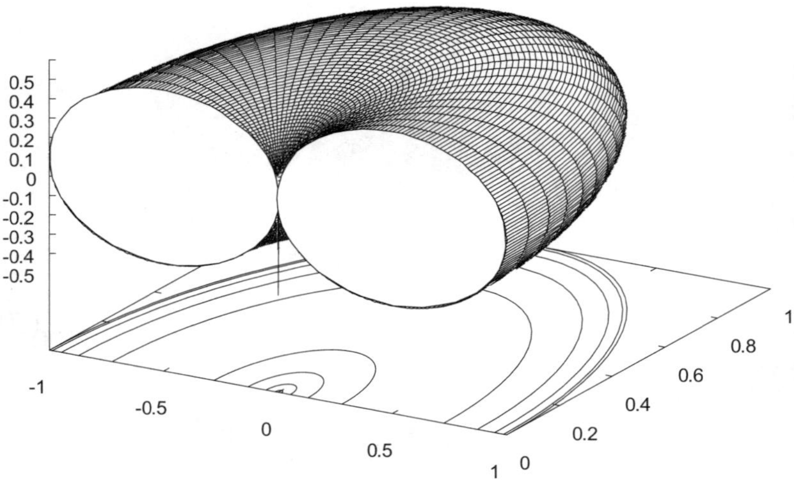

Fig. 3.12 Planar cut from the three-dimensional radiation pattern of a half-wave dipole antenna as shown in figure 3.6.

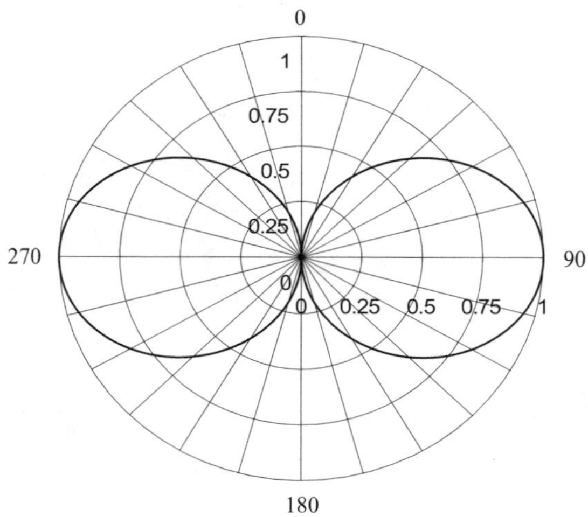

Fig. 3.13 Polar plot of the electric field radiation pattern of a half-wave dipole antenna.

Normally, the azimuth angle φ where the cut is taken should be specified, but since - in this particular case - we are dealing with a radiation pattern that is rotationally symmetric, all cuts are identical.

A planar cut of the (artificial) radiation pattern of the pyramidal horn antenna is shown in figure 3.14. This figure should be compared with figure 3.7.

This planar cut, transformed to the two-dimensional domain, is shown in figure 3.15. Note the different amplitude scales in figures 3.7 and 3.15.

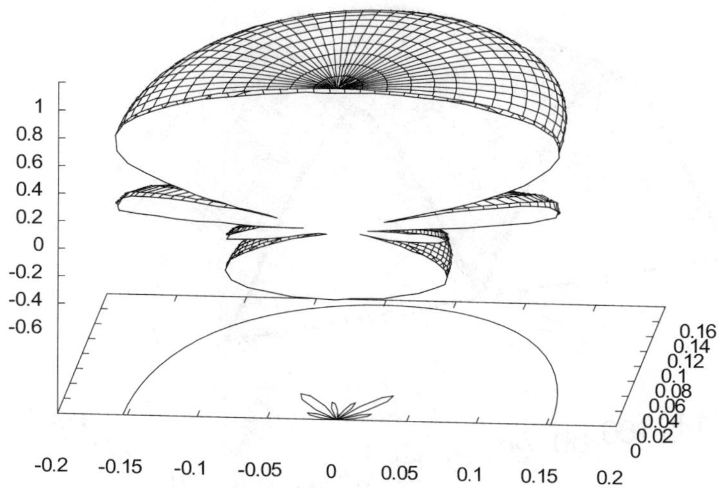

Fig. 3.14 Planar cut from the three-dimensional radiation pattern of a pyramidal horn antenna as shown in figure 3.7.

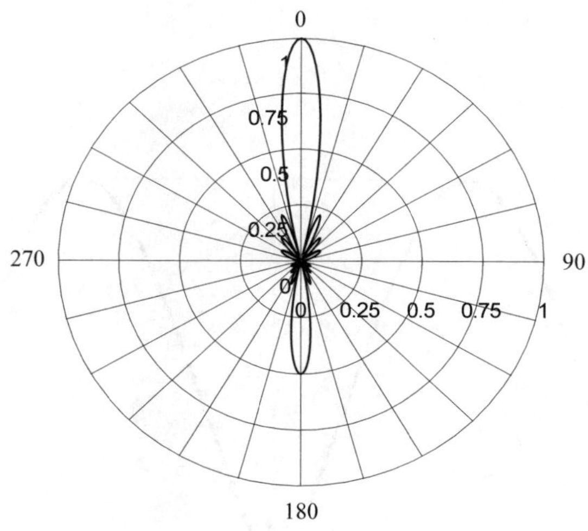

Fig. 3.15 Polar plot of the electric field radiation pattern of a pyramidal horn antenna.

Although, all information can be read from the polar plots and the polar plots correspond to our perception of the physical three-dimensional world, a rectangular plot may be found helpful in observing the details in the side lobes.

Rectangular plots are obtained by taking planar cuts from the three-dimensional radiation patterns of the form as shown in figures 3.8 to 3.11. A rectangular planar cut of the radiation pattern of a half-wave dipole antenna is shown in figure 3.16. The rectangular cut, transformed to the two-dimensional domain is shown in figure 3.17.

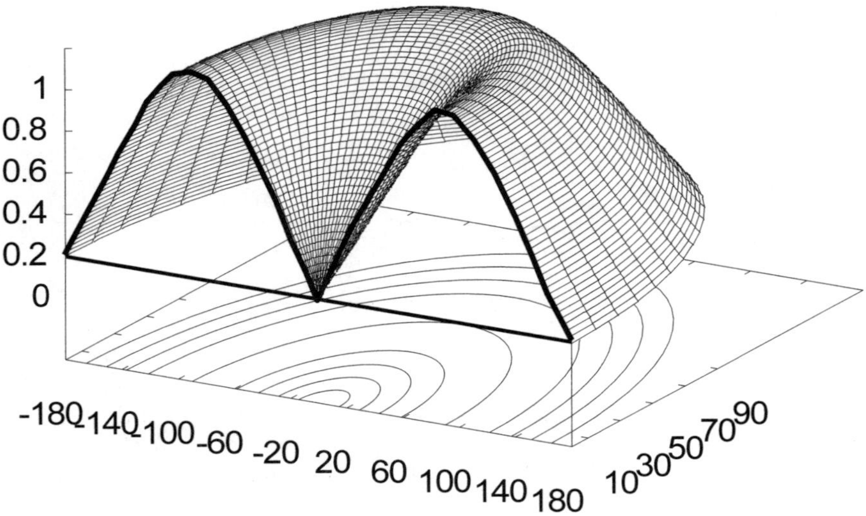

Fig. 3.16 Planar cut from the three-dimensional radiation pattern of a half-wave dipole antenna as shown in figure 3.8.

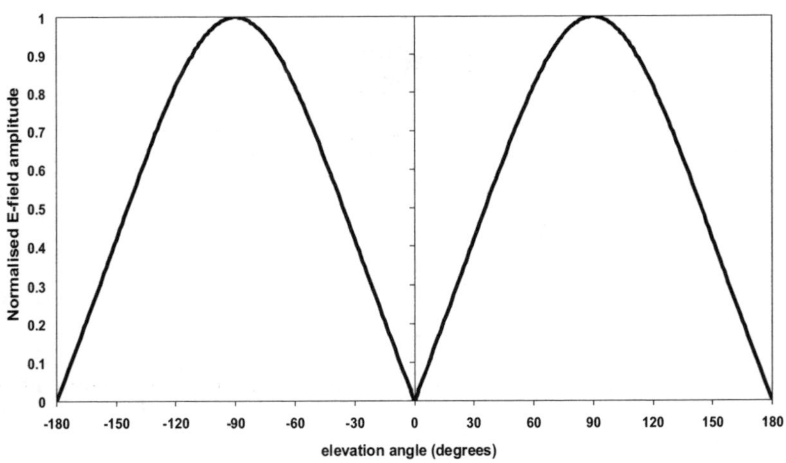

Fig. 3.17 Rectangular plot of the electric field radiation pattern of a half-wave dipole antenna.

A rectangular planar cut of the radiation pattern of a pyramidal horn antenna is shown in figure 3.18. The rectangular cut, transformed to the two-dimensional domain is shown in figure 3.19.

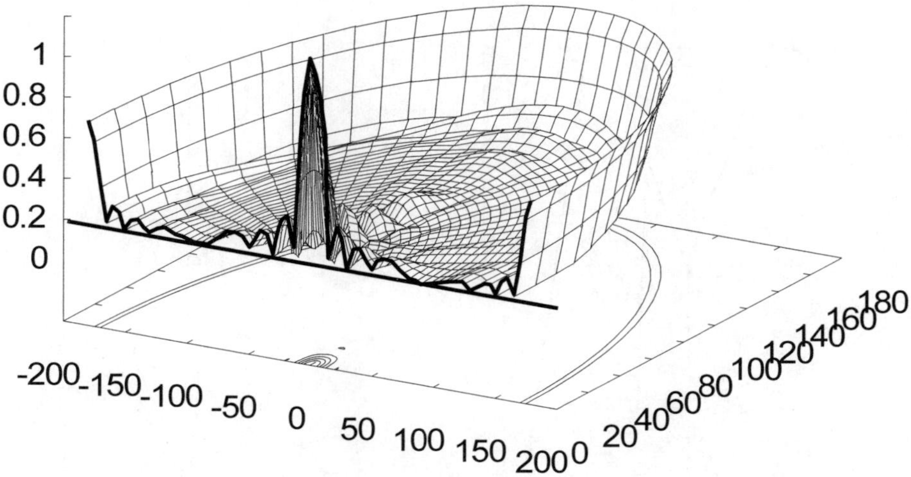

Fig. 3.18 Planar cut from the three-dimensional radiation pattern of a pyramidal horn antenna as shown in figure 3.10.

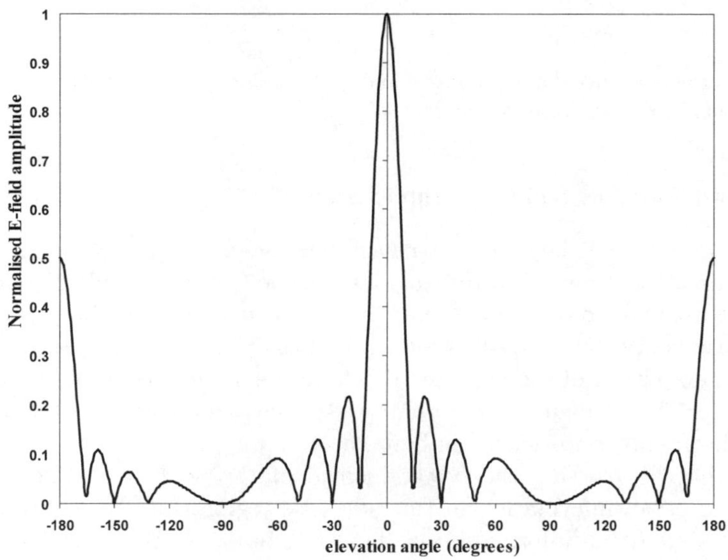

Fig. 3.19 Rectangular plot of the electric field radiation pattern of a pyramidal horn antenna.

The correspondence between polar and rectangular plots is shown in figure 3.20.

Whether a polar or a rectangular plot of the radiation pattern is being used, is for a big part a matter of taste. The author's taste is for rectangular patterns, especially for observing details in power patterns.

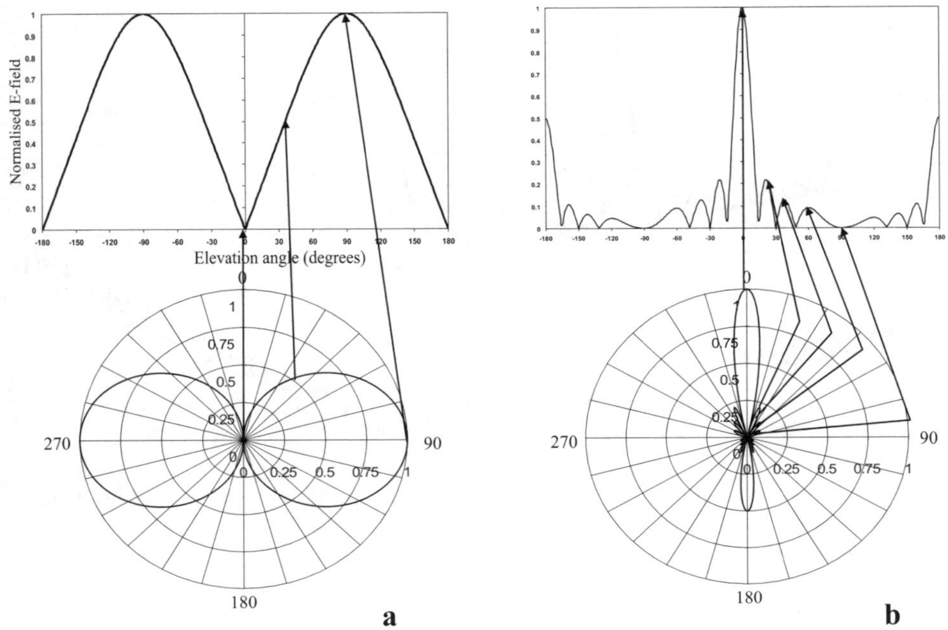

Fig. 3.20 Correspondence between polar and rectangular radiation patterns. a. Half-wave dipole antenna. b. Pyramidal horn antenna.

3.1.4 Power Patterns and Logarithmic Scale

The radiation patterns shown thus far have been so-called *field patterns.* In general, one is not directly interested in the angular electric field amplitude distribution, but more in the angular power distribution. In the far-field region, the power is related to the electric field through a square-law relationship. So by taking the square of the electric field amplitude of the normalised field pattern, we may obtain the normalised *power pattern.* The normalised power pattern (in the plane $\varphi = 0$) for the hypothetical pyramidal horn antenna is shown in figure 3.21.

We see that by plotting the power instead of the field amplitude, we seem to have lost the detail information in the side lobe region. Ideally, we would want to observe the same detail information in the side lobes as in the main lobe. This may be accomplished by plotting the angular power distribution not on a linear scale (as we have done in figure 3.21) but on a *logarithmic* scale.

3.1.4.1 Logarithms If

$$a^x = b, \tag{3.5}$$

where a and b are positive numbers and $a \neq 1$, then x is the *logarithm* of b to the base a:

$$y = \log_a b. \tag{3.6}$$

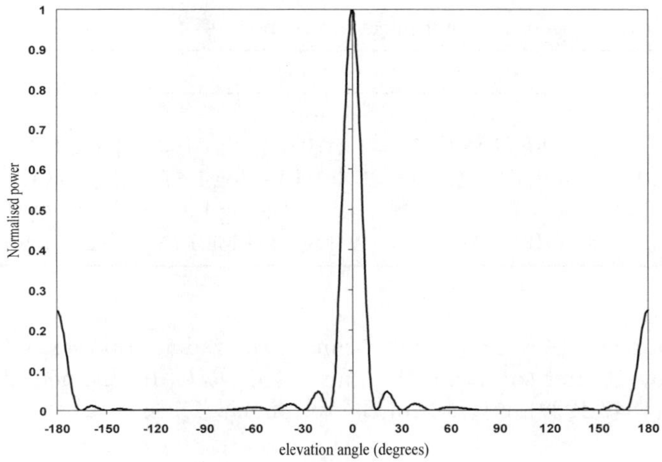

Fig. 3.21 Power radiation pattern for a pyramidal horn antenna.

We will only look at the base $a = 10$, so

$$y = \log_{10} b = \log b, \tag{3.7}$$

means that

$$10^y = b. \tag{3.8}$$

Logarithms to the base 10 are called *common logarithms.*
It can be easily verified then that $\log x$ relates to x as stated in table 3.1.

Table 3.1 $\log x$ vs. x for some values of x.

x	$\log x$
0.001	−3
0.01	−2
1	0
10	1
100	2
1000	3

One useful property of logarithms is that

$$\log(x \cdot y) = \log x + \log y. \tag{3.9}$$

This means that the relations as shown in table 3.2 for the multiplications by powers of 10 of the - arbitrarily chosen - number 1.87 hold.

From table 3.1 and table 3.2 we may conclude that in every decade (0.01 to 0.1, 0.1 to 1, 1 to 10, 10 to 100 and so on), the resolution is equal. This is exactly the property we were looking for in representing radiation power plots!

Table 3.2 Logarithmic values of multiplications by powers of 10 of the number 1.87.

x	$\log x$
1.87	$\log 1.87$
18.7	$\log(10 \cdot 1.87) = \log 10 + \log 1.87 = 1 + \log 1.87$
187.0	$\log(10^2 \cdot 1.87) = 2 \cdot \log 10 + \log 1.87 = 2 + \log 1.87$
0.187	$\log(10^{-1} \cdot 1.87) = -\log 10 + \log 1.87 = -1 + \log 1.87$
0.0187	$\log(10^{-2} \cdot 1.87) = -2 \cdot \log 10 + \log 1.87 = -2 + \log 1.87$

Instead of directly plotting the logarithm of the radiated power, we first multiply this number by 10 and so obtain the units of *decibel* (dB) named after *Alexander Graham Bell* (1847-1922), inventor of the telephone. Thus:

$$1dB = 10 \cdot \log(x) \longrightarrow x = 10^{\frac{1}{10}} = 1.26. \qquad (3.10)$$

The radiated power, expressed in decibels is shown in figure 3.22. This figure should be compared with figure 3.21.

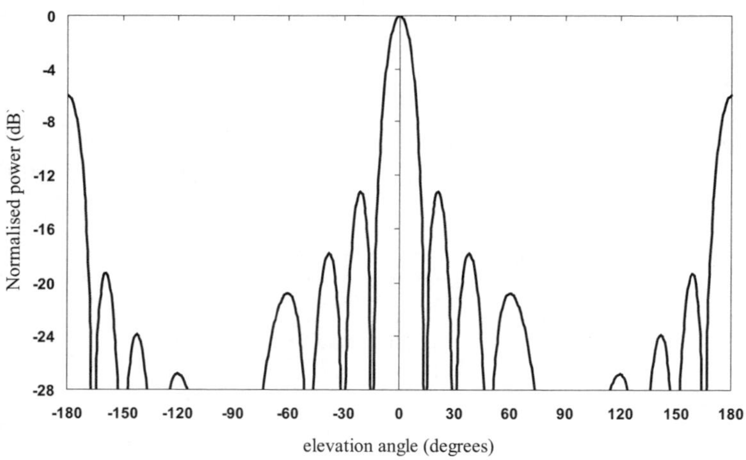

Fig. 3.22 Power radiation pattern for a pyramidal horn antenna on a logarithmic (decibel) scale.

3.1.5 Directivity and Gain

So far we have been comparing the radiated field and power only to the maximally radiated field or power of the same antenna. If we want to compare different antennas with each other, we need to have a reference to compare them to. This reference is taken to be the isotropic radiator.

Although we know that the isotropic radiator is a physical abstraction,[3] it still may serve as a reference for real life antennas. The (hypothetical) isotropic radiator, radiates equally in all directions. Its normalised, three-dimensional electric field or power pattern, on a linear scale, therefore is a sphere with radius *one*, see figure 3.23.

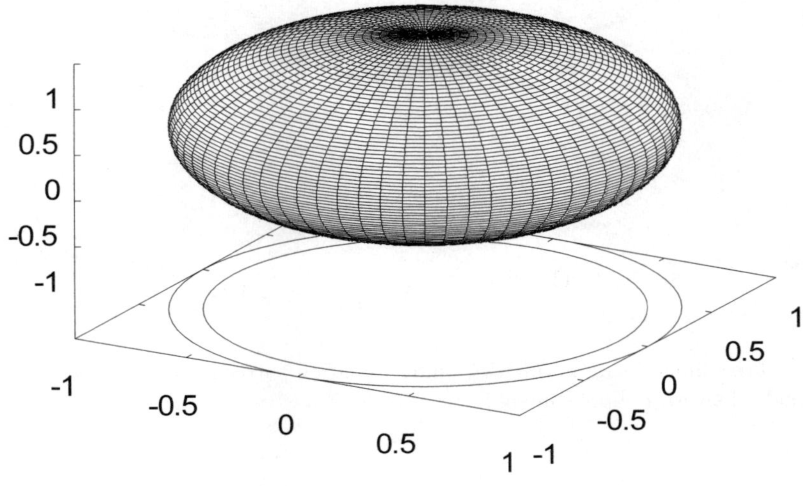

Fig. 3.23 Three-dimensional, normalised electric field or power radiation pattern of a (hypothetical) isotropic radiator.

The *directivity function*, $D(\vartheta, \varphi)$, is defined as the power radiated by an antenna into a direction (ϑ, φ), compared to the power radiated into that same direction by an isotropic radiator.

The power radiated into a certain direction by an isotropic radiator is equal to the amount radiated into every other direction and is equal to $P_t/4\pi$, where P_t is the totally transmitted power and $P_t/4\pi$ therefore is the radiation intensity of the isotropic radiator.[4] Thus

$$D(\vartheta, \varphi) = \frac{P(\vartheta, \varphi)}{P_t/4\pi}, \tag{3.11}$$

where $P(\vartheta, \varphi)$ is the power radiated by the actual antenna into the direction (ϑ, φ).

So, we compare the power radiated by the actual antenna to the power that would have been radiated by an isotropic radiator, radiating the same total amount of power.

In figure 3.24 we show both the three-dimensionally radiated power by a half-wave dipole and the radiated power by an isotropic radiator. Both antennas - the real half-wave dipole antenna and the hypothetical isotropic radiator - have the same totally transmitted power.

[3]Electromagnetic radiation only exists by virtue of accelerating charge and thus by displacement of charge. This acceleration and displacement may be modelled by assuming the existence of *more than one* isotropic radiator, physically displaced from one another. An *isolated* isotropic radiator therefore cannot exist.

[4]4π is the value of the solid angle of a complete sphere. Compare this three-dimensional angle with an angle in the two-dimensional domain, where 2π is the angle of a complete circle.

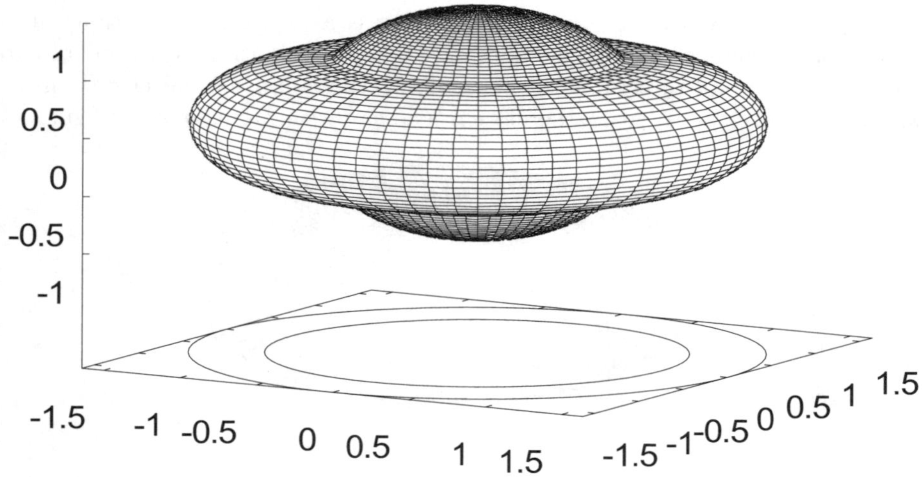

Fig. 3.24 Three-dimensional, normalised power radiation pattern of a (hypothetical) isotropic radiator and a half-wave dipole antenna.

The figure clearly shows the value of using the isotropic radiator as a reference. We see that in some directions, the half-wave dipole antenna radiates more power than the isotropic radiator while for other directions the opposite is true. Since both antennas have the same amount of totally transmitted power, we may transform the three-dimensional radiation pattern of one antenna into that of the other by reducing power in certain directions and increasing power into other directions. This process (transforming the isotropic radiation pattern into that of a half-wave dipole antenna) is shown - in a planar cut - in figure 3.25.

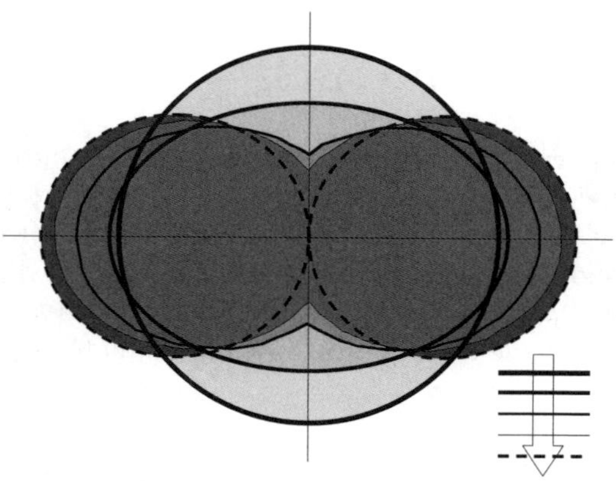

Fig. 3.25 Transforming the isotropically radiated power into the power radiated by the half-wave dipole, planar cut: reducing transmitted power in one direction means increasing transmitted power in another direction.

The *directivity*, D is defined as the maximum of the directivity function:

$$D = \max[D(\vartheta, \varphi)]. \tag{3.12}$$

For the half-wave dipole antenna, the directivity is $D = 1.64$ or $D = 2.15$dB, meaning that the half-wave dipole antenna at maximum radiates 1.64 times as much power as an isotropic radiator would do, transmitting the same total amount of power.

In general though, the total transmitted power of an antenna is not known, or difficult to assess. Therefore, a second function exists: the *gain function*, $G(\vartheta, \varphi)$. The gain function resembles the directivity function, except for the totally radiated power that has been replaced by the totally accepted power, P_{in},

$$G(\vartheta, \varphi) = \frac{P(\vartheta, \varphi)}{P_{in}/4\pi}. \tag{3.13}$$

P_{in} is easier to assess than the totally radiated power.

The gain function does *not* take impedance mismatch on the antenna terminals into account. If 99% of the power delivered to the antenna terminals is reflected, the gain function tells us how the remaining 1% of this power (i.e. the accepted power) is distributed in space. The *gain*, G, is the maximum of the gain function,

$$G = \max[G(\vartheta, \varphi)]. \tag{3.14}$$

The quotient of gain and directivity equals the quotient of totally radiated power and totally accepted power and is called the *radiation efficiency*, η [2]

$$\eta = \frac{G}{D} = \frac{P_t}{P_{in}}. \tag{3.15}$$

The efficiency is smaller than one due to ohmic and/or dielectric losses in the antenna.

Before moving on with the description of the parameters that describe an antenna, something should be said about the fact that so far we have been talking only about antennas transmitting.

3.1.6 Reciprocity

To say something about an antenna receiving, we take a look at the two-antenna system of figure 3.26 [4].

The two-antenna system may be considered as a two-port network exhibiting reciprocity.

This means that: the voltage at II, due to a current source at I, is equal to the voltage at I, due to the same current source at II. If we further assume that the two antennas are placed in each other's far-field regions, the voltage at the antenna terminals is - by definition [4] - equal to the receiving pattern of that antenna, the fields generated by the other antenna forming a plane wave at the receiving antenna position.

Therefore the reciprocity theorem, stated for antennas, reads: *the receiving pattern of an antenna is identical to its transmitting pattern.*

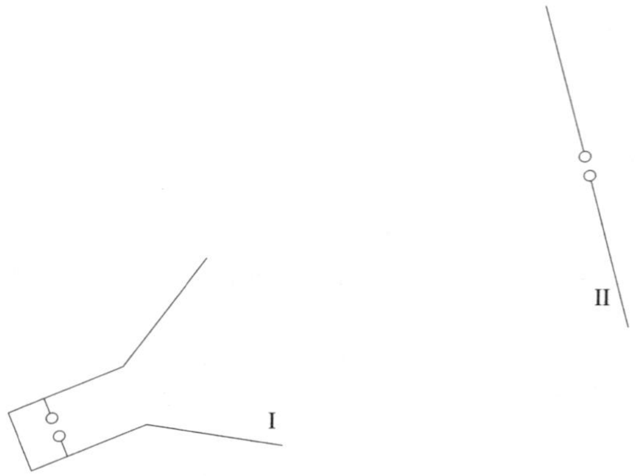

Fig. 3.26 Two-antenna system.

Thus, everywhere an antenna forms a lobe in transmission, it will be sensible for reception from that same direction and everywhere an antenna forms a 'null' in transmission, it will be 'blind' for that direction in reception.

3.1.7 Antenna Beamwidth

In comparing antennas we may use the gain we just discussed. But the gain is a number that only tells us about the maximum radiation. Often we want to know the shape of the area of maximum radiation. For that purpose we use the *beamwidth.* The beamwidth tells us about the shape of the main lobe. Different beamwidth definitions exist. The two most often used beamwidths are shown in figure 3.27.

The *half-power beamwidth*, ϑ_{HP}, is the angular separation between the points on a cut of the main lobe where the transmitted (received) power is half that of the maximum transmitted (received) power. This is shown in figure 3.27a. Since $10\log(0.5) = -3.01\text{dB} \approx -3\text{dB}$, ϑ_{HP} is found on a logarithmic scale on the interception points where the main lobe is 3dB under the maximum value, see figure 3.27b. The half-power beamwidth is therefore also known as the *3dB beamwidth.*

Other, less commonly used definitions for beamwidth are the *first null beamwidth*, ϑ_{FN}, indicated in figure 3.27a, b and the *10dB beamwidth.* In general, when the term *beamwidth* is used, the *3dB beamwidth* is meant.

Also indicated in figure 3.27b is the level of the first and highest side lobe. This level is known as the *side lobe level* (SLL).

The cuts are normally taken in the so-called *E-plane* and *H-plane*, also known as the *principal planes.* The E-plane is the plane that contains the electric field vector and the direction of maximum radiation. The H-plane is the plane that contains the magnetic field vector and the direction of maximum radiation. It is common practice to orient an antenna so that at least one of the principal plane patterns coincides with one of the geometrical principal planes [1]. As an example, figure 3.28 shows the normalised *field* patterns in the principal planes of a half-wave dipole antenna.

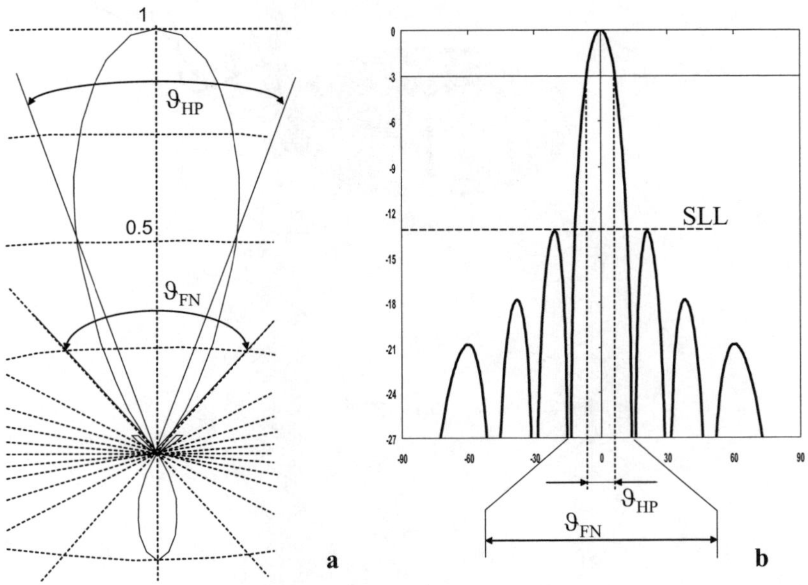

Fig. 3.27 Antenna beamwidth definitions. a. Polar plot, linear scale. b. Rectangular plot, logarithmic scale.

Note that for the z-directed half-wave dipole antenna, any plane $\varphi = constant$ is an E-plane.

3.2 ANTENNA IMPEDANCE AND BANDWIDTH

Antennas will never be used as standalone devices; we will have to consider antennas as being part of a system. Therefore, besides the radiation characteristics, we also need to know the impedance characteristics.

When looking into the antenna, at the antenna terminals, we may regard the antenna as a complex impedance Z_A,

$$Z_A = R_A + jX_A, \tag{3.16}$$

where R_A is the real or resistive part of the antenna impedance and X_A is the imaginary or reactive part of the antenna impedance. The real part accounts for the dissipation and consists of two parts, the ohmic losses (which may be considerable in small antennas), R_L and the (wanted) radiation losses, R_r,

$$R_A = R_L + R_r. \tag{3.17}$$

R_r is known as the *antenna radiation resistance.*

The reactive part of the antenna impedance accounts for the reactive near-field region of the antenna, where energy is being stored.

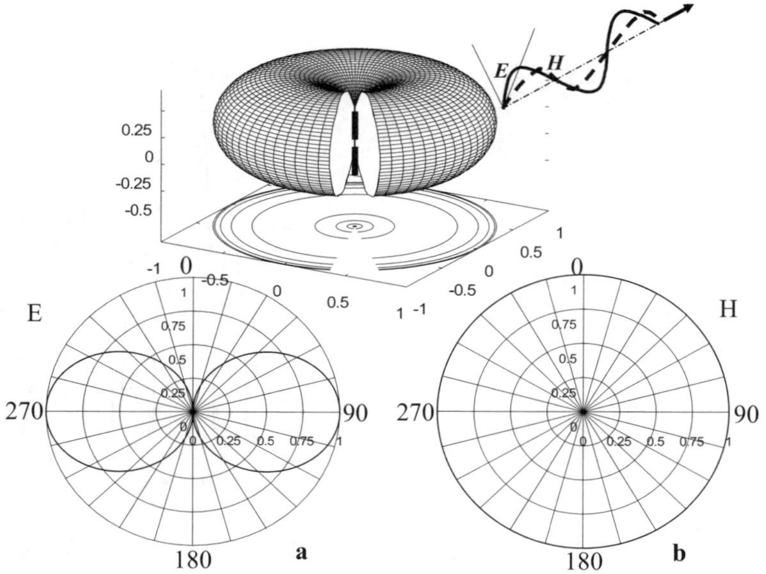

Fig. 3.28 Normalised field patterns in the principal planes of a z-directed half-wave dipole antenna. a. E-plane, containing the electric field vector and the direction of maximum radiation (eight-figure shape). b. H-plane, containing the magnetic field vector and the direction of maximum radiation (circle with unit radius).

We assume for the moment that the reactive and ohmic loss part of the antenna impedance may be neglected ($Z_A = R_r$) and that the antenna is connected to a generator with (real) impedance R_g,[5] see figure 3.29.

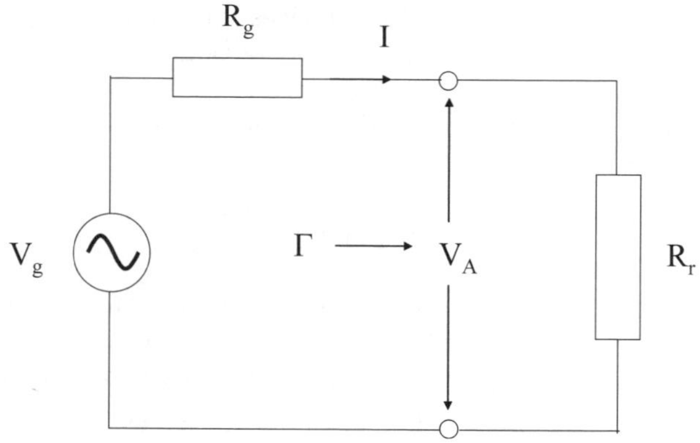

Fig. 3.29 Equivalent circuit for matching the lossless antenna to the generator.

[5] The principle of the so-called *conjugate matching* should be explained - in the most general form - for complex generator and antenna impedances, see for example [1] or [5]. For reasons of clarity, an explanation using real impedances is chosen here.

The (time average) power delivered by the generator to the antenna is given by

$$P = \frac{1}{2} Re \{V_A \cdot I^*\}, \tag{3.18}$$

where $Re\{x\}$ means the real part of the complex number x and I^* is the complex conjugate of I.

Since we are dealing with real quantities only, the power delivered to the antenna is

$$P = \frac{1}{2} |V_g|^2 \left| \frac{R_g}{R_g + R_r} \right|^2 \frac{1}{R_r}, \tag{3.19}$$

wherein use is made of

$$V_A = \frac{R_g}{R_g + R_r} V_g, \tag{3.20}$$

and

$$I^* = I = \frac{V_A}{R_r}. \tag{3.21}$$

To find the value of the antenna impedance that results in a maximum power transfer from the generator to the antenna, the following condition should be fulfilled

$$\frac{\partial P}{\partial R_r} = 0. \tag{3.22}$$

Upon substitution of equation (3.19) into equation (3.22), we finally find that the condition for maximum power transfer results in

$$R_r = R_g, \tag{3.23}$$

meaning that the radiation resistance should be equal to the generator resistance.

For the real generator and antenna impedance, this is also the condition for zero reflections at the antenna terminal. From (microwave) network theory, the reflection coefficient, Γ, looking into the antenna terminals, see figure 3.29, is found to be

$$\Gamma = \frac{R_r - R_g}{R_r + R_g}. \tag{3.24}$$

When we design the antenna such that $R_r = R_g$, we see that the reflection becomes zero.

The impedance of an antenna normally varies as function of the frequency and therefore, the matching also varies as function of the frequency. This means that an antenna will only operate efficiently within a restricted band of frequencies. The width of this band of frequencies is called the *bandwidth*. Normally, at the centre frequency (middle of the frequency band) the impedance matching will be best and going to lower or higher frequencies results in a degradation of impedance matching up to a level where matching has become unacceptably poor. These levels determine the boundaries of the frequency band. Reflection coefficient levels of -10dB ($|\Gamma|^2 \leq 0.1$) or -15dB ($|\Gamma|^2 \leq 0.03$) are commonly employed to determine the *impedance bandwidth*. For relatively small bandwidth antennas, bandwidth is expressed in percentages of the centre frequency [1]. If f_0 is the centre frequency, f_l is the lower boundary of the frequency band and f_u is the upper boundary of the frequency band, the bandwidth is given by:

$$BW = \frac{f_u - f_l}{f_0} \cdot 100\%. \tag{3.25}$$

Bandwidths expressed in percentages of the centre frequency are used up to a few decades. For larger bandwidth antennas the ratio of the upper and lower frequency boundaries is used (like 10 : 1 or 30 : 1) [1].

Next to the *impedance bandwidth* a *radiation pattern bandwidth* may also exist. Both bandwidths do not need to be identical. It depends on the antenna and the application to which of the two bandwidths is most critical.

Example The reflection coefficient of a certain antenna is measured as function of frequency and plotted (in dBs) in figure 3.30.

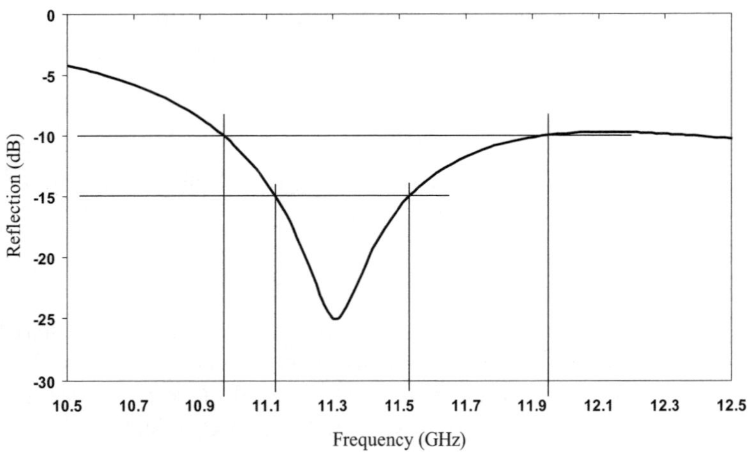

Fig. 3.30 Measured reflection as function of frequency.

The reflection is at minimum at the centre frequency, $f_0 = 11.3$GHz.
The -10dB-bandwidth then follows from the lower frequency limit $f_l = 10.97$GHz and upper frequency limit $f_u = 11.93$GHz, both read from the graph, and equation (3.25)

$$BW_{-10dB} = \frac{11.93 - 10.97}{11.3} \cdot 100\% = 8.50\%. \tag{3.26}$$

The -15dB-bandwidth follows from the graph and equation (3.25)

$$BW_{-15dB} = \frac{11.53 - 11.12}{11.3} \cdot 100\% = 3.63\%. \tag{3.27}$$

3.3 POLARISATION

The electric field in the far-field region of an antenna will in general possess two spherical coordinate components, E_ϑ and E_φ, see figure 3.31.

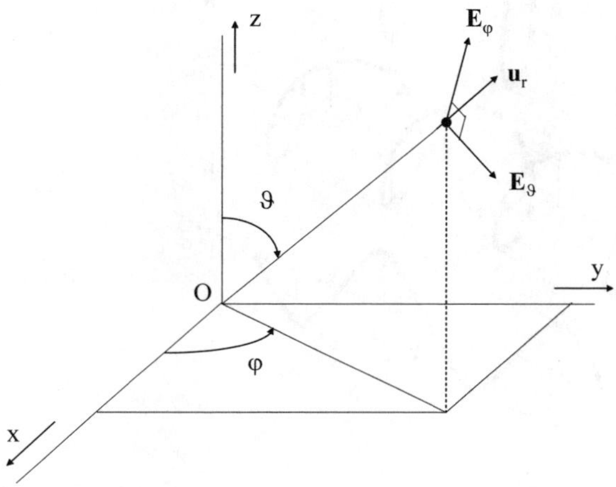

Fig. 3.31 Electric field components E_ϑ and E_φ in the far-field region of an antenna placed in the origin O. $\mathbf{u}_r$ is the unit vector in the direction of wave propagation.

In general, a phase difference will exist between these two field components. Therefore, the electric field *vector*, as function of time, t, will - for an arbitrary phase difference - describe an ellipse in the ϑ,φ-plane. The electric field and the antenna are called *elliptically polarised*, see figure 3.32a.

When the phase difference is plus or minus 90° and the field components are equal in amplitude, the ellipse becomes a circle and the polarisation is called *circular polarisation.* The antenna is called *circularly polarised.* When the phase difference is 0° or 180°, the ellipse becomes a line and the polarisation is called *linear polarisation.* The antenna is called *linearly polarised.*

3.3.1 Elliptical Polarisation

We can write the electric far-field as

$$\mathbf{E} = E_\vartheta \mathbf{u}_\vartheta + E_\varphi \mathbf{u}_\varphi, \tag{3.28}$$

where E_ϑ is the complex amplitude of the ϑ-component of the electric field and E_φ is the complex amplitude of the φ-component of the electric field. The vectors $\mathbf{u}_\vartheta$ and $\mathbf{u}_\varphi$ are unit vectors in respectively ϑ- and φ-directions. A brief overview of complex algebra and vector mathematics may be found in, respectively, appendices A and B.

The complex amplitudes E_ϑ and E_φ can be written as,

$$E_\vartheta = |E_\vartheta|\, e^{j\Psi_\vartheta}, \tag{3.29}$$

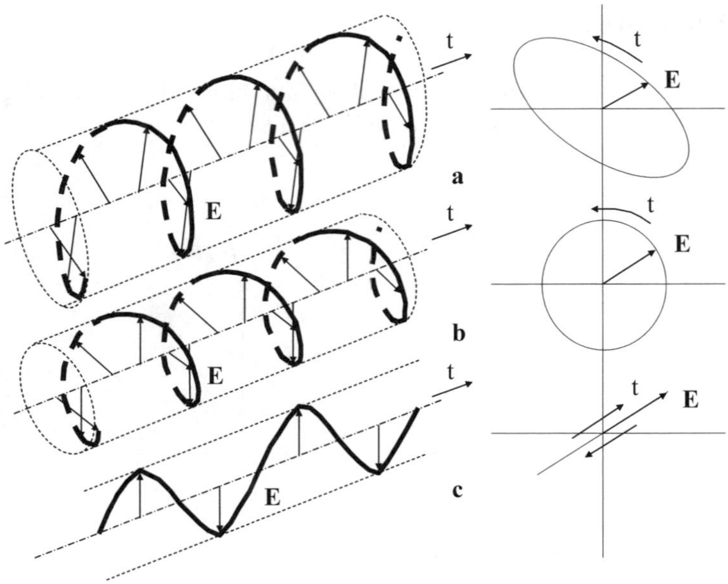

Fig. 3.32 Polarisation states. a. Elliptical polarisation. b. Circular polarisation. d. Linear polarisation.

and

$$E_\varphi = |E_\varphi| \, e^{j\Psi_\varphi}, \tag{3.30}$$

where Ψ_ϑ and Ψ_φ represent the phase of E_ϑ and E_φ, respectively.

The electric field vector therefore may be written as

$$\mathbf{E} = |E_\vartheta| \, e^{j\Psi_\vartheta} \left(\mathbf{u}_\vartheta + \rho \mathbf{u}_\varphi\right), \tag{3.31}$$

where

$$\rho = \frac{|E_\varphi|}{|E_\vartheta|} e^{j(\Psi_\varphi - \Psi_\vartheta)}. \tag{3.32}$$

To trace the extremity of the electric field in the ϑ,φ-plane, the real part of the electric field needs to be taken

$$\mathsf{E} = Re\left\{\mathbf{E}e^{j2\pi ft}\right\} = \\ |E_\vartheta| \left\{\cos\left(2\pi ft + \Psi_\vartheta\right)\mathbf{u}_\vartheta + |\rho| \cos\left(2\pi ft + \Psi_\vartheta + \Psi_\rho\right)\mathbf{u}_\varphi\right\}, \tag{3.33}$$

where f is the used frequency, t is the time and

$$\Psi\rho = \Psi_\varphi - \Psi_\vartheta. \tag{3.34}$$

Elimination of the time t results in a description of the trace of the extremity of the electric field vector in the ϑ,φ-plane. After some mathematical manipulations we get

$$\left(\frac{\mathsf{E}_\vartheta}{|E_\vartheta|}\right)^2 + \left(\frac{\mathsf{E}_\varphi}{|E_\varphi|}\right)^2 - \frac{2\mathsf{E}_\vartheta \mathsf{E}_\varphi}{|E_\vartheta|\,|E_\varphi|} \cos\left(\Psi_\rho\right) = \sin^2\left(\Psi_\rho\right). \tag{3.35}$$

Dividing this equation by $\sin^2(\Psi_\rho)$, results in

$$\left[\frac{1}{|E_\vartheta|^2\sin^2(\Psi_\rho)}\right]\mathsf{E}_\vartheta^2-\left[\frac{2\cos(\Psi_\rho)}{|E_\vartheta||E_\varphi|\sin^2(\Psi_\rho)}\right]\mathsf{E}_\vartheta\mathsf{E}_\varphi+\left[\frac{1}{|E_\varphi|^2\sin^2(\Psi_\rho)}\right]\mathsf{E}_\varphi^2=1, \quad (3.36)$$

which is the equation of an ellipse in the ϑ,φ-plane, see figure 3.33.

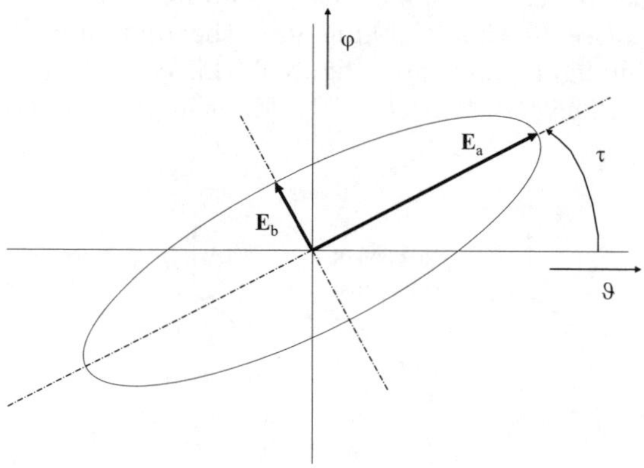

Fig. 3.33 Polarisation ellipse.

The tilt angle τ is given by

$$\tau = \arctan\left(\frac{|\mathbf{E}_\varphi|}{|\mathbf{E}_\vartheta|}\right). \quad (3.37)$$

The amount of ellipticity is expressed as the ratio of the semimajor and semiminor axes of the ellipse. This ratio is therefore known as the *axial ratio* and is usually expressed in decibels. With reference to figure 3.33

$$AR = 20\log\left(\frac{\mathbf{E}_a}{\mathbf{E}_b}\right). \quad (3.38)$$

We will now show that circular polarisation and linear polarisation are special cases of elliptical polarisation.

3.3.2 Circular Polarisation

For the special situation where $|\mathbf{E}_\vartheta| = |\mathbf{E}_\varphi| = |\mathbf{E}|$ and $\Psi_\rho = \pm\frac{\pi}{2}$, equation (3.36) simplifies to

$$\left(\frac{\mathsf{E}_\vartheta}{|E|}\right)^2+\left(\frac{\mathsf{E}_\varphi}{|E|}\right)^2=1, \quad (3.39)$$

which describes a circle in the ϑ,φ-plane.

We may further distinguish between the situations $\Psi_\rho = \frac{\pi}{2}$ and $\Psi_\rho = -\frac{\pi}{2}$. For the former situation we get, upon substitution in equation (3.33)

$$\mathsf{E}_L = |E|\left[\cos(2\pi f t')\,\mathbf{u}_\vartheta - \sin(2\pi f t')\,\mathbf{u}_\varphi\right], \quad (3.40)$$

where

$$2\pi f t' = 2\pi f t + \Psi_\rho. \tag{3.41}$$

Similarly, we get for the latter situation

$$\mathsf{E}_R = |E| \left[\cos\left(2\pi f t'\right) \mathbf{u}_\vartheta + \sin\left(2\pi f t'\right) \mathbf{u}_\varphi \right]. \tag{3.42}$$

For the first situation ($\Psi_\rho = \frac{\pi}{2}$) we have obtained an electric field vector that rotates counterclockwise in time, looking into the direction of propagation. This situation is depicted in figure 3.32 and in figure 3.34a. We call this circular polarisation state *left-hand circular polarisation* (LHCP), following the direction of a left-handed screw.

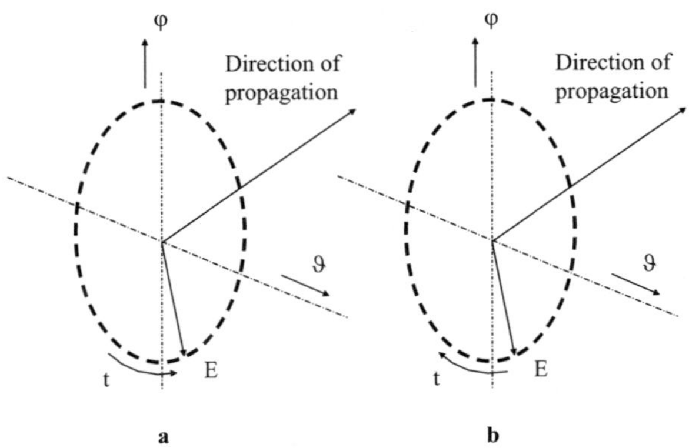

Fig. 3.34 Circular polarisation. a. Left-hand circular polarisation (LHCP). b. Right-hand circular polarisation (RHCP).

For the second situation ($\Psi_\rho = -\frac{\pi}{2}$) we have obtained an electric field vector that rotates clockwise in time, looking into the direction of propagation. This situation is depicted in figure 3.34b. We call this circular polarisation state *right-hand circular polarisation* (RHCP), following the direction of a right-handed screw.

3.3.3 Linear Polarisation

For the special situation where $\Psi_\rho = \pm\pi$ and $|\mathbf{E}_\vartheta|$ not necessarily equal to $|\mathbf{E}_\varphi|$, equation (3.35) simplifies to

$$\frac{\mathsf{E}_\vartheta}{|E|} + \frac{\mathsf{E}_\varphi}{|E|} = 0, \tag{3.43}$$

which describes a straight line in the ϑ,φ-plane, see also figure 3.32c. The *axial ratio* (AR) is infinite for a linearly polarised wave.

3.3.4 Axial Ratio

The axial ratio of a perfect circularly polarised wave is equal to unity. In practice, perfect circular polarisation is seldom encountered; a certain amount of ellipticity will always exist. This ellipticity is expressed in the axial ratio, the ratio of the semimajor to the semiminor axis lengths of the polarisation ellipse. These lengths may be expresses in terms of amplitudes of LHCP and RHCP components. The elliptical polarisation may be seen as consisting of a combination of LHCP and RHCP polarisation, the dominant part of these two determining the direction of rotation of the elliptical polarisation.

To show this we start by decomposing the electric field in LHCP and RHCP components:

$$\mathbf{E} = E_\vartheta \mathbf{u}_\vartheta + E_\varphi \mathbf{u}_\varphi = E_L \mathbf{u}_L + E_R \mathbf{u}_R, \tag{3.44}$$

where

$$\mathbf{u}_L = \frac{1}{\sqrt{2}} \left\{ \mathbf{u}_\vartheta + j\mathbf{u}_\varphi \right\}, \tag{3.45}$$

$$\mathbf{u}_R = \frac{1}{\sqrt{2}} \left\{ \mathbf{u}_\vartheta - j\mathbf{u}_\varphi \right\}, \tag{3.46}$$

and

$$E_L = \frac{1}{\sqrt{2}} \left\{ E_\vartheta - jE_\varphi \right\}, \tag{3.47}$$

$$E_R = \frac{1}{\sqrt{2}} \left\{ E_\vartheta + jE_\varphi \right\}. \tag{3.48}$$

E_L is the LHCP component of the wave, E_R is the RHCP component of the wave. The relation between an elliptically polarised wave and both circular polarisation components is graphically represented in figure 3.35.

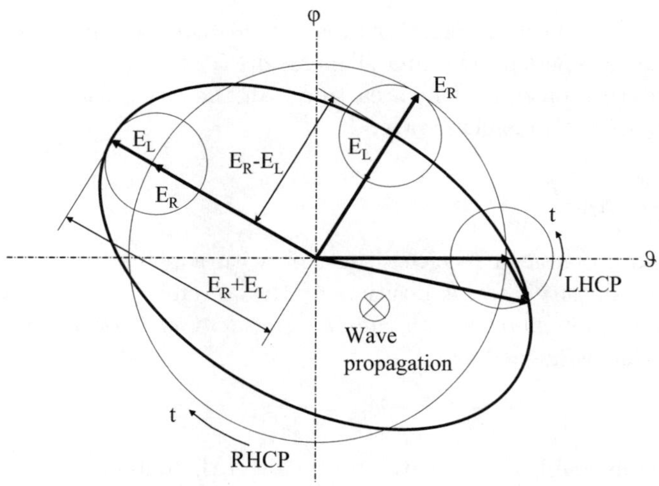

Fig. 3.35 Right-hand elliptically polarised wave, decomposed in RHCP and LHCP components.

In a predominantly RHCP wave (as shown in figure 3.35), the RHCP component is called the *co-polarisation* and the LHCP component is called the *cross-polarisation.* The opposite is true for a predominantly LHCP wave.

The axial ratio is given by

$$AR = 20\log\left|\frac{|E_L| + |E_R|}{|E_L| - |E_R|}\right|. \tag{3.49}$$

Circular polarisation may be beneficial both in mobile satellite communications and in radar applications.

At L-band frequencies, the ionosphere acts as a so-called *Faraday rotator* [6], meaning that a linearly polarised wave undergoes a rotation upon passing through the ionosphere. The use of circularly polarised transmit and receive antennas will annihilate negative effects of this rotation (co-polarised signal attenuation and increase of cross-polarised signal level).

In radar, circular polarisation may be employed to 'see through' rain [7]. A right-hand (left-hand) circularly polarised wave, incident upon a (near) spherical rain drop will be reflected as a left-hand (right-hand) circularly polarised wave, for which the receiving antenna is insensible and thus this rain scatter will be rejected, while reflections from a complicated structure, like an aircraft, will possess circularly polarised components with the right rotation direction that will be accepted by the antenna. A linearly polarised wave, reflected from a raindrop, would be accepted by the antenna since the antenna will accept linearly polarised waves, 180° shifted in phase, equally well as in-phase components.

3.4 ANTENNA EFFECTIVE AREA AND VECTOR EFFECTIVE LENGTH

Any antenna, be it a horn antenna or even a half-wave (wire) dipole antenna, may be considered as an aperture antenna. This means that we may associate with every antenna an aperture or equivalent area that - in the case of a receiving antenna - extracts energy from an incident wave.

3.4.1 Effective Area

Let's assume an antenna in receive situation[6] with a plane wave incident upon it, having a power density at the position of the antenna of S ($\mathrm{Wm^{-2}}$). We may characterise the antenna by an *maximum equivalent area* or *maximum equivalent aperture*, A_{em} that is defined as

$$A_{em} = \frac{P_T}{S}, \tag{3.50}$$

where P_T is the available power at the terminals of the antenna.

The effective area is, strictly speaking, a direction-dependent quantity, but, if no direction is specified - like in the above equation - the direction is assumed to be that of maximum directivity. Furthermore, without further specification, we assume that polarisation of antenna and the impinging plane wave are lined up and that the

[6]We have already seen that, by virtue of reciprocity, transmit and receive properties of antennas are identical.

antenna does not introduce dielectric and ohmic losses. Under these conditions the effective area as defined in the above equation is the *maximum effective area.*

Real aperture antennas, like electromagnetic horns, have effective apertures which are smaller than the physical ones. For electromagnetic horns, the effective aperture is in the order of 0.5 to 0.7 times the value of the physical aperture [8].

The maximum effective aperture of an antenna may be related to its directivity, this relation is derived in appendix C and results in

$$D = \frac{4\pi A_{em}}{\lambda^2}. \tag{3.51}$$

We see that an increase in (effective) area leads to an increase in directivity. Antenna beamwidth therefore is inversely proportional to aperture size.

The *effective aperture*, A_e, is related to the *maximum effective aperture*, A_{em}, through the radiation efficiency, η that accounts for ohmic and dielectric losses in the antenna

$$A_e = \eta A_{em}. \tag{3.52}$$

With equation (3.15) we then find the following relation between effective aperture and gain

$$G = \frac{4\pi A_e}{\lambda^2}. \tag{3.53}$$

Example To show that the concept of effective area is a mathematical abstraction that does not need to be related to physical area, we will calculate the effective area of a short dipole antenna of length $l \ll \lambda$ and negligible diameter [1, 2].

The antenna and its equivalent circuit are shown in figure 3.36. For educational reasons we assume antenna and load impedances to be real. Furthermore, we assume the short dipole to be lossless (which is not true in practice!).

The open circuit voltage of the short dipole antenna is

$$V = |\mathbf{E}_i|\, l = E_i l, \tag{3.54}$$

where E_i is the amplitude of the incoming (lined up) linearly polarised electric field. The maximum available power, P_a, realised when $R_L = R_A$, is given by

$$P_a = \frac{|V|^2}{8R_A} = \frac{|E_i|^2 l^2}{8R_A}, \tag{3.55}$$

where the short dipole radiation resistance, R_A may be calculated as [1, 2, 8]

$$R_A = 80\pi^2 \left(\frac{l}{\lambda}\right)^2, \tag{3.56}$$

where λ is the wavelength used.

The power density in the far-field region of the source, S, may be calculated as

$$S = \frac{1}{2} |\mathbf{E}_i \times \mathbf{H}_i^*| = \frac{1}{2} \frac{|E_i|^2}{Z_0}, \tag{3.57}$$

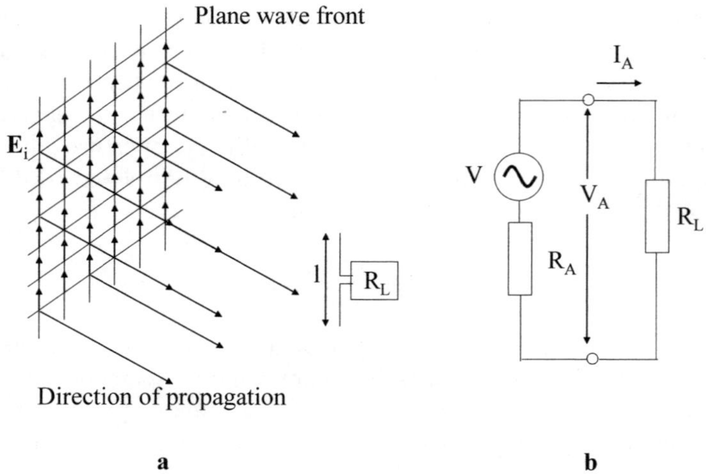

Fig. 3.36 Receiving short dipole antenna and equivalent circuit. a. Antenna connected to receiver with impedance R_L. b. Equivalent circuit.

where Z_0 is the intrinsic impedance of free space, which is equal to $Z_0 = 120\pi$ (Ω).

With equations (3.55), (3.56) and (3.57), we then find for the maximum effective area, A_{em}, of a short dipole (that has a negligible physical area!)

$$A_{em} = \frac{P_a}{S} = \frac{3}{8\pi}\lambda^2. \tag{3.58}$$

An effective dimension associated with straight-wire antennas, that appeals more to our intuition is the *vector effective length.*

3.4.2 Vector Effective Length

The *vector effective length* or *vector effective height* is used to determine the open circuit voltage induced on the antenna terminals, when a plane wave is incident on the antenna. When polarisations of plane wave and antenna are lined up, the effective length of a short dipole antenna is identical to its physical length, l, see equation (3.54) in the example above.[7] In general, the effective length is a direction-dependent quantity [1].

$$\mathbf{l}_e(\vartheta, \varphi) = l_\vartheta(\vartheta, \varphi)\mathbf{u}_\vartheta + l_\varphi(\vartheta, \varphi)\mathbf{u}_\varphi. \tag{3.59}$$

The open circuit voltage of an antenna is then obtained by projecting the incident electric field vector on the complex conjugate of the vector effective length [2]

$$V_{oc} = \mathbf{E}_i \cdot \mathbf{l}_e^*, \tag{3.60}$$

[7] For the short dipole in the example it was implicitly assumed that the current over the short dipole antenna is uniform. For any current distribution other than uniform, physical and effective length are not identical anymore.

where the complex conjugate is used to correct for the fact that the vector effective length is associated with the transmitting case, while the open circuit voltage is obtained in the receiving case.

The incident field vector - radiated by a transmitting antenna - may be expressed in terms of vector effective length according to [1, 9]

$$\mathbf{E}_i = -jZ_0 \frac{kI_{in}}{4\pi r} \mathbf{l}_e e^{-jkr}, \tag{3.61}$$

where I_{in} is the current at the input terminals of the transmitting antenna, r is the distance between transmitting and receiving antenna and $k = 2\pi/\lambda$.

Example The radiated electric far-field of a vertically oriented half-wave dipole antenna is given by [2]

$$\mathbf{E}_i = jZ_0 \frac{kI_{in}}{4\pi r} \left(\frac{\lambda}{\pi}\right) e^{-jkr} \sin(\vartheta) \frac{\cos\left[\left(\frac{\pi}{2}\right)\cos(\vartheta)\right]}{\sin^2(\vartheta)} \mathbf{u}_\vartheta \tag{3.62}$$

Substitution of equation (3.62) for $\vartheta = \frac{\pi}{2}$ into equation (3.61) gives for the effective length of a half-wave dipole antenna

$$|\mathbf{l}_e| = l_\vartheta = \frac{\lambda}{\pi} = 0.32\lambda. \tag{3.63}$$

The effective length of a half-wave dipole antenna is smaller than its physical length, just like we have seen that the effective aperture of an electromagnetic horn antenna is smaller than its physical size.

The concept of effective aperture or length is based on the principle that an imaginary aperture or wire antenna is conceived that intercepts the same amount of power as the original antenna does, but it does so *uniformly*, unlike the original antenna.[8] Only when the original antenna is very small, the power interception approximately takes place uniformly and effective and physical dimensions get close to one another. The concept of effective aperture is not restricted to physical aperture antennas and, likewise, the concept of effective length may also be applied to non-wire (aperture) antennas.

3.5 RADIO EQUATION

As stated before, an antenna is never used as a stand-alone component, but will always be part of a communication or radar system. When employed in a communication system, we have to deal with at least two antennas: a transmitting antenna and a receiving antenna. We assume that both antennas are lined up with respect to polarisation and maximum directivity and that they are positioned in each other's far-field regions, see figure 3.37.

[8] We have seen that the half-wave dipole has a sinusoidal current distribution over the wire. For the determination of the effective length, a constant current is assumed. Therefore, the effective length (0.32λ) is shorter than the physical length (0.50λ).

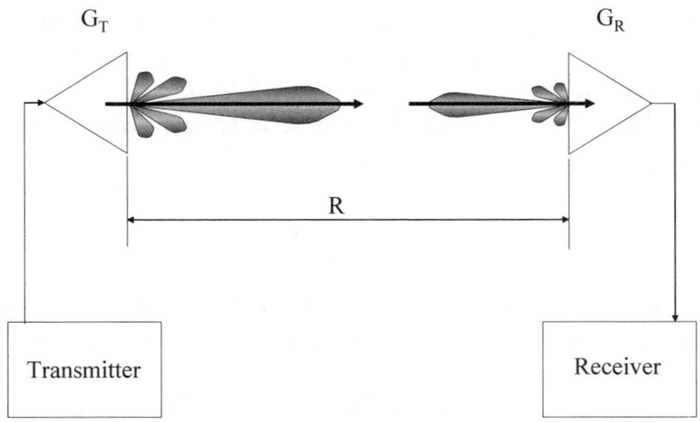

Fig. 3.37 Communication system consisting of a transmitter having an antenna with gain G_T and a receiver having an antenna with gain G_R. The antennas are displaced a distance R and assumed to be in each other's far-field regions and lined up with respect to polarisation and directivity.

The gain of the transmitting antenna is G_T, the gain of the receiving antenna is G_R.

The power density S at distance R from the transmitting antenna is

$$S = G_T \frac{P_T}{4\pi R^2}, \tag{3.64}$$

where P_T is the *input* power at the terminals of the transmitting antenna. The factor $1/(4\pi R^2)$ accounts for the spherical spreading of the energy. Since the transmit antenna is non-isotropic, this factor is multiplied with the gain of the antenna.

The amount of power intercepted by the receiving antenna, P_R is this power density, multiplied with the effective area of the receive antenna, A_{eR}. This effective area is directly related to the gain of the receive antenna

$$P_R = S A_{eR} = G_T \frac{P_T}{4\pi R^2} G_R \frac{\lambda^2}{4\pi}. \tag{3.65}$$

After rearranging the terms of this equation, we may relate the power received at the terminals of the receive antenna, P_R to the power delivered at the terminals of the transmit antenna, P_T

$$\frac{P_R}{P_T} = \left(\frac{\lambda}{4\pi R}\right)^2 G_R G_T. \tag{3.66}$$

This equation is known as the *radio equation* or *Friis transmission equation*. The term $(\lambda/(4\pi R))^2$ is known as the *free space loss factor*.

Example Assume a broadcasting system, operating at 100MHz, employing a half-wave dipole antenna, having a gain of 2.15dBi.[9] The power accepted by the transmit antenna is 1kW.

[9] The units of dBi refer to a gain relative to an isotropic radiator. Another unit sometimes encountered is the dBd, the gain relative to an elementary dipole radiator (2.15dBi = 0.0dBd).

The minimum required power delivered by the receiving antenna is 1nW. When the maximum range is 500km, what should be the minimum gain of the receiving antenna?

The required gain is given by

$$G_R = \frac{16\pi^2 R^2 P_R}{\lambda^2 P_T G_T}. \tag{3.67}$$

The wavelength is given by $\lambda = c_0/f$, where c_0 is the speed of light in vacuum ($c_0 \approx 3 \cdot 10^8 \text{ms}^{-1}$) and f is the frequency. The wavelength therefore is 3.0m. The gain of the transmitting antenna is $G_T = 10^{2.15/10} = 1.64$.[10] The desired minimum gain is then found to be

$$G_R = \frac{16 \cdot 9.87 \cdot 2.5 \cdot 10^{11} \cdot 10^{-9}}{9 \cdot 10^3 \cdot 1.64} = 2.67, \tag{3.68}$$

or $G_R = 10\log(2.67) = 4.27\text{dB}_i$.

Mind that the equation is solved for the situation where both antennas are lined up with respect to gain and polarisation.

Example Consider a mobile communication system consisting of two identical transmitter-receiver sets operating at 1GHz. The same half-wave dipole antenna is used both for transmission and for reception. $G_T = G_R = G = 2.15$dBi. The power delivered to the antenna in transmission is 1W. The minimum power at the antenna terminals in reception is −65dBm. Find the maximum allowable distance, R, between the two sets.

The maximum allowable distance is given by

$$R = \frac{\lambda}{4\pi} G \sqrt{\frac{P_T}{P_R}}. \tag{3.69}$$

The wavelength is $c_0/f = 3 \cdot 10^8/10^9$ =0.30m. $G = 1.64$. P_T =1W. The minimum power at reception is expressed in dBm, meaning decibels with respect to 1mW. Therefore, $P_R = 10^{-65/10}\text{mW} = 10^{-65/10} \cdot 10^{-3}\text{W} = 3.16 \cdot 10^{-10}\text{W}$. The maximum distance is then found to be

$$R = \frac{0.30}{12.57} 1.64 \sqrt{\frac{1}{3.16 \cdot 10^{-10}}} = 2.20\text{km}. \tag{3.70}$$

3.6 RADAR EQUATION

A radar system is very much like the communication system described in the previous section. The difference is that instead of a direct link, the electromagnetic waves, emitted by the transmitter T in figure 3.38a, now reach the receiver, R, via a reflection against a target.

[10]The product $P_T G_T$ is known as the *effective isotropic radiated power* (EIRP). The EIRP is the power intensity that could have been obtained from an isotropic radiator if it had an input power, P_{in} equal to the EIRP.

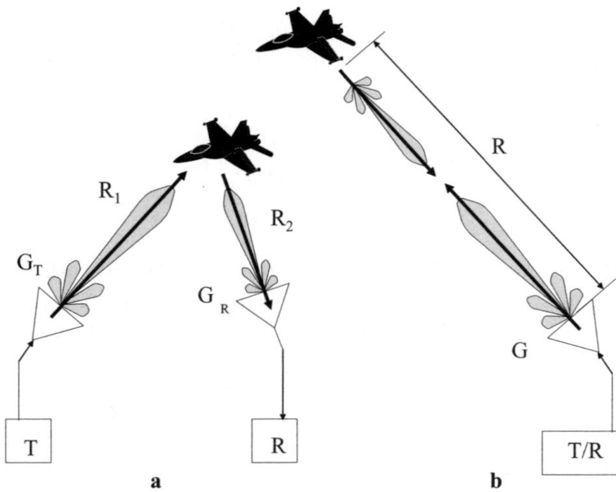

Fig. 3.38 Radar system a. Bistatic radar. The transmitter is connected to an antenna with gain G_T and is at a distance R_1 from the reflecting target. The receiver has an antenna with gain G_R and is at a distance R_2 from the target. b Monostatic radar. The same antenna, having gain G, is used for transmission and reception. The distance to the target is R. The target is assumed to be in each antenna's far-field region.

The power density incident on the target, S_i, is given by

$$S_i = G_T \frac{P_T}{4\pi R_1^2}, \tag{3.71}$$

where G_T is the gain of the transmit antenna, P_T is the power at the input of the transmit antenna and R_1 is the distance between transmit antenna and target. It is implicitly assumed that the beam of the transmit antenna is directed to the target.

The power intercepted by the target, P_i, is proportional to the power density, S_i and is given by

$$P_i = \sigma S_i, \tag{3.72}$$

where σ (m^2) is known as the *radar cross-section* (RCS) of the target.

3.6.1 Radar Cross-Section

The radar cross-section of a target is the equivalent area intercepting that amount of power that, when scattered equally in all directions, produces an echo at the radar equal to that coming from the target [7].

There is in general not a simple rule, relating the physical size of a target to its radar cross-section, although, in general, larger targets exhibit larger RCSs. Table 3.3 [1, 7] gives an overview of some typical RCS values.

We now return to the radar equation.

Table 3.3 Typical RCS values.

Object	RCS (m^2)
Conventional, winged missile	0.5
Small fighter or 4-passenger jet	2
Large fighter	6
Medium bomber or medium jet airliner	20
Large bomber or large jet airliner	40
Jumbo jet	100
Small pleasure boat	2
Cabin cruiser	10
Pickup truck	200
Automobile	100
Bicycle	2
Man	1
Bird	0.01
Insect	0.00001

The power density at the position of the receiver after scattering from the target, S_s is given by

$$S_s = \frac{P_i}{4\pi R_2^2} = \frac{\sigma G_T P_T}{(4\pi)^2 R_1^2 R_2^2}. \tag{3.73}$$

The power available at the receiver, P_R, is

$$P_R = A_{er} S_s, \tag{3.74}$$

where A_{er} is the effective aperture of the receiving antenna. With use of equation (3.53), we find

$$P_R = P_T \frac{\sigma G_T G_R \lambda^2}{(4\pi)^3 R_1^2 R_2^2}. \tag{3.75}$$

This equation, relating received and transmitted power, is known as the *radar equation*.

The setup with a separate transmit and receive antenna, physically displaced, see figure 3.38a, is called *bistatic radar*. The more common *monostatic radar*, see figure 3.38b, uses the same antenna both for transmission and reception. The consequences for the radar equation are that $G_T = G_R = G$ and $R_1 = R_2 = R$, leading to

$$P_R = P_T \frac{\sigma G^2 \lambda^2}{(4\pi)^3 R^4}. \tag{3.76}$$

Example It is necessary to detect a target with a RCS of 1m^2 at a range of 150km. A monostatic radar is used. The gain of the antenna employed is 40dB at a frequency of 3GHz. The minimum power at the terminals of the antenna in receiving mode is −100dBm.

What is the transmitting power needed and what is the size of the effective antenna aperture?

The transmitting power is given by

$$P_T = P_R \frac{(4\pi)^3 R^4}{\sigma G^2 \lambda^2}. \tag{3.77}$$

$G = 10^{40/10} = 10^4$ and $\lambda = 3 \cdot 10^8 / 3 \cdot 10^9 = 10^{-1}\text{m}$. $P_R = 10^{-100/10}\text{mW} = 10^{-13}\text{W}$. The transmitted power is then found to be

$$P_T = 10^{-13} \frac{1984.40\left(150 \cdot 10^3\right)^4}{10^8 \cdot 10^{-2}} \text{W} = 100.46\text{kW}. \tag{3.78}$$

The effective aperture is

$$A_e = \frac{G\lambda^2}{4\pi} = \frac{10^4 \cdot 10^{-2}}{12.57} \text{m}^2 = 7.96\text{m}^2. \tag{3.79}$$

Example Answer the same questions if the frequency is now 1GHz. We assume that the antenna is replaced by an antenna that has a gain of 40dBi at 1GHz.

The wavelength has changed to $\lambda = 0.30\text{m}$. The transmitted power is now found to be

$$P_T = 10^{-13} \frac{1984.40\left(150 \cdot 10^3\right)^4}{10^8 \cdot 0.30^2} \text{W} = 11.16\text{kW}. \tag{3.80}$$

The effective aperture of the antenna is

$$A_e = \frac{G\lambda^2}{4\pi} = \frac{10^4 \cdot 0.30^2}{12.57} \text{m}^2 = 71.60\text{m}^2. \tag{3.81}$$

Thus, at a lower frequency we may detect targets using less power, but we need larger antennas.

REFERENCES

1. Contantine E. Balanis, *Antenna Theory, Analysis and Design, second edition*, John Wiley & Sons, New York, 1997.

2. Warren L. Stutzman and Gary A. Thiele, *Antenna Theory and Design*, John Wiley & Sons, New York, 1998.

3. R.C. Johnson (ed.) *Antenna Engineering Handbook, third edition*, McGraw-Hill, New York, 1993.

4. Roger F. Harrington, *Time-Harmonic Electromagnetic Fields*, John Wiley & Sons, New York, 2001.

5. David M. Pozar, *Microwave Engineering, second edition*, John Wiley & Sons, New York, 1998.

6. G. Murakami and G.S. Wickizer, Ionosphere Phase Distortion and Faraday Rotation and Radio Waves, *RCA Review*, Vol. 30, 1969, pp. 488–491.

7. Merill I. Skolnik, *Introduction to Radar Systems*, McGraw-Hill, Auckland, 1981.

8. John D. Kraus, *Antennas*, McGraw-Hill, New York, 1950.

9. R.E. Collin and F.J. Zucker, *Antenna Theory, Part I*, McGraw-Hill, New York, 1969.

4

The Linear Broadside Array Antenna

With the just gained physical, or qualitative, understanding of electromagnetic radiation and antennas - including array and phased array antennas - we are now capable of developing a quantitative understanding of these antennas. We will start with the relatively easy-to-understand linear broadside array antenna, i.e. a system of radiators positioned along a straight line, producing a beam perpendicular to this line. The specifics of the linear array antenna will be outlined, making use of the antenna parameters described in the previous chapter. The mathematics needed for the derivation of these parameters for the linear array rely mostly on straightforward goniometrics.

4.1 A LINEAR ARRAY OF NON-ISOTROPIC POINT-SOURCE RADIATORS

Let's assume a system of identical radiators, placed at equidistant positions along a straight line, see figure 4.1. This system is a so-called *linear array antenna.* Neither the identicalness of the radiators, nor the equidistant positions are prerequisites for a linear array antenna. They are introduced though to avoid obscuring the explanation of the linear array antenna basics. Besides, in practical situations, linear array antennas are often realised with identical radiators that are equidistantly positioned.

Also for clarity reasons, we assume that the individual radiators that make up the array antenna do not occupy any volume, but we allow them to have a non-isotropic radiation pattern. In other words, we assume the radiators to be - physically not realisable - *non-isotropic point sources.* Thereby we allow the radiators to have directivity, i.e. have a non-trivial radiation pattern, but for the moment we do not have to bother with restrictions in positioning the elements due to their own physical dimensions since we simply assume these physical dimensions to be non-existent.

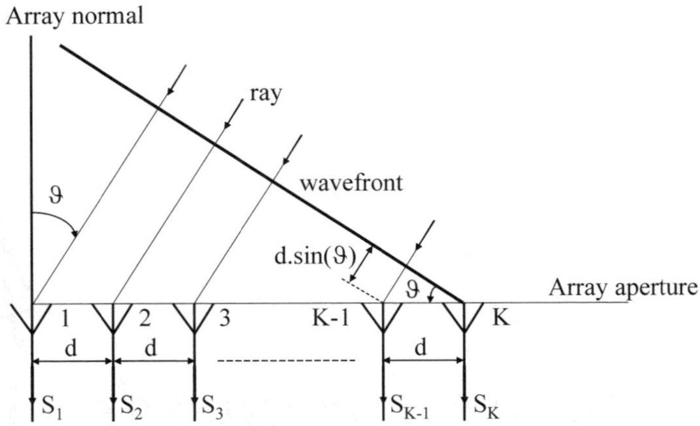

Fig. 4.1 A linear array of K radiators, equidistantly positioned along a straight line, where a plane wave is incident under an angle ϑ with respect to the array normal.

Furthermore, for explaining the array antenna basics, we consider the array antenna to be a receiving antenna. This is not a restriction, since by virtue of the *reciprocity theorem* [1, 2], we know that the characteristics of an antenna when used as a transmitting antenna are identical to those when used as a receiving antenna.

Accepting these restrictions, we assume, see figure 4.1, that the wavefront of a plane wave is incident upon the linear array antenna under an angle ϑ with the array aperture. The wavefront is perpendicular to the direction of the plane wave. This direction is indicated by *rays* in the figure.

A wavefront is defined by the characteristic that all points on the wavefront have equal amplitude and phase values.

A point source emits waves having spherical wavefronts. Also the wavefronts emitted by a real-life antenna, when observed at a large distance from this antenna, may be considered as being spherical. When observed over a finite area - as is usually the case since our receiving antenna is of finite size - the wavefronts may be regarded as being locally planar, see figure 4.2.

4.2 PLANE WAVES

We have already seen that a time harmonic signal, $s(t)$, may be represented as

$$s(t) = A\cos(\omega t) = A\Re\left\{e^{j\omega t}\right\}. \tag{4.1}$$

Herein, $\Re\{x\}$ is the real part of the complex argument x. The angular frequency ω is related to the frequency f through $\omega = 2\pi f$.

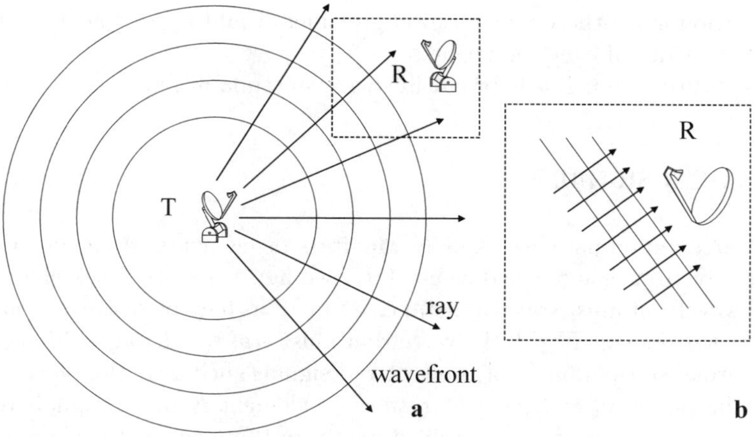

Fig. 4.2 Wavefronts. a. Spherical wavefronts at a large distance from a transmitting antenna T, received by a receiving antenna R. b. Locally planar wavefronts.

A is the amplitude of signal $s(t)$. The signal $s(t)$ is periodic, which means that the signal value at instant $t = t_0$ will be repeated at instances of time $t = t_0 + n\frac{1}{f}$, where n is an integer number.

A wave is a disturbance in both *time* and *space*. A wave function therefore is a time harmonic function that not only repeats itself in time, but also in space:

$$s'(t) = A' \cos\left(\omega t \pm \frac{2\pi}{\lambda} l\right) = A' \Re\left\{e^{j\left(\omega t \pm \frac{2\pi}{\lambda} l\right)}\right\} = A' \Re\left\{e^{j(\omega t \pm kl)}\right\}, \tag{4.2}$$

where l is the distance along the propagation path and A' is the amplitude of signal $s'(t)$.

We see that for l being an integer multiple of *wavelengths* λ, the signal $s'(t)$ repeats itself. The parameter $k = \frac{2\pi}{\lambda}$ is called the *wave number*.

The phase of signal $s'(t)$ - denoted $\Phi'(t)$ - is given by

$$\Phi'(t) = \omega t \pm kl. \tag{4.3}$$

The minus sign indicates a wave travelling away from the source, while the plus sign indicates a wave travelling towards the source. This can be easily understood by keeping the phase, $\Phi'(t)$, constant - as should be done for following a wavefront - while increasing the time, t. The phase can only be kept constant by increasing the distance l if the minus sign applies and by decreasing the distance l if the plus sign applies.

It is customary in technical publications to mention the time dependency $e^{j\omega t}$ once and suppress is further on.[1] Adding the time dependency in every equation would not

[1] Also a time dependency according to $e^{-j\omega t}$ is used sometimes. Therefore the time periodicity needs to be mentioned once *explicitly*.

add information since the exponential expressions would appear as a multiplication factor on both sides of every equation.

We now return to our linear broadside array antenna of figure 4.1.

4.3 RECEIVED SIGNAL

Our linear array antenna consists of K radiators or elements. At a certain moment, the planar wavefront has reached element K, see figure 4.1. To reach element $K-1$, the planar wavefront must travel a distance $d\sin(\vartheta)$, as may be distilled from the same figure. To reach element $K-2$, the wavefront must travel a distance $2d\sin(\vartheta)$, and so on. If we normalise the phases of the received signals such that the phase at element K is zero, the phase *differences* with respect to element K of the signals received by the other elements represent the received phases of these elements. These phases, Φ_i, are obtained by multiplying the path lengths with the free space wave number, k_0, as follows from equation (4.3) after suppression of the $e^{j\omega t}$ time-dependency

$$\Phi_i = k_0(K-i)d\sin(\vartheta) \text{ for } i = 1, 2, \ldots, K, \tag{4.4}$$

where

$$k_0 = \frac{2\pi}{\lambda_0}. \tag{4.5}$$

Herein, λ_0 is the wavelength in free space.

With the term *received signal*, the current flowing through the clamps of the element or the amplitude of the guided wave travelling through the waveguide connected to the element is meant.[2] The exact nature of this signal (current, voltage or current or voltage wave amplitude) is of no concern for the explanation of the array antenna basics.

The *complex* signals received by the elements of the array antenna, $S_i(\vartheta)$, may be written as

$$S_i(\vartheta) = S_e(\vartheta)a_i e^{jk_0(K-i)d\sin(\vartheta)} \text{ for } i = 1, 2, \ldots, K, \tag{4.6}$$

where $S_e(\vartheta)$ represents the complex radiation pattern of one - isolated - individual radiator and a_i is the amplitude received by the ith element. For the moment we will assume that all amplitudes received by the elements are equal and normalised to one, i.e.

$$a_i = 1 \text{ for } i = 1, 2, \ldots, K. \tag{4.7}$$

We call this a *uniform aperture distribution.*

If we combine all received signals without introducing additional phase differences between the elements, we may simply add the received signals described by equation (4.6) for all elements i. The total received signal, $S(\vartheta)$, is then found to be

$$S(\vartheta) = \sum_{i=1}^{K} S_i(\vartheta) = S_e(\vartheta) \sum_{i=1}^{K} e^{jk_0(K-i)d\sin(\vartheta)}. \tag{4.8}$$

[2]The waveguide may be a hollow rectangular waveguide, a two-wire or a coaxial transmission line, for example.

Combining the received signal without introducing additional phase differences may be accomplished by using a feeding or summing network (in the appropriate waveguide technology) that ensures equal path lengths to all elements of the array. Such a feeding network is schematically shown in figure 4.3.[3]

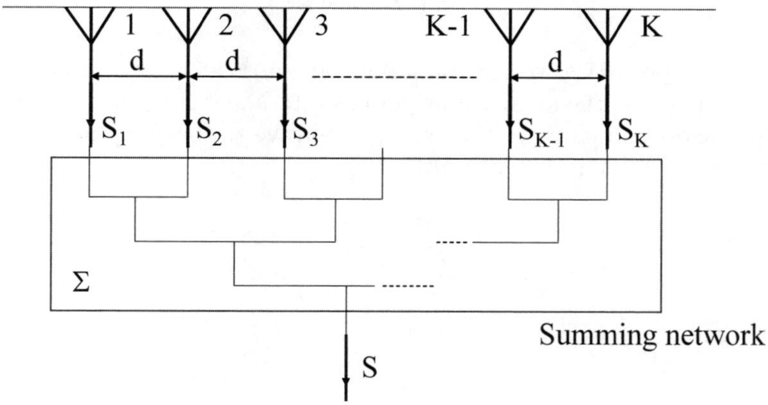

Fig. 4.3 Linear array antenna with equal path length summing network.

In writing down equation (4.8), we have implicitly assumed that the radiation pattern of an individual radiator remains the same upon placing this radiator in an array environment. Apart from the interaction introduced by the inter-element distances and the phase differences these distances cause, no further interaction between the radiators is assumed. In other words: 'we neglect *mutual coupling* between the radiators'. Although situations may arise where this assumption is valid, in general it is not true. For the explanation of array antenna basics though we neglect mutual coupling effects for the moment. Later on we will return to this phenomenon.

Having accepted this, we return to equation (4.8) and see that the received signal may be separated in a component due to a single radiator and in a component due to the *array configuration* only

$$S(\vartheta) = S_e(\vartheta)S_a(\vartheta), \tag{4.9}$$

where $S_e(\vartheta)$ is known as the *element factor* and

$$S_a(\vartheta) = \sum_{i=1}^{K} e^{jk_0(K-i)d\sin(\vartheta)} \tag{4.10}$$

is known as the *array factor*.

The element factor is the radiation pattern of a single radiator, the array factor is the radiation pattern of an array of K *isotropic* radiators. The radiation pattern of

[3]This type of feeding arrangement is known as a *corporate feeding network*.

the linear array antenna, $S(\vartheta)$, is obtained by multiplying the element factor, $S_e(\vartheta)$, with the array factor, $S_a(\vartheta)$. This operation is known as *pattern multiplication.*

Example Consider a linear array antenna consisting of eight elements. The element voltage radiation pattern is given by

$$S_e(\vartheta) = \cos(\vartheta). \tag{4.11}$$

Given this hypothetical voltage radiation pattern,[4] calculate and show the element factor power pattern, the array factor power pattern and the power radiation pattern of the total array as function of the angle ϑ relative to the array normal (broadside) for the following element distances d:

1. $d = \frac{\lambda_0}{4}$;

2. $d = \frac{\lambda_0}{2}$;

3. $d = \lambda_0$;

4. $d = \frac{5\lambda_0}{4}$.

Using equations (4.9), (4.10) and (4.11) results in the radiation power patterns shown in figures 4.4, 4.5, 4.6 and 4.7 for, respectively, $d = \frac{\lambda_0}{4}$, $d = \frac{\lambda_0}{2}$, $d = \lambda_0$ and $d = \frac{5\lambda_0}{4}$. The element power pattern is calculated as $20\log(|S_e(\vartheta)|)$, the normalised array factor power pattern is calculated as $20\log(|S_a(\vartheta)|/8)$ and the normalised power pattern of the total array is calculated as $20\log(|S_e(\vartheta)|\,|S_a(\vartheta)|/8)$.

The first thing that strikes us when taking a closer look at figures 4.4 to 4.7 is that our expectations, based on the physical reasoning developed in the preceding chapters, come true. We clearly see that the main beam of the linear array antenna gets smaller when the elements occupy a larger area, i.e. when the element distance increases. Furthermore, we see that after passing a critical element distance, a further increase of the element distance leads to the introduction of additional main beams.

We also see that the total linear array antenna behaviour is dominated by the array factor. The directive properties of the elements merely act as an angular filter that lowers the radiated power of the array antenna for angles getting closer to *endfire*, i.e. directions parallel to the linear array antenna. Due to the dominant character of the array factor we will discuss this array factor in more detail.

[4] Although the element radiation pattern is a hypothetical one, it bears a strong resemblance with the radiation pattern of a horizontal half-wave dipole antenna as may be seen by comparing the (power) radiation pattern with the ones shown in the previous chapter for a vertical half-wave dipole antenna. The pattern also resembles that of a slot in an infinite ground plane. Assuming these latter elements, we only need to consider the radiation pattern in the upper hemisphere, $-90° \leq \vartheta \leq 90°$.

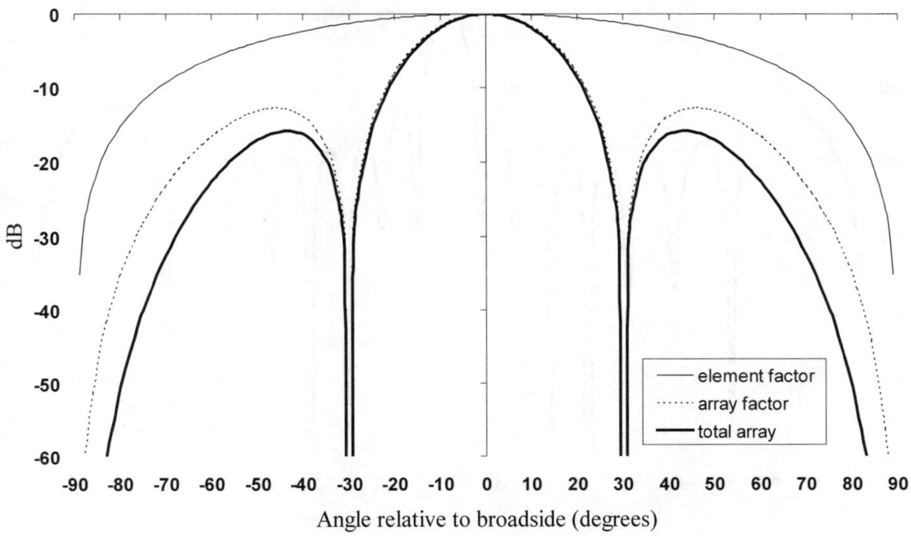

Fig. 4.4 Power radiation patterns of the element factor, the array factor and the total array of a linear 8-element broadside array with element distance $d = \frac{\lambda_0}{4}$.

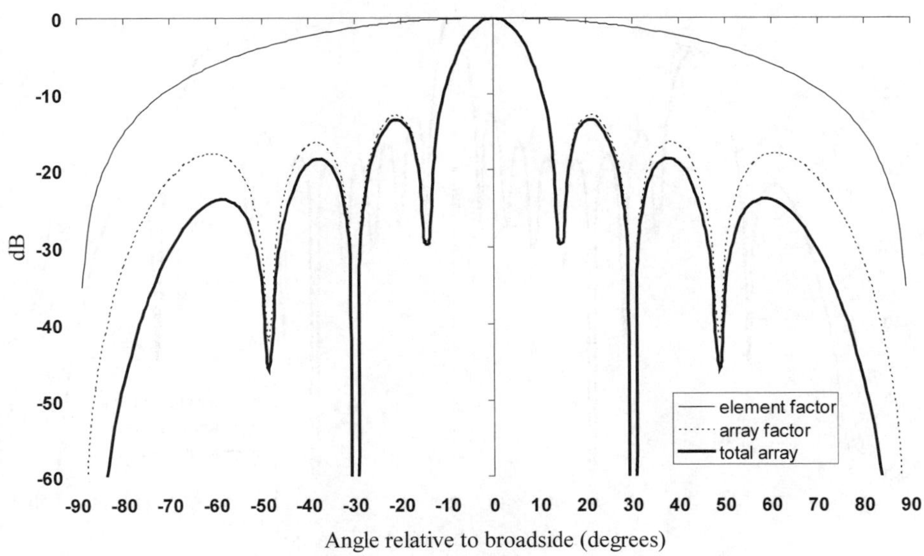

Fig. 4.5 Power radiation patterns of the element factor, the array factor and the total array of a linear 8-element broadside array with element distance $d = \frac{\lambda_0}{2}$.

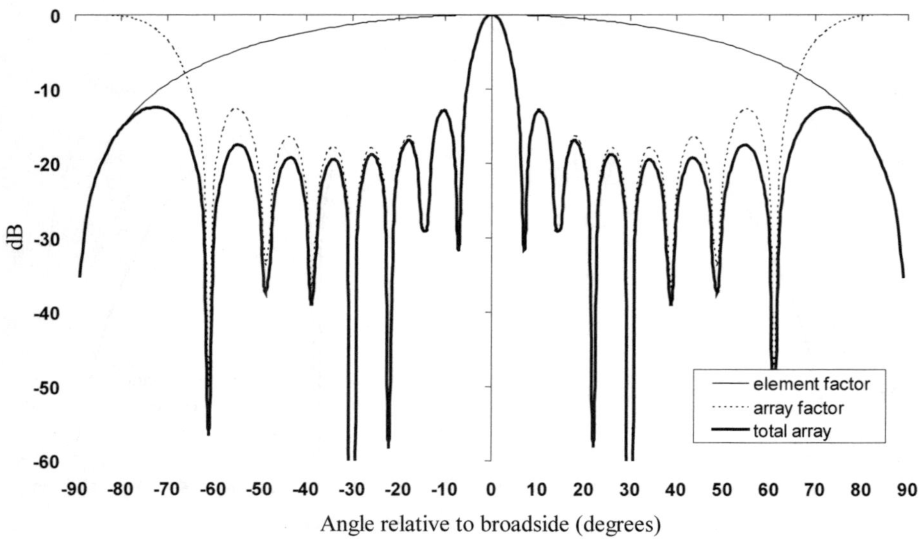

Fig. 4.6 Power radiation patterns of the element factor, the array factor and the total array of a linear 8-element broadside array with element distance $d = \lambda_0$.

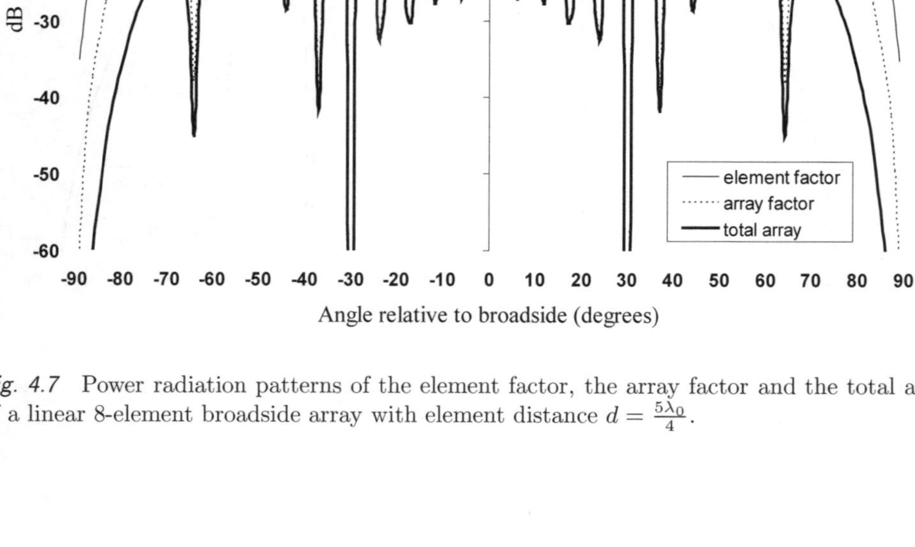

Fig. 4.7 Power radiation patterns of the element factor, the array factor and the total array of a linear 8-element broadside array with element distance $d = \frac{5\lambda_0}{4}$.

4.4 ARRAY FACTOR

The array factor is given by equation (4.10), which can be rewritten - in more compact form - into

$$S_a(\vartheta) = \sum_{i=1}^{K} e^{jk_0(K-i)d\sin(\vartheta)} = \sum_{i=1}^{K} e^{j(K-i)T}, \tag{4.12}$$

where

$$T = k_0 d \sin(\vartheta). \tag{4.13}$$

In equation (4.12) we may recognise a finite geometric series.
If we multiply both sides of this equation with e^{jT}, we get

$$S_a(\vartheta)e^{jT} = e^{jKT} + e^{j(K-1)T} + \cdots + e^{j2T} + e^{jT}. \tag{4.14}$$

Next, we subtract equation (4.12) from equation (4.14) and thus obtain

$$S_a(\vartheta)\left(e^{jT} - 1\right) = \left(e^{jKT} - 1\right), \tag{4.15}$$

which may be written, after splitting and rearranging the exponential terms, as

$$S_a(\vartheta) = \frac{e^{j\frac{KT}{2}}\left(e^{j\frac{KT}{2}} - e^{-j\frac{KT}{2}}\right)}{e^{j\frac{T}{2}}\left(e^{j\frac{T}{2}} - e^{-j\frac{T}{2}}\right)} = e^{j\frac{K-1}{2}T}\frac{\sin\left(\frac{K}{2}T\right)}{\sin\left(\frac{1}{2}T\right)}, \tag{4.16}$$

so that, finally

$$|S_a(\vartheta)| = \left|\frac{\sin\left(\frac{K}{2}k_0 d\sin(\vartheta)\right)}{\sin\left(\frac{1}{2}k_0 d\sin(\vartheta)\right)}\right| = \left|\frac{\sin\left(\pi\frac{Kd}{\lambda_0}\sin(\vartheta)\right)}{\sin\left(\pi\frac{d}{\lambda_0}\sin(\vartheta)\right)}\right|. \tag{4.17}$$

4.5 SIDE LOBES AND GRATING LOBES

With the thus found expression for the radiation pattern of a linear array antenna consisting of isotropic radiators, we may analyse some characteristics of *side lobes* and *grating lobes*.

4.5.1 First Side-Lobe Level

Equation (4.17) shows that the maximum of the (voltage) array factor occurs for $\vartheta = 0$ and is equal to K. For angles ϑ close to broadside, i.e. around the main beam, we may approximate the array factor of equation (4.17) using an approximation for the sine function for small arguments, $\sin(x) \approx x$. This leads to the following expression for the absolute value of the array factor near broadside

$$|S_a(\vartheta)| \approx K\left|\frac{\sin(Kx)}{Kx}\right|, \tag{4.18}$$

where

$$x = \pi \frac{d}{\lambda_0} \sin(\vartheta). \tag{4.19}$$

We have specifically chosen for this representation, with a factor K in both the numerator and denominator.

It is easily verified, see for example [2], appendix I, that the second maximum of $\left|\frac{\sin(Kx)}{Kx}\right|$ appears for the argument being equal to approximately $Kx \approx 4.5$ and that the amplitude of this maximum is equal to 0.21723.

Therefore the level of the radiation power pattern's first side lobe will be approximately $20 \log 0.21723 = -13.26\text{dB}$, down relative to the pattern maximum.

Comparing this value with the first side lobe levels of the array factor as shown in figures 4.4, 4.5, 4.6 and 4.7 indicates the validity of our approximation for the array factor in the vicinity of broadside.

4.5.2 Grating Lobes

Equation (4.17) not only shows that the maximum of the (voltage) array factor (K) occurs for $\vartheta = 0$. The equation also shows that the array factor is a periodic function of ϑ and that multiple maxima occur whenever

$$\pi \frac{d}{\lambda_0} \sin(\vartheta) = m\pi \text{ for } m = 1, 2, 3, \ldots, \tag{4.20}$$

i.e. for the argument of the sine function being an integer multiple of π. Note that due to the *absolute* value of the sine function appearing in the expression for the radiation pattern, the periodicity has become π instead of 2π.

From the above equation, a restriction for the element distance follows that ensures having only one maximum ($m = 1$) within the range $-\vartheta_{max} \leq \vartheta \leq \vartheta_{max}$

$$\frac{d}{\lambda_0} \leq \frac{1}{|\sin(\vartheta_{max})|}. \tag{4.21}$$

If we do not want to have secondary maxima or lobes within the whole angular range, $-\frac{\pi}{2} \leq \vartheta \leq \frac{\pi}{2}$, the restriction becomes more severe:

$$\frac{d}{\lambda_0} \leq \frac{1}{\left|\sin\left(\frac{\pi}{2}\right)\right|} = 1. \tag{4.22}$$

If this condition is not met, more than one maximum will occur. These extra maxima are known as *grating lobes*, a term originating from optics.

In our example, grating lobes are occurring for the situation where $d = \lambda_0$, see figure 4.6 and for the situation where $d = \frac{5\lambda_0}{4}$, see figure 4.7. Figure 4.6 reveals that keeping too strict to the grating lobe condition as stated in equation (4.22) does not prevent grating lobe effects to be not present at all in the radiation pattern. Choosing an inter-element distance too close to one wavelength, still results in unwanted effects in the radiation pattern due to the *slopes* of the first grating lobes. Thus, keeping the grating lobe *maximum* out of visible space ($-\frac{\pi}{2} \leq \vartheta \leq \frac{\pi}{2}$) alone is not enough to ensure not observing grating lobe effects. Therefore, the inter-element distance should be chosen with great care and should be *smaller* than one wavelength.

One should be aware of the fact that a *grating lobe* differs from an ordinary *side lobe.* Side lobes are the result of constructive and destructive interference from different radiating parts of the antenna. The level of a side lobe is always below that of the main beam. A grating lobe is due to the *periodicity* in the radiation pattern and is formed in directions where a maximum in-phase addition of radiated fields occur. A grating lobe should be compared with the main beam instead of to an ordinary side lobe. The level of a grating lobe is equal to that of the main beam, since a grating lobe is a repeated main beam. As can be seen from the array factor in figures 4.6 and 4.7, each grating lobe is accompanied by its own set of side lobes that are - just like the grating lobe itself - copies of the radiation pattern around broadside.

Due to the earlier mentioned 'angular filtering characteristics' of the element factor, the grating lobe phenomenon seems to be less severe in figure 4.6 than in figure 4.7. In the first figure, the element factor greatly reduces the amplitude of the grating lobe. As we can see in both figures, also the side lobe levels are affected (lowered) by the element radiation pattern.

The element radiation pattern ensures that more power is radiated in the forward direction at the expense of less power being radiated into other directions. If the element did not have these directive properties, i.e. if the element were an isotropic radiator, this favourism for the forward direction would not exist. We may employ another mechanism of broadside favourism in the array antenna to reduce the side lobe levels. Since we cannot control the element radiation pattern, we have to manipulate the array factor to create additional directive properties in there. We may accomplish this by changing our summation network in such a way that the contributions of the elements are not equal anymore but weighted.

4.6 AMPLITUDE TAPER

We start with a uniformly excited linear array, for which the array factor is given by

$$S_a(\vartheta) = \sum_{i=1}^{K} a_i e^{jk_0(K-i)d\sin(\vartheta)}, \tag{4.23}$$

where

$$a_i = 1 \text{ for } i = 1, 2, \ldots, K. \tag{4.24}$$

To avoid the results of the amplitude weighting or *tapering* being obscured by grating lobe effects, we choose an inter-element distance equal to half a wavelength, well beyond the limit as stated in equation (4.22).

The number of elements we take to be equal to eight and the element factor is given by

$$S_e(\vartheta) = \cos(\vartheta). \tag{4.25}$$

This choice will allow us to compare newly derived results with results already obtained in our earlier example.

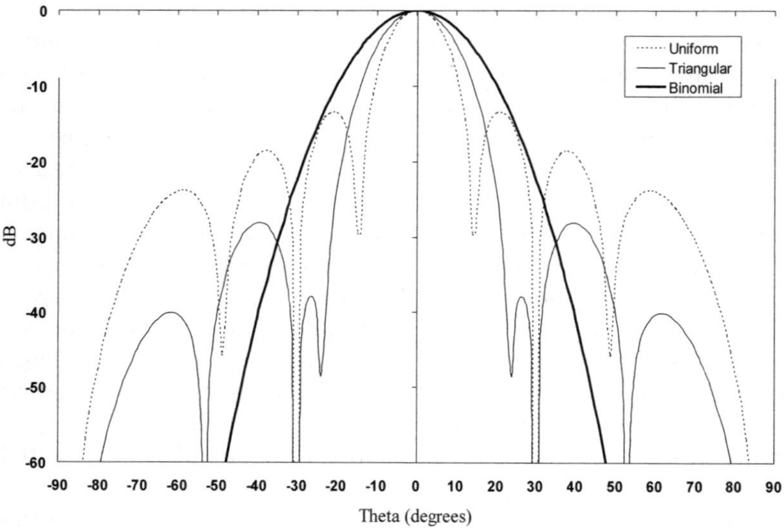

Fig. 4.8 Power radiation patterns of a linear 8-element broadside array with element distance $d = \frac{\lambda_0}{2}$ for a uniform, a triangular and a binomial amplitude taper.

The radiation pattern for the uniformly excited linear array is shown in figure 4.5 and is repeated in figure 4.8. The uniform amplitude taper is shown in figure 4.9.

The easiest way to apply an amplitude taper now, favouring centre elements over edge elements, is to linearly decrease the contribution of the array elements going from the centre to the edges of the array. A *triangular amplitude taper*, for which the a_i in equation (4.24) are - arbitrarily - chosen to be

- $a_1 = a_8 = 0.05$;
- $a_2 = a_7 = 0.10$;
- $a_3 = a_6 = 0.15$;
- $a_4 = a_5 = 0.20$,

see figure 4.9, results in the radiation pattern show in figure 4.8.

We see that this triangular amplitude taper indeed leads to a lower side lobe level, at the expense of a broader beam compared to the one obtained for a uniform amplitude taper.

It appears to be even possible to remove the side lobes all together, by choosing the amplitude coefficients equal to the coefficients of a binomial series [3]. The *binomial amplitude taper*

- $a_1 = a_8 = 0.0078$;
- $a_2 = a_7 = 0.0547$;

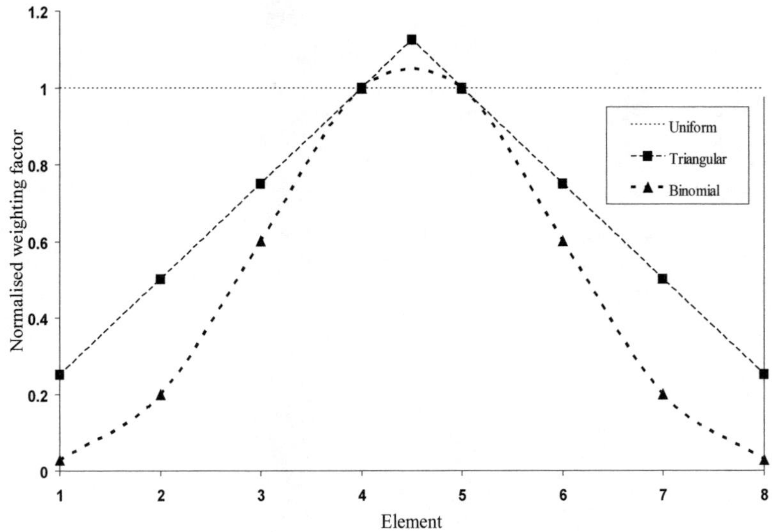

Fig. 4.9 Amplitude weighting or taper for a linear 8-element array antenna. Uniform, triangular and binomial amplitude taper, normalised to unity at the centre elements.

- $a_3 = a_6 = 0.1641$;
- $a_4 = a_5 = 0.2734$,

is shown in figure 4.9 and the resulting radiation pattern is shown in figure 4.8.

We observe a further broadening of the beam to be paid as price for the removal of the side lobes.

Having reached this point, it seems to be worthwhile to put into practice what we have learned thus far before moving on. To that purpose, we will, in the next chapter, treat the design of a linear array of microstrip patch radiators in great detail.

REFERENCES

1. Roger F. Harrington, *Time-Harmonic Electromagnetic Fields*, John Wiley & Sons, New York, 2001.
2. Contantine E. Balanis, *Antenna Theory, Analysis and Design, second edition*, John Wiley & Sons, New York, 1997.
3. Warren L. Stutzman and Gary A. Thiele, *Antenna Theory and Design, second edition*, John Wiley & Sons, New York, 1998.

5

Design of a 4-Element, Linear, Broadside, Microstrip Patch Array Antenna

In this chapter we will put the linear, broadside array antenna theory of the previous chapter into practice. We will go into every detail concerning the analysis, design and realisation of a 4-element, linear, broadside, microstrip patch array antenna, including its corporate feed network. We will develop an approximate mathematical model for calculating the radiation pattern of a microstrip patch radiator as well as its input impedance and we will develop expressions for the elements of the scattering matrix of a split-T power divider. These models will be employed in the design of the 4-element, linear, broadside, microstrip patch array antenna. When necessary, additional theoretical background material will be provided, either in the text or in appendices at the end of this book.

5.1 INTRODUCTION

Excellent full-wave electromagnetic simulation software is commercially available today. It is still necessary though for the user to have a basic understanding of the structures to analyse. Since full-wave simulation times for complicated structures - like, for example, array antennas - may become considerable, the first stages in the design process of an array antenna should be performed using basic knowledge and maybe simplified models. In this chapter, these first design stages will be discussed in detail for a 4-element, linear, broadside array antenna consisting of microstrip patch elements. We specifically choose a 4-element array antenna, operating at a frequency of 2.4GHz, since this array antenna will make it possible to demonstrate array antenna characteristics, while at the same time being of limited size making indoor radiation pattern measurements feasible (far-field condition!). The frequency is chosen to be

Array and Phased Array Antenna Basics Hubregt J. Visser

in a licence-free frequency-band and the choice for microstrip patch elements at this frequency will allow for ease in realisation.

In the following we will adopt a modular approach in which we will first design the subsystems like antenna elements, power dividers and subarrays. In the design process, full-wave analysis computer codes will be employed occasionally to verify these subsystems and eventually the complete antenna will be measured and compared with expectations. The array antenna will be excited by means of a corporate feed network, designed for a uniform aperture distribution.

In the array antenna design process we will assume that mutual coupling effects are negligible. Therefore we will ensure that the distance between the array antenna elements is sufficiently large.

5.2 RECTANGULAR MICROSTRIP PATCH ANTENNA

One of the key parameters, characterising any antenna - besides the radiation pattern - is the *input impedance*, which determines the matching bandwidth of the antenna when connected to the front-end (transmitter/receiver).

A control over the input impedance of a microstrip patch antenna is available by using the closed-form equations that were derived for these antennas in the 1970s and 1980s, see for example [1, 2, 4, 5]. The closed-form equations are characterised by resulting in fast calculations when implemented in computer code, while at the same time providing a high degree of accuracy. Since - in this section - we are dealing with a single radiator,[1] we can use these equations to accurately predict both the input impedance as well as the radiation behaviour of the rectangular microstrip patch antenna.

5.2.1 Cavity Model

The microstrip patch antenna (see figure 5.1) consists, in its basic form, of a rectangular, electrically conducting element that is photoetched from one side of a printed circuit board (PCB). The element may be fed with a microstrip transmission line, with a coaxial feed or by means beyond the scope of this chapter.

The feed introduces a (time-varying) electric field between ground-layer and patch. The antenna may be regarded as a cavity with electrically conducting bottom and top walls and magnetically conducting side walls. This cavity imposes boundary conditions on the electromagnetic field components inside the cavity, such that only discrete solutions, the so-called *modes*, can exist. The dimensions of the cavity determine the resonance frequency of the antenna.

[1]For an array antenna, in general we would no longer be able to employ closed-form equations. Due to mutual electromagnetic interactions between the array elements, we would need to resort to (lengthy and complicated) numerical methods for calculating the element - and inter-element - characteristics. However, for a fixed-beam array antenna - the subject of this chapter - a careful choice of array element positions may result in mutual coupling effects being of secondary order, justifying leaving out mutual coupling effects all together in the analysis.

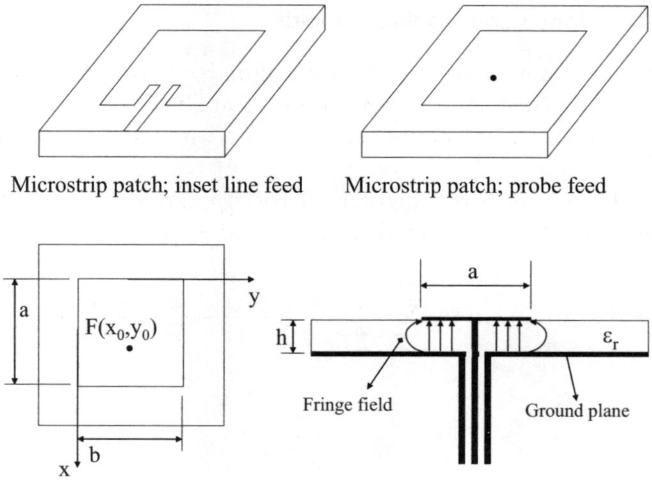

Fig. 5.1 Microstrip patch antenna.

Since the side walls are not perfectly magnetic conducting, the electric field lines are not completely perpendicular to the bottom and top layer at the position of the side walls, as shown in figure 5.1. These so-called *fringe fields* give rise to radiation leaking from the cavity, thus making the antenna operation possible. The radiation may be considered as originating from two slots, separated by the length b of the patch, see figure 5.1. For the two slot fields to be in-phase, the slot separation needs to be half a wavelength (i.e. the wavelength in the dielectric substrate of the PCB). This is depicted in figure 5.2.

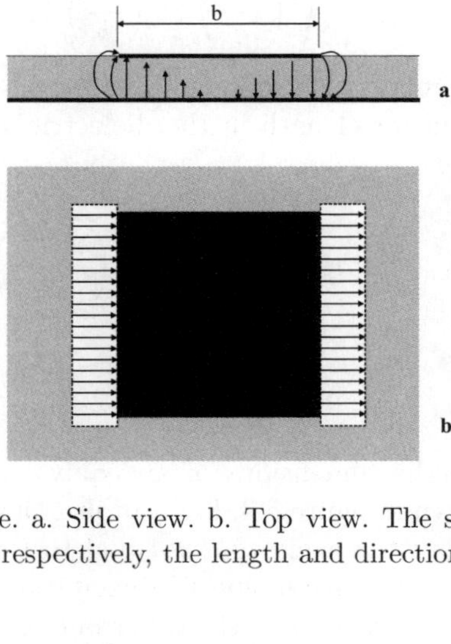

Fig. 5.2 Slot fields in phase. a. Side view. b. Top view. The strength and direction of the electric field is indicated by, respectively, the length and direction of the arrow.

5.2.2 Input Impedance and Radiated Fields

To calculate the input impedance of the rectangular microstrip patch antenna, only the key equations - needed for the realisation of analysis software - will be stated. Detailed derivations may be found in the literature listed at the end of this chapter.

5.2.2.1 Electric Field For a rectangular microstrip patch antenna with dimensions as shown in figure 5.1, the (z-directed) electric field is given by [1]:

$$E_z(x,y) = jI_0\eta_0 k_0 \sum_{m=0}^{\infty}\sum_{n=0}^{\infty} \frac{\psi_{mn}(x,y)\,\psi_{mn}(x_0,y_0)}{k^2 - k_{mn}^2} G_{mn}, \tag{5.1}$$

where

$$\eta_0 = \sqrt{\frac{\mu_0}{\varepsilon_0}}, \tag{5.2}$$

is the characteristic impedance of free space (μ_0 is the free space permeability and ε_0 is the free space permittivity) and

$$\psi_{mn}(x,y) = \frac{\chi_{mn}}{\sqrt{ab}} \cos\left(\frac{m\pi}{a}x\right) \cos\left(\frac{n\pi}{b}y\right), \tag{5.3}$$

is the modal field for mode indices m and n, where

$$\chi_{mn} = \begin{cases} 1 & \text{if } m = 0 \wedge n = 0 \\ \sqrt{2} & \text{if } m = 0 \vee n = 0 \\ 2 & \text{if } m \neq 0 \wedge n \neq 0 \end{cases}. \tag{5.4}$$

Further

$$k_{mn} = \sqrt{\left(\frac{m\pi}{a}\right)^2 + \left(\frac{n\pi}{b}\right)^2}, \tag{5.5}$$

is the wave number of the mode with mode indices m and n. $k = \omega\sqrt{\varepsilon\mu}$ is the wave number associated with the wavelength in the dielectric substrate. $k_0 = \omega\sqrt{\varepsilon_0\mu_0}$ is the free space wave number.

Finally,

$$G_{mn} = sinc\left(\frac{m\pi d_x}{2a}\right) sinc\left(\frac{n\pi d_y}{2b}\right), \tag{5.6}$$

where

$$sinc(x) = \frac{\sin(x)}{x}, \tag{5.7}$$

and where d_x and d_y are the dimensions of an equivalent rectangular, current I_0 carrying excitation, as shown in figure 5.3. For a probe, the dimensions d_x and d_y are chosen such that the area of the excitation is equal to that of the (circular) probe cross-section. For a microstrip transmission line excitation, one of the dimensions is set to the width of the transmission line, the other one is set to zero.

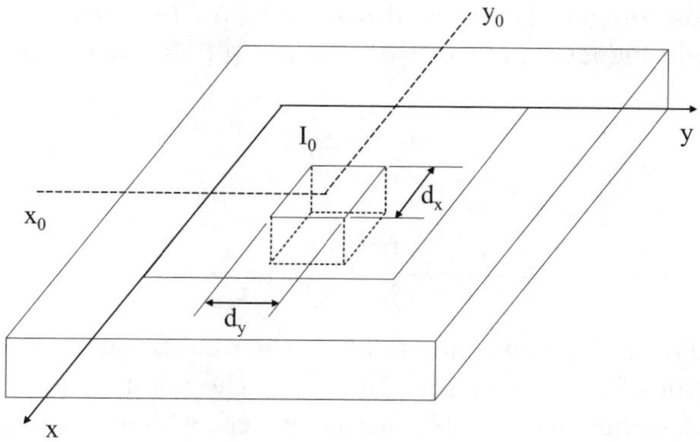

Fig. 5.3 Microstrip patch antenna equivalent excitation at $(x, y) = (x_0, y_0)$.

The input impedance Z_{in} is found as the ratio of the input voltage over the excitation current I_0. The input voltage is $-hE_z(x_0, y_0)$, yielding

$$Z_{in} = -j\eta_0 k_0 h \sum_{m=0}^{\infty} \sum_{n=0}^{\infty} \frac{\psi_{mn}^2(x_0, y_0)}{k^2 - k_{mn}^2} G_{mn}. \tag{5.8}$$

Whenever the wave number in the dielectric corresponds to a modal wave number (i.e. at a modal resonance), the impedance reaches infinity, a situation not encountered in real life. The reason for this non-realistic behaviour is found in the neglecting of the losses, including the radiation (a desired loss).

5.2.2.2 Losses To obtain a more realistic model, all losses are lumped into one effective loss tangent δ_{eff} [2].[2] Thus the wave number is replaced by a complex wave number, yielding

$$E_z(x, y) = jI_0\eta_0 k_0 \sum_{m=0}^{\infty} \sum_{n=0}^{\infty} \frac{\psi_{mn}(x, y)\,\psi_{mn}(x_0, y_0)}{\varepsilon_r(1 - j\delta_{eff})k_0^2 - k_{mn}^2} G_{mn}, \tag{5.9}$$

and

$$Z_{in} = -j\eta_0 k_0 h \sum_{m=0}^{\infty} \sum_{n=0}^{\infty} \frac{\psi_{mn}^2(x_0, y_0)}{\varepsilon_r(1 - j\delta_{eff})k_0^2 - k_{mn}^2} G_{mn}, \tag{5.10}$$

where k_0 is the wave number in free space.

[2]Lossy dielectric materials are characterised by a complex relative permittivity, $\varepsilon_r = \varepsilon_r' - j\varepsilon_r''$. The loss tangent is defined as $\delta_d = \tan(\delta) = \varepsilon_r''/\varepsilon_r'$. Waves propagate through the dielectric proportional to $e^{\gamma l}$, where l is the distance and γ is the propagation constant. The propagation constant is given by $\gamma = -j\omega\sqrt{\mu_0\varepsilon_0(\varepsilon_r' - j\varepsilon_r'')}$. In practice $\varepsilon_r''/\varepsilon_r' \ll 1$, and upon propagating through the dielectric, the waves undergo an exponential decay according to $e^{\alpha l}$, where $\alpha \approx \frac{1}{2}\omega\sqrt{\varepsilon_r'\varepsilon_0\mu_0}\delta_d$. Hereby the term *loss tangent* is explained.

The effective loss tangent is obtained from the quality factor of the cavity formed by the electric and magnetic walls of the microstrip patch antenna:

$$\delta_{\mathit{eff}} = \frac{1}{Q_0}, \tag{5.11}$$

where

$$\frac{1}{Q_0} = \frac{1}{Q_r} + \frac{1}{Q_c} + \frac{1}{Q_d}. \tag{5.12}$$

Herein is Q_0 the total quality factor, Q_r is the quality factor of the radiation, Q_c is the quality factor of the conductors and Q_d is the quality factor of the dielectric. The occurrence of surface waves has been neglected, which is a valid assumption for thin dielectrics [1]. The quality factors are given by [1–3]

$$Q_c = h\sqrt{\pi f \mu_0 \sigma}, \tag{5.13}$$

$$Q_d = \frac{1}{\tan\delta}, \tag{5.14}$$

$$Q_r = \frac{2\omega W_E}{P_r}. \tag{5.15}$$

In the above equations, σ is the conductivity of the electric conductors, $\tan\delta$ is the loss tangent of the dielectric, P_r is the total radiated power and W_E is the stored electric energy.

The stored electric energy, for mode (m,n) is calculated to be, with use of equation (5.9)

$$W_{E_{mn}} = \frac{\varepsilon_0\varepsilon_r}{4}\iiint_{cavity}\left|\vec{E}\right|^2 dV = |E_{0_{mn}}|^2\,\frac{\varepsilon_0\varepsilon_r hab}{4\chi^2_{mn}}, \tag{5.16}$$

where

$$E_{0_{mn}} = \frac{jI_0\chi^2_{mn}\omega\mu_0\cos\left(\frac{m\pi}{a}x_0\right)\cos\left(\frac{n\pi}{b}y_0\right)sinc\left(\frac{m\pi d_x}{2a}\right)sinc\left(\frac{n\pi d_y}{2b}\right)}{ab\left[\varepsilon_r\varepsilon_0\mu_0\omega^2\left(1-j\delta_{\mathit{eff}}\right)-\left(\frac{m\pi}{a}\right)^2-\left(\frac{n\pi}{b}\right)^2\right]}. \tag{5.17}$$

The total radiated power is

$$P_r = \frac{1}{2\eta_0}\int_0^{\frac{\pi}{2}}\int_0^{2\pi}\left(|E_\vartheta|^2 + |E_\varphi|^2\right)r^2\sin(\vartheta)d\vartheta d\varphi. \tag{5.18}$$

The calculation of the effective loss tangent may be simplified by substituting $k_{\mathit{eff}} = k$ in Q_r. For a moderately high Q cavity this will lead to acceptable results, the cavity being dominated by the resonant (m,n)-mode [2].

5.2.2.3 Far Field To calculate the total radiated power - also needed for the determination of the equivalent loss tangent - we need to determine the far-field components $E_{\vartheta_{mn}}$ and $E_{\varphi_{mn}}$ as a function of mode amplitude $E_{0_{mn}}$, for mode (m,n).

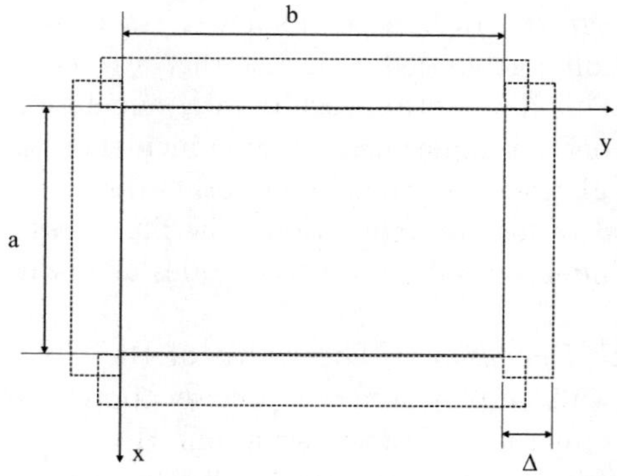

Fig. 5.4 Rectangular microstrip patch antenna and radiating slots.

For the far-field components, we employ the radiating slot model of [7]. The far field is considered to originate from four slots with width Δ, surrounding the patch antenna, shown in figure 5.4.

The radiated far field is given by [7]

$$E_{mn}(r) = \frac{jk_0 e^{-jk_0 r}}{2\pi r}\left(\hat{i}_\vartheta E_{\vartheta_{mn}} + \hat{i}_\varphi E_{\varphi_{mn}}\right), \tag{5.19}$$

where

$$E_{\vartheta_{mn}} = \left[\widetilde{E}_{x_{mn}}(\xi,\eta)\cos(\varphi) + \widetilde{E}_{y_{mn}}(\xi,\eta)\sin(\varphi)\right], \tag{5.20}$$

$$E_{\varphi_{mn}} = \left[-\widetilde{E}_{x_{mn}}(\xi,\eta)\sin(\varphi)\cos(\vartheta) + \widetilde{E}_{y_{mn}}(\xi,\eta)\cos(\varphi)\cos(\vartheta)\right]. \tag{5.21}$$

In the above equations,

$$\widetilde{E}_{x_{mn}}(\xi,\eta) = \left[(-1-(-1)^m)\, j\sin\left(\xi\frac{a'}{2}\right) + (1-(-1)^m)\cos\left(\xi\frac{a'}{2}\right)\right],$$
$$hE_{0_{mn}}\frac{b'}{2} sinc\left(\frac{\Delta}{2}\xi\right) j^n \left[sinc\left(\eta\frac{b'}{2}+\frac{n\pi}{2}\right) + (-1)^n sinc\left(\eta\frac{b'}{2}-\frac{n\pi}{2}\right)\right], \tag{5.22}$$

and

$$\widetilde{E}_{y_{mn}}(\xi,\eta) = \left[(-1-(-1)^n)\, j\sin\left(\eta\frac{b'}{2}\right) + (1-(-1)^n)\cos\left(\eta\frac{b'}{2}\right)\right],$$
$$hE_{0_{mn}}\frac{a'}{2} sinc\left(\frac{\Delta}{2}\eta\right) j^m \left[sinc\left(\xi\frac{a'}{2}+\frac{m\pi}{2}\right) + (-1)^m sinc\left(\xi\frac{a'}{2}-\frac{m\pi}{2}\right)\right], \tag{5.23}$$

where

$$a' = a+\Delta, b' = b+\Delta, \tag{5.24}$$
$$\xi = k_0\sin(\vartheta)\cos(\varphi), \eta = k_0\sin(\vartheta)\sin(\varphi), \tag{5.25}$$

and Δ is the width of the slot, see figure 5.4.

The far field is integrated over a semi-sphere, since we assume the microstrip patch to be positioned on a grounded dielectric that stretches out to infinity in both transverse directions. This assumption results in a calculated radiation pattern that differs from the actual one for angles near grazing incidence. Since the radiated power in these directions is negligible in comparison to other directions, the infinite grounded dielectric is a good first-order approximation. The finiteness effects of the grounded dielectric could be included by diffraction techniques as discussed in [6].

5.2.2.4 Effective Patch Dimensions The width of the slot is approximated in [7] as $\Delta = h$. Here, we will employ a more precise approximation, based upon the capacitance of an open microstrip transmission line [1].

In the theoretical derivation thus far, we implicitly assumed that the electric field in the cavity is perpendicular to the electrical conductors everywhere in the cavity. As shown in figure 5.1, this assumption is not valid at the edges of the patch. However, we can still work with the assumption of a perpendicular field, if we change the transverse dimensions a and b into effective dimensions a_{eff} and b_{eff}. These effective dimensions are lengthened with respect to the physical dimensions, to account for the fringe field. The effective dimensions are given by

$$a_{eff} = a + \Delta, \tag{5.26}$$

$$b_{eff} = b + \Delta, \tag{5.27}$$

where [9]

$$\Delta = 0.824h \frac{(\varepsilon_{eff} + 0.3)\left(\frac{b}{h} + 0.262\right)}{(\varepsilon_{eff} - 0.258)\left(\frac{b}{h} + 0.813\right)}, \tag{5.28}$$

$$\varepsilon_{eff} = \frac{\varepsilon_r + 1}{2} + \frac{\varepsilon_r - 1}{2}\left(1 + 10\frac{h}{b}\right)^{-\frac{1}{2}}. \tag{5.29}$$

In the above we have assumed the TM_{01}-mode to be dominant, meaning that the length b of the microstrip patch radiator determines the resonance frequency. We have assumed all slot widths to be equal and have referred this width to the parameter b.

5.2.3 Rectangular Microstrip Patch Antenna Design

For our linear array antenna we need a rectangular microstrip patch radiator that is excited by a microstrip transmission line. This will allow us to construct both the radiator and the array antenna feed network in the same technology. Our antenna thus will consist of a grounded dielectric slab with on top of this slab the radiators and feed network shaped in a single copper layer that may be photoetched from the dielectric slab or constructed as shown later on in this chapter.

All the subsystems, including the rectangular microstrip patch radiator, will be designed for connection to a 50Ω transmission line. 50Ω is nowadays the standard characteristic impedance for RF transmission lines. This *chosen* value is a compromise between a characteristic impedance of 30Ω that will lead to a maximum power-carrying capability - for physically realisable coaxial transmission lines - and a characteristic impedance of 77Ω that will lead to a minimum attenuation.

For low budget reasons, we will realise our linear array antenna not on a microwave laminate substrate, but on (lossy) standard FR4 PCB material. Since the relative permittivity of this material is in general not known to a great accuracy, we will construct a piece of microstrip transmission line of known width and measure the scattering parameters using a vector network analyser (see also chapter 11). From these measurement results we calculate the characteristic impedance and next determine the relative permittivity of the substrate, using [10]

$$Z_0 = \begin{cases} \dfrac{60}{\sqrt{\varepsilon_{eff}}} \ln\left(\dfrac{8h}{W} + \dfrac{W}{4h}\right) & \text{for } \dfrac{W}{h} \leq 1 \\ \dfrac{120\pi}{\sqrt{\varepsilon_{eff}}\left[\frac{W}{h} + 1.393 + 0.667\ln\left(\frac{W}{h} + 1.444\right)\right]} & \text{for } \dfrac{W}{h} \geq 1 \end{cases}, \tag{5.30}$$

where W is the width of the transmission line and h is the thickness of the substrate.

Thus, we found that our specific sample of FR4 material (h=1.6mm) was characterised by a relative permittivity of $\varepsilon_r = 4.28$ and a loss tangent of $\tan\delta = 0.016$. The width of a 50Ω transmission line is then found with [10]

$$\frac{W}{h} = \begin{cases} \dfrac{8e^A}{e^{2A} - 2} & \text{for } \dfrac{W}{h} < 2 \\ \dfrac{2}{\pi}\left[B - 1 - \ln(2B-1) + \dfrac{\varepsilon_r - 1}{2\varepsilon_r}\left\{\ln(B-1) + 0.39 - \dfrac{0.61}{\varepsilon_r}\right\}\right] & \text{for } \dfrac{W}{h} > 2 \end{cases}, \tag{5.31}$$

where

$$A = \frac{Z_0}{60}\sqrt{\frac{\varepsilon_r + 1}{2}} + \frac{\varepsilon_r - 1}{\varepsilon_r + 1}\left(0.23 + \frac{0.11}{\varepsilon_r}\right), \tag{5.32}$$

and

$$B = \frac{377\pi}{2Z_0\sqrt{\varepsilon_r}}. \tag{5.33}$$

Substitution of the found value for the relative permittivity and the height of the substrate as well as the desired 50Ω characteristic impedance value leads to a microstrip transmission line width of 3.12mm.

Having found the transmission line width and the relative permittivity of the substrate, this now allows us to design a rectangular microstrip patch radiator that is resonant at 2.4GHz. Thereto we could employ equation (5.10) in an iterative way. Although this equation, used to calculate the input reflection as function of frequency, requires at least an order of magnitude less in CPU time than a general full-wave method, simulation times can still become relatively long, due to the slow convergence of the series in this equation. Although the computation times may be acceptable for analysis purposes, we would like to speed up the calculation times for synthesis (iterative analysis) purposes.

Keeping this in mind and knowing that the microstrip patch antenna characteristics are mainly (but not completely!) determined by the dominant mode, i.e. the mode with the lowest resonance frequency, in our case the $(m, n) = (0, 1)$ mode, we can approximate the contribution of all higher order modes into a single inductive component [1]. The expression for the input impedance then becomes

$$Z_{in} = -j\eta_0 k_0 h \sum_{m=0}^{\circ} \sum_{n=0}^{1} \frac{\psi_{mn}^2(x_0, y_0)}{\varepsilon_r(1 - j\delta_{\mathit{eff}})k_0^2 - k_{mn}^2} G_{mn} + j\sqrt{\frac{\mu_0}{\varepsilon_r \varepsilon_0}} \tan\left(2\pi h f \sqrt{\varepsilon_r \varepsilon_0 \mu_0}\right). \tag{5.34}$$

We will use this equation - that generates the input impedance very fast - to find the dimensions of the rectangular microstrip patch radiator that is excited by a (3.12mm wide) microstrip transmission line, see figure 5.5.

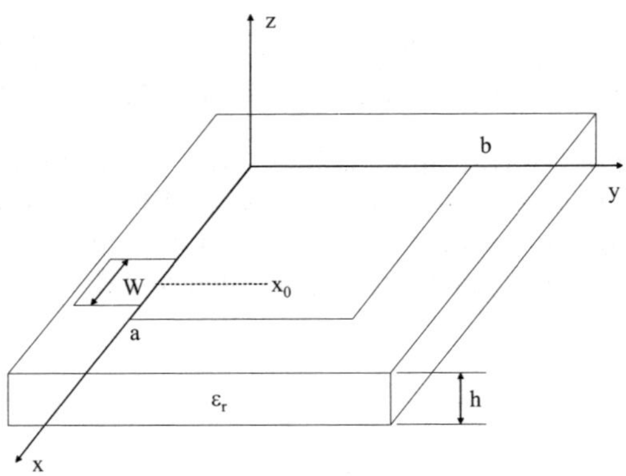

Fig. 5.5 Geometry of a microstrip line fed rectangular microstrip patch antenna.

We start with a length b equal to half the wavelength in the substrate and then vary this length to obtain a minimum reflection at 2.4GHz. Next we change the width a to further minimise this reflection. We choose to keep the position of the feed line fixed at half the patch width ($x_0 = \frac{a}{2}$).

We thus find: a=58.00mm, b=29.30mm, h=1.60mm, $\varepsilon_r = 4.28$, $\tan\delta = 0.0016$, d_x=3.12mm, d_y=0, x_0=29.00mm, y_0=0 and $\sigma = 5.8 \cdot 10^7$. The reflection coefficient as a function of frequency for these parameters is shown in figure 5.6, calculated with the cavity model equation shown above and using a full-wave method.

We see that the (simple) cavity model predicts the resonance of the patch radiator reasonably well. Upon a closer inspection, we see that the full-wave simulation results agree with those of the cavity model but that for frequencies just above 2.4GHz something else is happening that is not predicted by the (simple) cavity model.

In order to analyse these differences further, we have calculated the input impedance of the rectangular patch antenna using the slowly converging series expression for the cavity model and compared real and imaginary parts with those calculated with the full-wave method. The results are shown in figure 5.7.

The figure reveals that the cavity model predicts the first resonance (imaginary part of the input impedance equal to zero) very accurately compared with the full-wave calculations. The cavity model also shows the occurrence of higher order modes, but fails in predicting the correct behaviour.

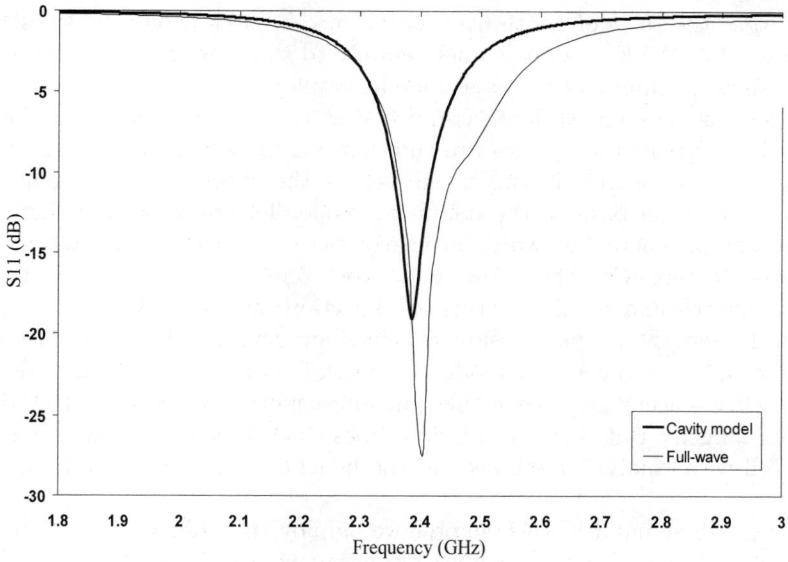

Fig. 5.6 Reflection coefficient as a function of frequency, calculated for a rectangular microstrip patch radiator.

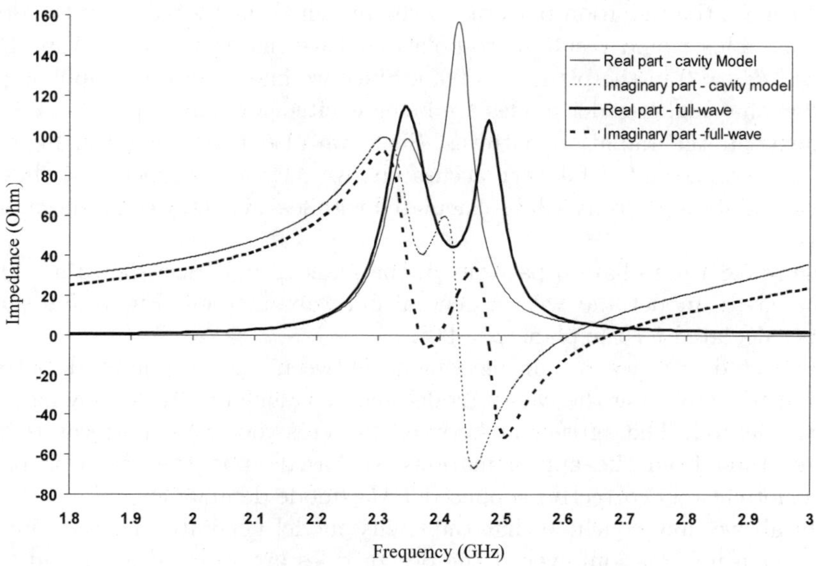

Fig. 5.7 Real and imaginary part of the input impedance as a function of frequency, calculated for a rectangular microstrip patch radiator.

In designing the rectangular patch radiator we have violated the rule-of-thumb design rule that the ratio of width over length of the patch should be smaller than approximately 1.5. We have done so deliberately to show the reason for this design rule and to show the limits of the cavity model employed.

In our analysis thus far we have assumed that the TM_{01}-mode is the dominant mode and that higher order modes are sufficiently suppressed to be ignored in the analysis. We have done this in the calculation of the effective loss tangent that is determined entirely on basis of the dominant mode. Therefore, we see that for our rectangular patch, where the width is nearly twice the length and where higher order modes, starting with the TM_{10}-mode *are* excited, the analysis starts to fail at higher order resonances. By putting a restriction on the width of the patch in relation to the length, we may assure a behaviour dominated by the TM_{01}-mode and for these radiators our cavity model will work fine. If we want to include higher order mode effects accurately, we should put more effort in correctly calculating the effective loss tangent. But then, calculation times could become comparable to those of general full-wave analysis methods and the benefit of the cavity model no longer is present.

In the example at hand in this chapter we employ the philosophy to employ the cavity model as is for obtaining a starting design very fast, accepting the limited accuracy and then validate and maybe fine-tune this design by employing a general full-wave analysis method. For the moment we are happy with the return loss as predicted by the cavity model as compared with the full-wave analysis results.

It remains to evaluate the radiation patterns as predicted by the cavity model. We could calculate the radiation patterns in the principal planes based on the dominant mode only. This would result in co-polarised patterns only ($E_\vartheta = 0$ in the plane $\varphi = 0$ and $E_\varphi = 0$ in the plane $\varphi = 90°$). Since we know from our input impedance evaluation that higher order modes are being excited, we may expect cross-polarised components in the radiation patterns. Since we also know that the higher order modes are not accounted for very accurately, we may also expect that these cross-polarised radiation patterns will be calculated with less accuracy than the co-polarised patterns.

In figure 5.8 the radiation patterns in the plane $\varphi = 0°$ are shown as calculated with the cavity model and with a general full-wave method. Figure 5.9 shows the patterns calculated for the plane $\varphi = 90°$.

As expected, we see a fair agreement between the co-polarised patterns as calculated by employing the cavity model and as calculated by employing a general full-wave method. The agreement between the cross-polarised patterns is not that good, resulting from the approximations we based upon the (for this particular radiator not entirely correctly) supposed TM_{01}-mode dominance.

All in all we may conclude that the cavity model generates accurate impedance and radiation information, even if the design rules are violated a bit and that this information can be calculated very fast. This makes the cavity model a valuable tool for generating designs or pre-designs very fast. Pre-designs may be fine-tuned - if necessary - using (commercially available) full-wave methods.

With a design tool for the radiator thus being developed, we may try to do the same for the components of the corporate feed network.

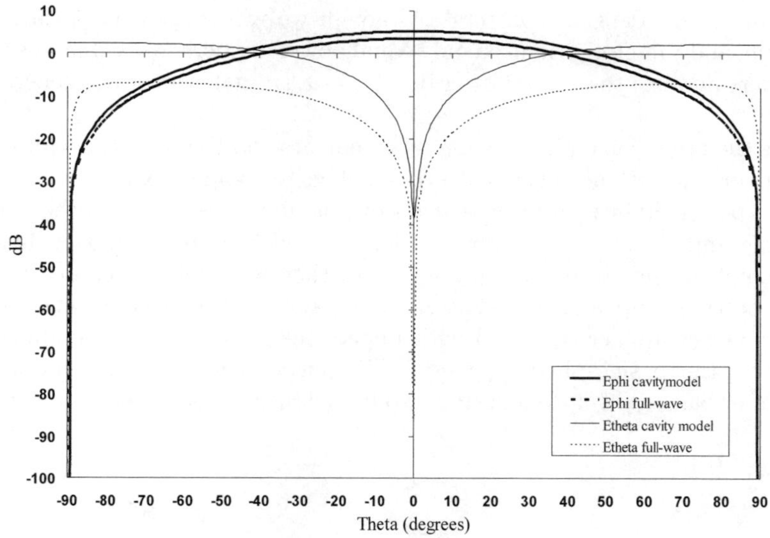

Fig. 5.8 Co- and cross-polarised power radiation pattern in the plane $\varphi = 0°$, calculated for a rectangular microstrip patch radiator.

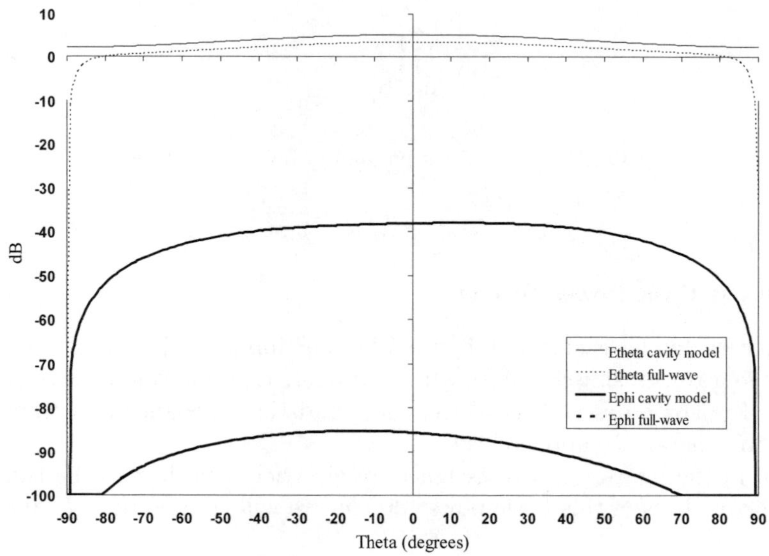

Fig. 5.9 Co- and cross-polarised power radiation pattern in the plane $\varphi = 90°$, calculated for a rectangular microstrip patch radiator.

5.3 SPLIT-T POWER DIVIDER

The feed network of our 4-element array antenna will be of the corporate type. Since the array antenna elements are microstrip patch radiators, the feed network will be realised in microstrip technology. Within this microstrip feed network, we will need microstrip power dividers or 'splitters'. Although we will make a uniformly excited array antenna and therefore will need only equal power dividers, we will discuss general power dividers and use the equal power divider as a special case of the general power divider.

To assess the reflections from the antenna elements and the reflections internally in the feed network, we will need the scattering matrix (see appendix E) of a basic power divider. Our power divider and the analysis of this divider is based on the Wilkinson power divider and the analysis as given in [12]. To get a simple, compact divider, we have chosen not to use a resistive component. Furthermore, we have chosen to employ a 50Ω characteristic impedance level at every stage of the feed network and therefore, we have our power divider equipped with impedance transformers (see appendix D) immediately at the 'basic' splitter's junction. Therefore we may envisage the splitter as consisting of a 'basic' power divider and two impedance transformers, see figure 5.10.

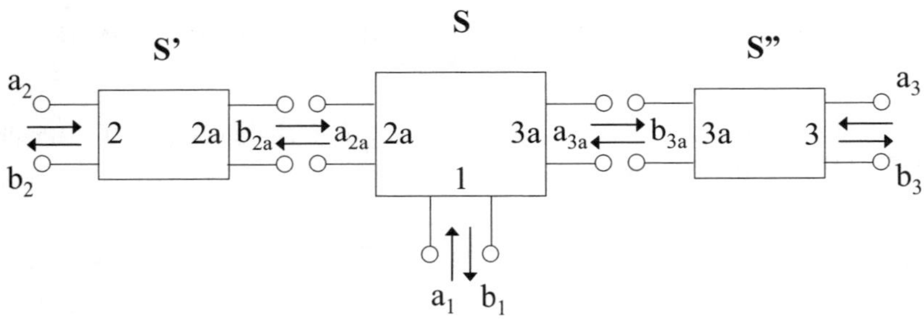

Fig. 5.10 Power divider separated in functional blocks.

5.3.1 Analysis Basic Power Divider

The basic power divider is shown in figure 5.11. The inputs and outputs are assumed to be connected to transmission lines with characteristic impedances of, respectively, Z_0, Z_2 and Z_3. For a description of the characteristic impedance and transmission line theory in general, see appendix D.

Although in the figure we have assigned an electrical length ϑ to the two output ports, for the analysis of this basic power divider we will assume that $\vartheta \downarrow 0$.

5.3.1.1 Impedance Levels The input power (at port 1) is equal to P_1. The output powers at, respectively, ports 2 and 3 are denoted P_2 and P_3. The power division is such that

$$P_3 = K^2 P_2, \tag{5.35}$$

where K is provided by the designer.

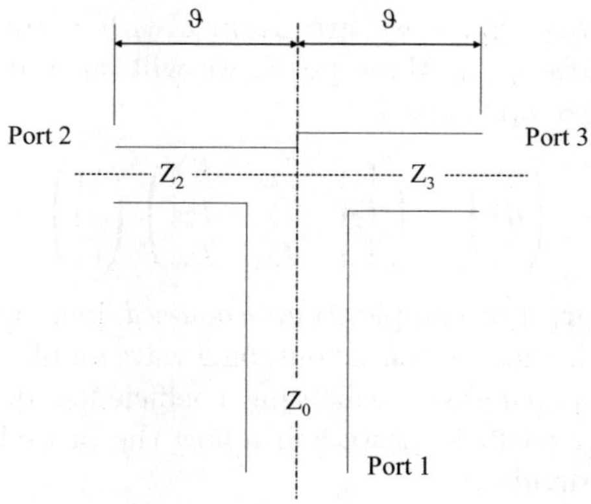

Fig. 5.11 Basic power divider. The electric length ϑ is vanishing.

Since $P_1 = P_2 + P_3$, it may be easily validated that

$$P_2 = \frac{1}{1+K^2} P_1, \tag{5.36}$$

$$P_3 = \frac{K^2}{1+K^2} P_1. \tag{5.37}$$

If we denote the voltage from transmission line 1 at the position of the junction as V_0^+, the voltage from transmission line 2 at the junction as V_2^- and the voltage from transmission line 3 at the junction as V_3^-, then the input and output powers may be written as

$$P_1 = \frac{1}{2}\frac{|V_0^+|^2}{Z_0}, \tag{5.38}$$

$$P_2 = \frac{1}{2}\frac{|V_2^-|^2}{Z_2}, \tag{5.39}$$

and

$$P_3 = \frac{1}{2}\frac{|V_3^-|^2}{Z_3}. \tag{5.40}$$

Since, at the junction, $V_0^+ = V_2^- = V_3^-$, the characteristic impedances are related to the power levels through

$$\frac{Z_2}{Z_3} = \frac{P_3}{P_2} = K^2, \tag{5.41}$$

$$\frac{Z_2}{Z_0} = \frac{P_1}{P_2} = 1 + K^2, \tag{5.42}$$

so that we can relate the impedances Z_2 and Z_3 to Z_0 by

$$Z_2 = (1 + K^2) Z_0, \tag{5.43}$$

$$Z_3 = \frac{1+K^2}{K^2} Z_0. \tag{5.44}$$

5.3.1.2 Scattering Matrix Since we are dealing with a three-port circuit having unequal impedance levels at the three ports, we will make use of the *unnormalised scattering matrix* [11], see appendix E.

$$\begin{pmatrix} d_1 \\ d_2 \\ d_3 \end{pmatrix} = \begin{pmatrix} T_{11} & T_{12} & T_{13} \\ T_{21} & T_{22} & T_{23} \\ T_{31} & T_{32} & T_{33} \end{pmatrix} \begin{pmatrix} c_1 \\ c_2 \\ c_3 \end{pmatrix} \tag{5.45}$$

where c_i, $i = 1, 2, 3$, are the complex *unnormalised* ingoing wave amplitudes, d_i, $i = 1, 2, 3$, are the complex *unnormalised* outgoing wave amplitudes and T_{ij}, $i = 1, 2, 3$, $j = 1, 2, 3$, are the *unnormalised* scattering coefficients. In the derivation of the unnormalised scattering coefficients we will follow the procedure as outlined in [12] for a Wilkinson power divider.[3]

We start with terminating the two outputs in their characteristic impedances, see figure 5.12.

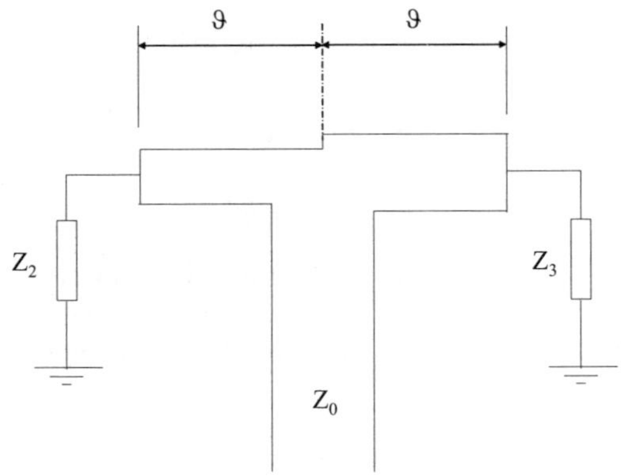

Fig. 5.12 Basic power divider terminated in characteristic impedances.

With the three-port network of figure 5.12, driven from port 1 and (this is very important!) $\vartheta \downarrow 0$, we experience equal voltages at port 2 and 3 and therefore we may apply a short circuit between these ports and take the parallel circuit of the two transmission lines of electrical length ϑ. Also the terminating impedances need to be taken in parallel then. The equivalent circuit and signals at the junctions are depicted in figure 5.13.

The equations governing this network are, see also [12]

$$d_1 = \Gamma_1 c_1 + (1 - \Gamma_1) c_a, \tag{5.46}$$

$$d_a = (1 + \Gamma_1) c_1 - \Gamma_1 c_a, \tag{5.47}$$

$$c_a = \Gamma_2 e^{-j2\vartheta} d_a, \tag{5.48}$$

$$d_2 = (1 + \Gamma_2) e^{-j\vartheta} d_a. \tag{5.49}$$

[3] A Wilkinson power divider is a three-port circuit that divides the power delivered at its input-port to the two output-ports and that is characterised by the property that its three ports are all matched. This matching is accomplished by the incorporation of a resistive element in the three-port that dissipates mismatches.

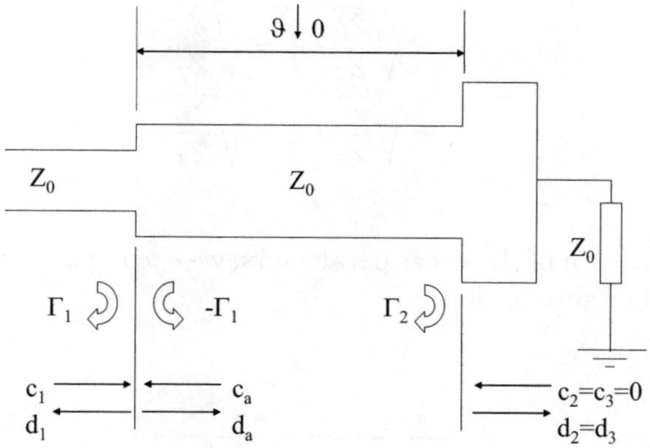

Fig. 5.13 Equivalent circuit for basic power divider, driven at port 1 and terminated in characteristic impedances.

Now, using $\vartheta \downarrow 0$ and $\Gamma_1 = \Gamma_2 = 0$, reduces these equations to

$$d_1 = c_a, \tag{5.50}$$
$$d_a = c_1, \tag{5.51}$$
$$c_a = 0, \tag{5.52}$$
$$d_2 = d_a, \tag{5.53}$$

so that, see [12],

$$T_{21} = T_{31} = \left. \frac{d_2}{c_1} \right|_{c_2=c_3=0} = 1, \tag{5.54}$$

$$T_{11} = \left. \frac{d_1}{c_1} \right|_{c_2=c_3=0} = 0. \tag{5.55}$$

On the basis of reciprocity (see appendix E) we find

$$T_{12} = \frac{Z_0}{Z_2} T_{21}, \tag{5.56}$$

and

$$T_{13} = \frac{Z_0}{Z_3} T_{31}. \tag{5.57}$$

The corresponding *normalised* scattering coefficients are found to be (see appendix E)

$$S_{11} = T_{11} = 0, \tag{5.58}$$

$$S_{12} = \sqrt{\frac{Z_2}{Z_0}} T_{12} = \sqrt{\frac{Z_0}{Z_2}}, \tag{5.59}$$

$$S_{13} = \sqrt{\frac{Z_3}{Z_0}} T_{13} = \sqrt{\frac{Z_0}{Z_3}}, \tag{5.60}$$

$$S_{21} = \sqrt{\frac{Z_0}{Z_2}} T_{21} = \sqrt{\frac{Z_0}{Z_2}}, \tag{5.61}$$

$$S_{31} = \sqrt{\frac{Z_0}{Z_3}} T_{31} = \sqrt{\frac{Z_0}{Z_3}}. \tag{5.62}$$

For the determination of the other parameters, we excite the three-port from ports 2 and 3 as shown in figure 5.14.

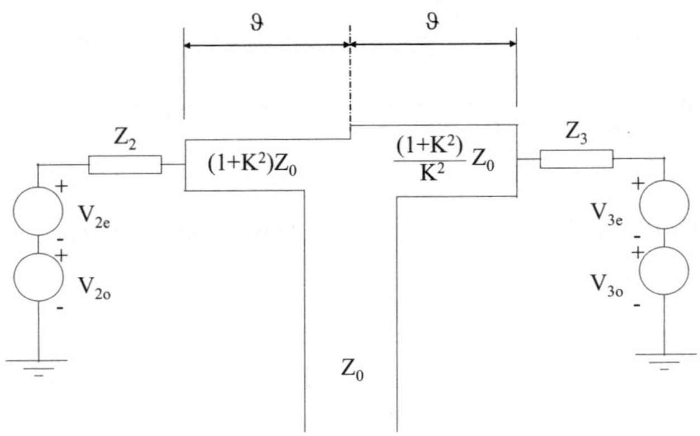

Fig. 5.14 Basic power divider, driven at ports 2 and 3 for even or odd excitation.

We start with the so-called *even excitation* for which $V_{2e} = V_{3e}$ and $V_{2o} = V_{3o} = 0$. For this situation, again bearing in mind that $\vartheta \downarrow 0$, the circuit of figure 5.14 may be represented by the circuits as shown in figure 5.15.

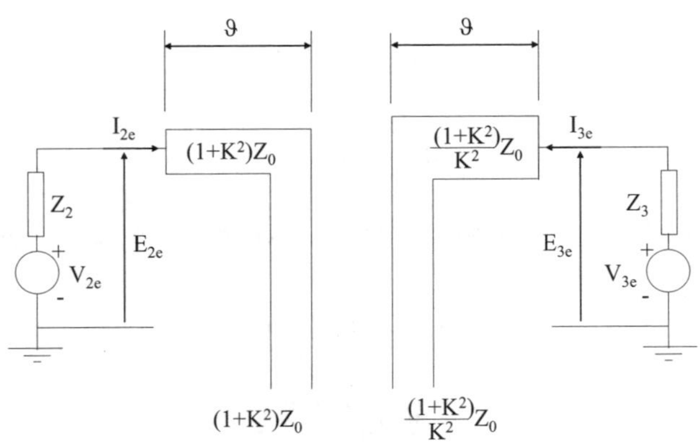

Fig. 5.15 Equivalent circuit of basic power divider for even excitation.

The currents and voltages for the *even excitation* are related through, see figure 5.15

$$I_{2e} = \frac{V_{2e}}{Z_2 + Z_{2e}}, \tag{5.63}$$

$$E_{2e} = \frac{Z_{2e} V_{2e}}{Z_2 + Z_{2e}}, \tag{5.64}$$

where

$$Z_{2e} = (1 + K^2) Z_O, \tag{5.65}$$

is the impedance looking into the left arm shown in figure 5.15.

Thus

$$I_{2e} = \frac{V_{2e}}{2(1 + k^2) Z_O}, \tag{5.66}$$

$$E_{2e} = \frac{V_{2e}}{2}. \tag{5.67}$$

Further

$$I_{3e} = K^2 I_{2e}, \tag{5.68}$$

and

$$E_{3e} = E_{2e}. \tag{5.69}$$

For the *odd excitation*, $V_{2e} = V_{3e} = 0$ and $V_{2o} = -K^2 V_{3o}$. The equivalent circuits are shown in figure 5.16.

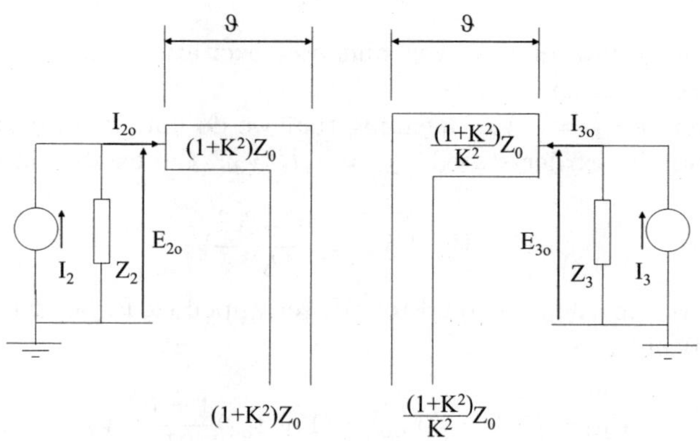

Fig. 5.16 Equivalent circuit of basic power divider for odd excitation.

The currents I_2 and I_3 in figure 5.16 are given by

$$I_2 = \frac{V_{2o}}{Z_2} = \frac{V_{2o}}{(1 + K^2) Z_O}, \tag{5.70}$$

$$I_3 = \frac{V_{3o}}{Z_3} = -\frac{V_{2o}}{(1 + K^2) Z_O}. \tag{5.71}$$

The voltages and currents for the *odd excitation* are related through, see figure 5.16

$$E_{2o} = (I_2 - I_{2o})Z_2 = V_{2o} - Y_{2o}(1 + K^2)Z_0E_{2o}, \tag{5.72}$$

where use has been made of

$$I_{2o} = Y_{2o}E_{2o}, \tag{5.73}$$

where Y_{2o} is the admittance looking into the left arm of the circuit shown in figure 5.16.
Thus

$$E_{2o} = \frac{V_{2o}}{1 + Y_{2o}(1 + K^2)Z_0}, \tag{5.74}$$

$$I_{2o} = \frac{V_{2o}Y_{2o}}{1 + Y_{2o}(1 + K^2)Z_0}. \tag{5.75}$$

Since $I_3 = -I_2$, we may envisage a short circuit at the end of the transmission lines of electrical length ϑ, looking from ports 2 and 3. This means that - since this derivation is only valid for $\vartheta \downarrow 0$ - $Y_{2o} \to \infty$ and therefore

$$E_{2o} = 0, \tag{5.76}$$

$$I_{2o} = \frac{V_{2o}}{(1 + K^2)Z_0}, \tag{5.77}$$

and

$$E_{3o} = 0, \tag{5.78}$$

$$I_{3o} = -I_{2o}. \tag{5.79}$$

Combining the results of the even and odd excitation will yield the remaining unnormalised scattering coefficients.

At first we choose $V_{2o} = -V_{2e}$, meaning that we do not excite port 2.

Using this and the earlier stated $V_{2o} = -K^2V_{3o}$, the excitation of port 3, V, is found to be

$$V = V_{3e} + V_{3o} = \frac{1 + K^2}{K^2}V_{2e}. \tag{5.80}$$

The incident voltage, V_{inc} is equal to $\frac{1}{2}V$, see appendix F. So, the total voltage at port 3, V_{T3} is given by

$$V_{T3} = (1 + T_{33})V_{inc} = (1 + T_{33})\frac{1 + K^2}{2K^2}V_{2e}. \tag{5.81}$$

This incident voltage also satisfies

$$V_{T3} = E_{3e} + E_{3o} = \frac{1}{2}V_{2e}, \tag{5.82}$$

where use has been made of equations (5.67), (5.69) and (5.78).

From equations (5.81) and (5.82) we find

$$T_{33} = -\frac{1}{1 + K^2}. \tag{5.83}$$

The coefficient T_{23} is found from

$$T_{23} = \frac{V_{T2}}{V_{inc}}, \tag{5.84}$$

where

$$V_{T2} = E_{2e} + E_{2o} = \frac{1}{2}V_{2e}, \tag{5.85}$$

and

$$V_{inc} = \frac{1+K^2}{2K^2}V_{2e}. \tag{5.86}$$

So

$$T_{23} = \frac{K^2}{1+K^2}. \tag{5.87}$$

As second choice we take $V_{3o} = -V_{3e}$, meaning that we do not excite port 3. The excitation of port 2, V', is given by

$$V' = V_{2e} + V_{2o} = (1+K^2)V_{2e}. \tag{5.88}$$

The incident voltage, V'_{inc}, at port 2 is then

$$V'_{inc} = \frac{1}{2}V' = \frac{1}{2}(1+K^2)V_{2e}. \tag{5.89}$$

The total voltage at port 2 therefore will be

$$V_{T2} = (1+T_{22})V'_{inc} = (1+T_{22})\frac{1}{2}(1+K^2)V_{2e}. \tag{5.90}$$

This total voltage also satisfies

$$V_{T2} = E_{2e} + E_{2o} = \frac{1}{2}V_{2e}, \tag{5.91}$$

so that we find

$$T_{22} = -\frac{K^2}{1+K^2}. \tag{5.92}$$

The scattering coefficient T_{32} is found with

$$T_{32} = \frac{V_{T3}}{V'_{inc}}, \tag{5.93}$$

where

$$V_{T3} = E_{3e} + E_{3o} = \frac{1}{2}V_{2e}, \tag{5.94}$$

$$V_{inc} = \frac{1}{2}(1+K^2)V_{2e}, \tag{5.95}$$

leading to

$$T_{32} = \frac{1}{1+K^2}. \tag{5.96}$$

The *normalised* scattering coefficients follow from, see appendix E

$$S_{22} = T_{22}, \tag{5.97}$$

$$S_{23} = \sqrt{\frac{Z_3}{Z_2}} T_{23}, \tag{5.98}$$

$$S_{32} = \sqrt{\frac{Z_2}{Z_3}} T_{32}, \tag{5.99}$$

$$S_{33} = T_{33}. \tag{5.100}$$

With use of the equations (5.43), (5.44), (5.58) to (5.62), (5.84), (5.87), (5.92), (5.96) and (5.97) to (5.100), we find for the normalised scattering coefficients

$$S_{11} = 0, \tag{5.101}$$

$$S_{12} = \frac{1}{\sqrt{1+K^2}}, \tag{5.102}$$

$$S_{13} = \frac{K}{\sqrt{1+K^2}}, \tag{5.103}$$

$$S_{21} = S_{12}, \tag{5.104}$$

$$S_{22} = -\frac{K^2}{1+K^2}, \tag{5.105}$$

$$S_{23} = \frac{K}{1+K^2}, \tag{5.106}$$

$$S_{31} = S_{13}, \tag{5.107}$$

$$S_{32} = S_{23}, \tag{5.108}$$

$$S_{33} = -\frac{1}{1+K^2}. \tag{5.109}$$

5.3.2 Analysis Impedance Transformer

For the quarter lambda impedance transformer we will make use of the ABCD matrix of a section of transmission line.

The ABCD matrix of a two-port network relates output voltage, V_2, and current, I_2, to input voltage V_1, and current, I_1, through

$$\begin{pmatrix} V_1 \\ I_1 \end{pmatrix} = \begin{pmatrix} A & B \\ C & D \end{pmatrix} \begin{pmatrix} V_2 \\ I_2 \end{pmatrix}. \tag{5.110}$$

The ABCD matrix of a section of transmission line of length ϑ is given by [13, 14]

$$\begin{pmatrix} A & B \\ C & D \end{pmatrix} = \begin{pmatrix} \cos(\vartheta) & jZ_c \sin(\vartheta) \\ j\frac{1}{Z_c}\sin(\vartheta) & \cos(\vartheta) \end{pmatrix}, \tag{5.111}$$

where Z_c is the characteristic impedance of the transmission line and

$$\vartheta = \frac{2\pi}{\lambda} l, \tag{5.112}$$

where l is the physical length of the transmission line section.

When the transmission line section is a quarter lambda impedance transformer (see appendix D), the physical length l is equal to a quarter of the centre wavelength, λ_0

$$l = \frac{\lambda_0}{4}. \tag{5.113}$$

The characteristic impedance Z_c is equal to the square root of the product of characteristic impedances encountered at the left and right of the transformer (see appendix D), as shown in figure 5.17.

$$Z_c = \sqrt{Z_{01} Z_{02}}. \tag{5.114}$$

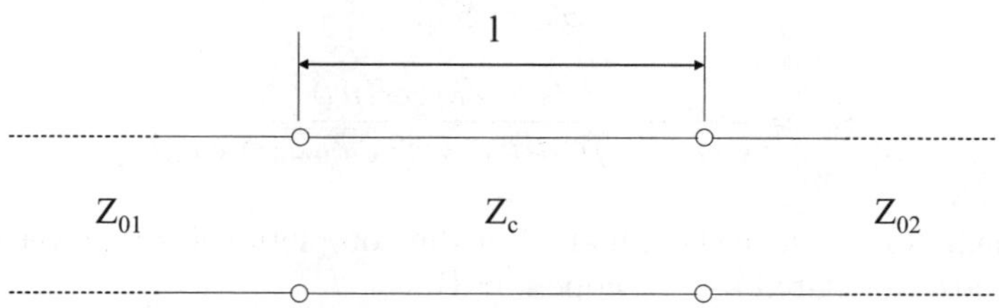

Fig. 5.17 Quarter lambda impedance transformator in between two transmission lines.

The normalised scattering matrix elements may be determined from the ABCD matrix elements according to [13, 14]

$$S_{11} = \frac{AZ_{02} + B - CZ_{01}Z_{02} - DZ_{01}}{AZ_{02} + B + CZ_{01}Z_{02} + DZ_{01}}, \tag{5.115}$$

$$S_{12} = \frac{2(AD - BC)\sqrt{Z_{01}Z_{02}}}{AZ_{02} + B + CZ_{01}Z_{02} + DZ_{01}}, \tag{5.116}$$

$$S_{21} = \frac{2\sqrt{Z_{01}Z_{02}}}{AZ_{02} + B + CZ_{01}Z_{02} + DZ_{01}}, \tag{5.117}$$

$$S_{22} = \frac{-AZ_{02} + B - CZ_{01}Z_{02} + DZ_{01}}{AZ_{02} + B + CZ_{01}Z_{02} + DZ_{01}}. \tag{5.118}$$

5.3.3 Power Divider Scattering Matrix

With reference to figure 5.10, the scattering matrix elements of the two impedance transformers are given by

$$S'_{11} = \frac{(Z_0 - Z_2)\cos(\vartheta_2)}{(Z_2 + Z_0)\cos(\vartheta_2) + j2\sqrt{Z_0 Z_2}\sin(\vartheta_2)}, \tag{5.119}$$

$$S'_{12} = \frac{2\sqrt{Z_0 Z_2}}{(Z_2 + Z_0)\cos(\vartheta_2) + j2\sqrt{Z_0 Z_2}\sin(\vartheta_2)}, \tag{5.120}$$

$$S'_{21} = S'_{12}, \tag{5.121}$$

$$S'_{22} = \frac{(Z_2 - Z_0)\cos(\vartheta_2)}{(Z_2 + Z_0)\cos(\vartheta_2) + j2\sqrt{Z_0 Z_2}\sin(\vartheta_2)}, \tag{5.122}$$

and

$$S''_{11} = \frac{(Z_0 - Z_3)\cos(\vartheta_3)}{(Z_3 + Z_0)\cos(\vartheta_3) + j2\sqrt{Z_0 Z_3}\sin(\vartheta_3)}, \tag{5.123}$$

$$S''_{12} = \frac{2\sqrt{Z_0 Z_3}}{(Z_3 + Z_0)\cos(\vartheta_3) + j2\sqrt{Z_0 Z_3}\sin(\vartheta_3)}, \tag{5.124}$$

$$S''_{21} = S''_{12}, \tag{5.125}$$

$$S''_{22} = \frac{(Z_3 - Z_0)\cos(\vartheta_3)}{(Z_3 + Z_0)\cos(\vartheta_3) + j2\sqrt{Z_0 Z_3}\sin(\vartheta_3)}. \tag{5.126}$$

The elements of the scattering matrix of the complete split-T power divider, S^T_{ij}, $i, j = 1, 2, 3$, are calculated as, see appendix G

$$S^T_{11} = S_{11} + S_{31}G + FC, \tag{5.127}$$

$$S^T_{12} = S_{12}S'_{12} + S_{32}S'_{12}G + FD, \tag{5.128}$$

$$S^T_{13} = S_{13}S''_{12} + S_{33}S''_{12}G + FE, \tag{5.129}$$

$$S^T_{21} = S'_{21}C, \tag{5.130}$$

$$S^T_{22} = S'_{22} + S'_{21}D, \tag{5.131}$$

$$S^T_{23} = S'_{21}E, \tag{5.132}$$

$$S^T_{31} = S''_{21}\left[\frac{S_{31}}{A} + HC\right], \tag{5.133}$$

$$S^T_{32} = S''_{21}\left[\frac{S_{32}S'_{12}}{A} + HD\right], \tag{5.134}$$

$$S^T_{33} = S''_{22} + S''_{21}\left[\frac{S_{33}S''_{12}}{A} + HE\right], \tag{5.135}$$

where

$$A = 1 - S_{33}S''_{11}, \tag{5.136}$$

$$B = A(1 - S_{22}S'_{11}) - S_{23}S''_{11}S_{32}S'_{11}, \tag{5.137}$$

$$C = \frac{S_{21}A + S_{23}S''_{11}S_{31}}{B}, \tag{5.138}$$

$$D = \frac{S_{22}S'_{12}A + S_{23}S''_{11}S_{32}S'_{12}}{B}, \tag{5.139}$$

$$E = \frac{S_{23}S''_{12}A + S_{23}S''_{11}S_{33}S''_{12}}{B}, \tag{5.140}$$

$$F = S_{12}S'_{11} + \frac{S_{13}S''_{11}S_{32}S'_{11}}{A}, \tag{5.141}$$

$$G = \frac{S_{13}S''_{11}}{A}, \tag{5.142}$$

$$H = \frac{S_{32}S'_{11}}{A}. \tag{5.143}$$

5.3.4 Split-T Power Divider Simulation

Before we compare the simulations according to the model presented in the previous sections with simulations performed using a general full-wave analysis method, we need to say a few words about (microstrip) transmission line discontinuities. In general, a transmission line discontinuity like a bend, a change in width, or - as in our case - a T-junction, will introduce parasitic reactances. These parasitic reactances may degrade phase and amplitude performance if not properly accounted for in the design of the discontinuity. Figure 5.18a shows a T-junction and its equivalent electrical circuit with parasitic reactances. One could try to find the values of these reactances and include the equivalent electrical circuit into the simulations.

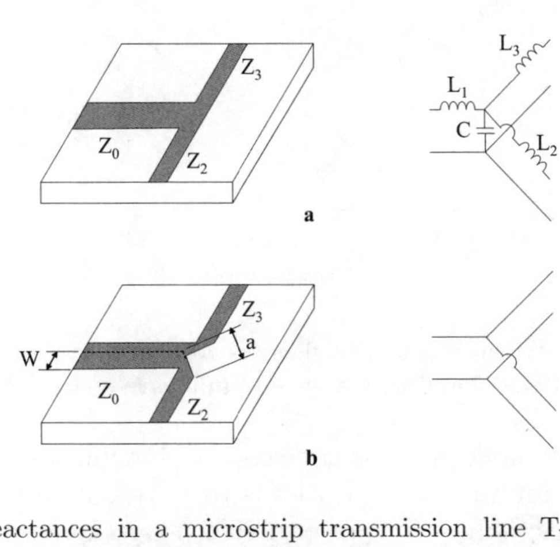

Fig. 5.18 Parasitic reactances in a microstrip transmission line T-junction. a. Equivalent electrical model. b. Mitering for compensating parasitic reactances.

A more pragmatic approach though is shown in figure 5.18b. The analysis will be performed on the model as developed in the previous sections only, but in the realisation of the junction, the corners will be *mitred*, thus compensating for the parasitic reactances. A practical value for the length of the mitre, a, is given by, [10]

$$a = 1.8W, \tag{5.144}$$

where W is the width of the microstrip transmission line, see figure 5.18.

The full-wave analysis results presented in the figures 5.19 and 5.20 are based on T-split power dividers with mitred corners. Once the power ratio K^2 has been chosen, the characteristic impedances Z_2 and Z_3 follow from, respectively, equations (5.43) and (5.44). The widths of the microstrip transmission lines for impedances $Z_0 = 50\Omega$ and $\sqrt{Z_2 Z_0}$ and $\sqrt{Z_3 Z_0}$ for the impedance transformers are calculated with the help of equations (5.31) to (5.33). The substrate is the same as that used for the microstrip patch radiators, so $\varepsilon_r = 4.28$ and h=1.6mm.

Figure 5.19 shows the reflection and transmission coefficients versus frequency for a T-split power divider that divides the input power equally over the two outputs, $K = 1$. The width of the input and output microstrip transmission lines is 3.12mm. The width of the two identical quarter lambda transformer sections is 1.66mm and their lengths are 17.73mm each.

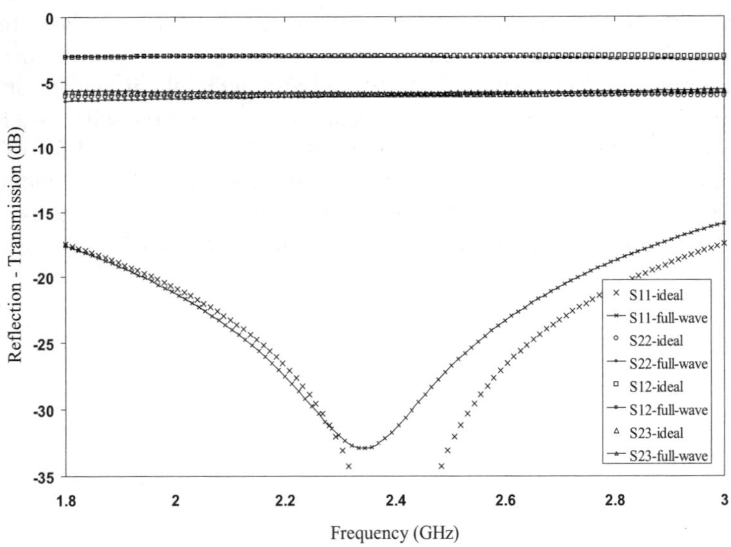

Fig. 5.19 Reflection and transmission coefficients for a $K = 1$ split-T power divider as calculated with the ideal model and with a general full-wave method for a mitred lay-out.

We see that our ideal model agrees very well with the full-wave results. Not mitering the corners appears - for this configuration - to have only a small influence on the amplitude characteristics. Since the power divider is symmetrical, we do not show the S_{33} and S_{13} results, since these are identical to, respectively, the S_{22} and S_{12} results.

As a final example, demonstrating the validity of our ideal model, figure 5.20 shows the simulated results for a $K = \sqrt{2}$ split-T power divider. Since this power divider is not symmetrical we have to display more scattering parameter vs. frequency results as for the $K = 1$ case.

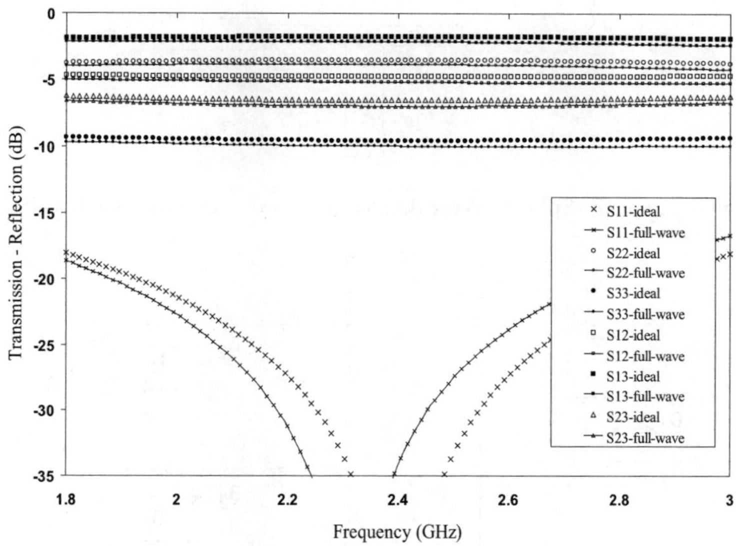

Fig. 5.20 Reflection and transmission coefficients for a $K = \sqrt{2}$ split-T power divider as calculated with the ideal model and with a general full-wave method for a mitred lay-out.

Again the results of the ideal model agree with the full-wave analysis results to a level that makes the ideal model feasible for design purposes. Having established this, we move on to the corporate feeding network. This network will consist of one or more split-T power dividers.

5.4 TRANSMISSION AND REFLECTION COEFFICIENTS FOR A CORPORATE FED ARRAY ANTENNA

With the analysis of the subcomponents of the linear array antenna (microstrip patch antenna and split-T power dividers) being completed, we can now turn to the analysis of the complete array antenna. The scattering matrix of the split-T power divider will play a central role in this analysis.

Let's start with the first combined subcomponent of our array antenna, i.e. a split-T power divider with two microstrip patch antennas, see figure 5.21.

This system may be represented by the circuit shown in figure 5.22.

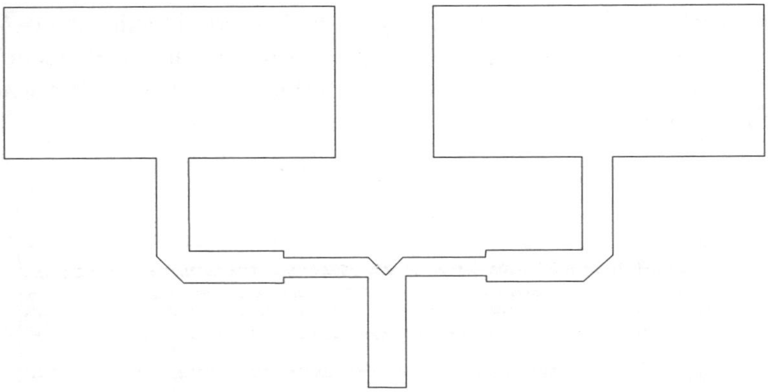

Fig. 5.21 Subcomponent: Split-T power divider with two microstrip patch radiators.

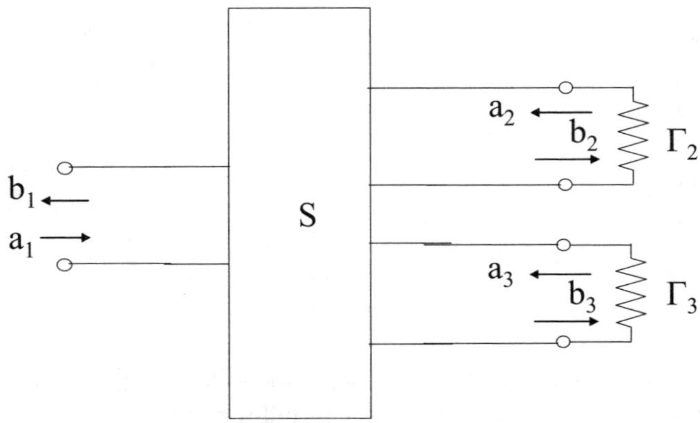

Fig. 5.22 Equivalent circuit of the subcomponent shown in figure 5.21.

The split-T power divider is characterised by its scattering matrix:

$$\begin{pmatrix} b_1 \\ b_2 \\ b_3 \end{pmatrix} = \begin{pmatrix} S_{11} & S_{12} & S_{13} \\ S_{21} & S_{22} & S_{23} \\ S_{31} & S_{32} & S_{33} \end{pmatrix} \begin{pmatrix} a_1 \\ a_2 \\ a_3 \end{pmatrix}. \tag{5.145}$$

The microstrip patch radiators are represented as loads that are characterised by the reflection coefficients Γ_2 and Γ_3 respectively. These reflection coefficients relate the amplitudes of the incoming and outgoing waves of the two output ports of the power divider according to

$$a_2 = \Gamma_2 b_2, \tag{5.146}$$

and

$$a_3 = \Gamma_3 b_3. \tag{5.147}$$

Upon substitution of equations (5.146) and (5.147) into equation (5.145), we may express all outgoing wave amplitudes in terms of the amplitude of the wave going into the subsystem a_1, [15]

$$b_2 = \frac{(1 - S_{33}\Gamma_3)\, S_{21} + S_{23}\Gamma_3 S_{31}}{(1 - S_{22}\Gamma_2)(1 - S_{33}\Gamma_3) - S_{23}S_{32}\Gamma_2\Gamma_3} a_1, \tag{5.148}$$

$$b_3 = \frac{S_{31}(1 - S_{22}\Gamma_2) + S_{32}S_{21}\Gamma_2}{(1 - S_{22}\Gamma_2)(1 - S_{33}\Gamma_3) - S_{23}S_{32}\Gamma_2\Gamma_3} a_1, \tag{5.149}$$

$$b_1 = S_{11}a_1 + S_{12}\Gamma_2 b_2 + S_{13}\Gamma_3 b_3. \tag{5.150}$$

The reflection coefficients Γ_i, $i = 2, 3$, are obtained from the microstrip patch radiator input impedance Z_{inp} and the characteristic impedance of the transmission line the patch radiator is connected to, Z_c

$$\Gamma_i = \frac{Z_{inp} - Z_c}{Z_{inp} + Z_c}. \tag{5.151}$$

Thus, we implicitly assume that the radiators are connected directly to the split-T power divider. In reality, finite length transmission lines are situated between power divider output ports and microstrip radiator input ports.

5.4.1 Two-Element Subarray

In the process of designing a four-element linear microstrip patch array antenna, we will first look at the two-element subarray. We have seen that the analysis is built around the scattering matrix of the split-T power divider. In the analysis, we need to incorporate the transmission lines that connect the components. The two-element subarray and its equivalent microwave network are shown in figure 5.23.

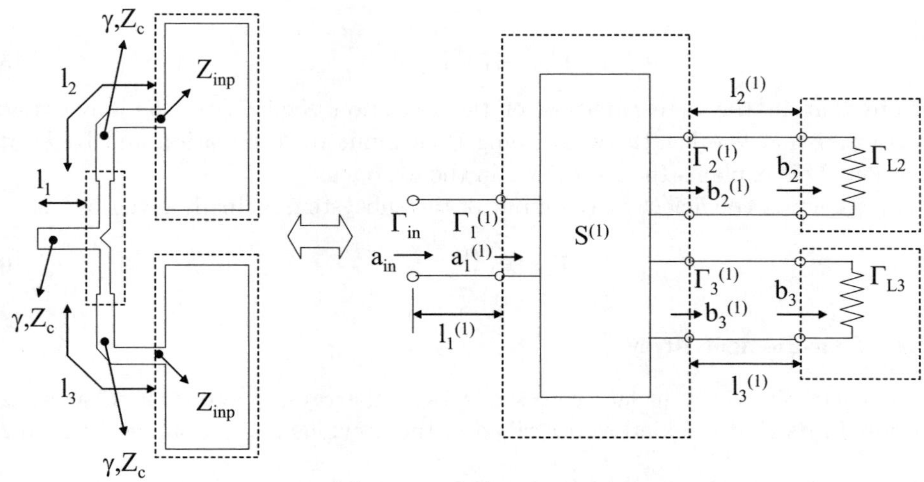

Fig. 5.23 Two-element subarray and equivalent microwave network.

The signals and parameters associated with the split-T power divider three-port are identified with a superscript $^{(1)}$. Transmission lines are connected to all three ports of the power divider.

Then, using equations (5.148) to (5.150)

$$b_2^{(1)} = \frac{\left(1 - S_{33}^{(1)}\Gamma_3^{(1)}\right) S_{21}^{(1)} + S_{23}^{(1)}\Gamma_3^{(1)} S_{31}^{(1)}}{\left(1 - S_{22}^{(1)}\Gamma_2^{(1)}\right)\left(1 - S_{33}^{(1)}\Gamma_3^{(1)}\right) - S_{23}^{(1)} S_{32}^{(1)}\Gamma_2^{(1)}\Gamma_3^{(1)}} a_1^{(1)}, \tag{5.152}$$

$$b_3^{(1)} = \frac{S_{31}^{(1)}\left(1 - S_{22}^{(1)}\Gamma_2^{(1)}\right) + S_{32}^{(1)} S_{21}^{(1)}\Gamma_2^{(1)}}{\left(1 - S_{22}^{(1)}\Gamma_2^{(1)}\right)\left(1 - S_{33}^{(1)}\Gamma_3^{(1)}\right) - S_{23}^{(1)} S_{32}^{(1)}\Gamma_2^{(1)}\Gamma_3^{(1)}} a_1^{(1)}, \tag{5.153}$$

$$b_1^{(1)} = S_{11}^{(1)} a_1^{(1)} + S_{12}^{(1)}\Gamma_2^{(1)} b_2^{(1)} + S_{13}^{(1)}\Gamma_3^{(1)} b_3^{(1)}. \tag{5.154}$$

Herein,

$$a_1^{(1)} = a_{in} e^{-\gamma l_1^{(1)}}, \tag{5.155}$$

where γ is the propagation constant of the transmission line.[4] The amplitudes of the waves entering the antennas, b_2 and b_3 are obtained in a similar way from $b_2^{(1)}$ and $b_3^{(1)}$ respectively:

$$b_2 = b_2^{(1)} e^{-\gamma l_2^{(1)}}, \tag{5.156}$$

$$b_3 = b_3^{(1)} e^{-\gamma l_3^{(1)}}. \tag{5.157}$$

The reflection coefficients $\Gamma_2^{(1)}$ and $\Gamma_3^{(1)}$, see figure 5.23, are obtained from the reflection coefficients at the radiator input ports according to

$$\Gamma_2^{(1)} = \Gamma_{L2} e^{-2\gamma l_2^{(1)}}, \tag{5.158}$$

and

$$\Gamma_3^{(1)} = \Gamma_{L3} e^{-2\gamma l_3^{(1)}}. \tag{5.159}$$

Note that in the determination of the reflection coefficients, the wave travels the transmission line length twice, going from input to termination and back after reflection. This explains the factor two in the exponent.

The reflection coefficient at the input of the subsystem is finally given by

$$\Gamma_{in} = \Gamma_1^{(1)} e^{-2\gamma l_1^{(1)}}. \tag{5.160}$$

5.4.2 Four-Element Array

The four-element array in fact consists of two subarrays, where each subarray is a subarray in itself of the kind as described in the previous paragraph, see figure 5.24.

[4] We do not use a superscript $^{(1)}$ for the propagation constant, since we have designed the power splitter such that transmission lines with equal characteristic impedance and propagation constant are connected to all ports.

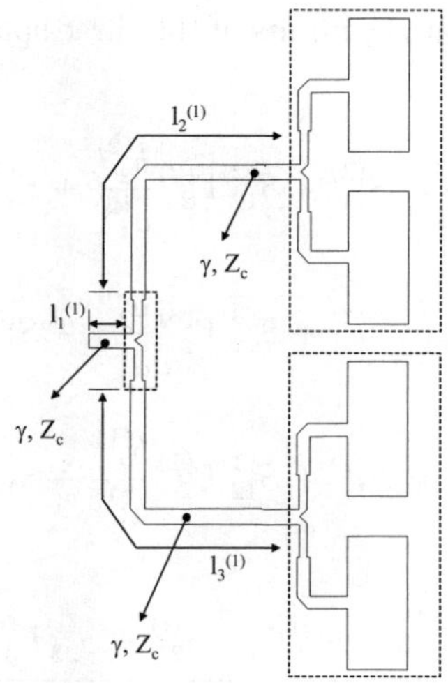

Fig. 5.24 Four-element array consisting of two two-element subarrays.

The naming convention is such that the first split-T power divider encountered from the input going to the radiators is designated (1), the other ones are numbered in increasing order. The equivalent microwave network is shown in figure 5.25.

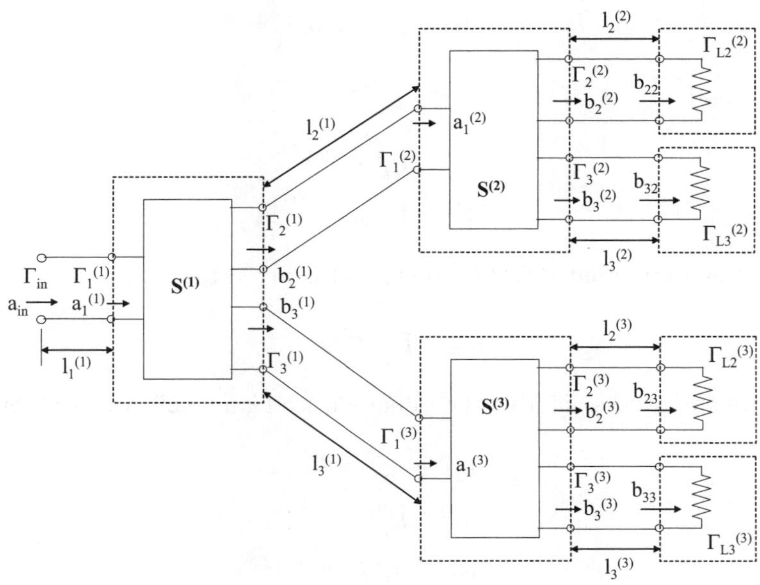

Fig. 5.25 Equivalent microwave network of the array antenna shown in figure 5.24.

The reflection coefficients at the inputs of the three split-T power dividers are given by

$$\Gamma_1^{(2)} = \frac{b_1^{(2)}}{a_1^{(2)}} = S_{11}^{(2)} + S_{12}^{(2)}\Gamma_2^{(2)}\frac{b_2^{(2)}}{a_1^{(2)}} + S_{13}^{(2)}\Gamma_3^{(2)}\frac{b_3^{(2)}}{a_1^{(2)}}, \tag{5.161}$$

$$\Gamma_1^{(3)} = \frac{b_1^{(3)}}{a_1^{(3)}} = S_{11}^{(3)} + S_{12}^{(3)}\Gamma_2^{(3)}\frac{b_2^{(3)}}{a_1^{(3)}} + S_{13}^{(3)}\Gamma_3^{(3)}\frac{b_3^{(3)}}{a_1^{(3)}}, \tag{5.162}$$

$$\Gamma_1^{(1)} = \frac{b_1^{(1)}}{a_1^{(1)}} = S_{11}^{(1)} + S_{12}^{(1)}\Gamma_2^{(1)}\frac{b_2^{(1)}}{a_1^{(1)}} + S_{13}^{(1)}\Gamma_3^{(1)}\frac{b_3^{(1)}}{a_1^{(1)}}, \tag{5.163}$$

where

$$\frac{b_2^{(i)}}{a_1^{(i)}} = \frac{\left[1 - S_{33}^{(i)}\Gamma_3^{(i)}\right] S_{21}^{(i)} + S_{23}^{(i)}\Gamma_3^{(i)} S_{31}^{(i)}}{\left[1 - S_{22}^{(i)}\Gamma_2^{(i)}\right]\left[1 - S_{33}^{(i)}\Gamma_3^{(i)}\right] - S_{23}^{(i)} S_{32}^{(i)}\Gamma_2^{(i)}\Gamma_3^{(i)}}, \tag{5.164}$$

and

$$\frac{b_3^{(i)}}{a_1^{(i)}} = \frac{S_{31}^{(i)}\left[1 - S_{22}^{(i)}\Gamma_2^{(i)}\right] + S_{32}^{(i)} S_{21}^{(i)}\Gamma_2^{(i)}}{\left[1 - S_{22}^{(i)}\Gamma_2^{(i)}\right]\left[1 - S_{33}^{(i)}\Gamma_3^{(i)}\right] - S_{23}^{(i)} S_{32}^{(i)}\Gamma_2^{(i)}\Gamma_3^{(i)}}, \tag{5.165}$$

for $i = 1, 2, 3$.

In these last five expressions,

$$\Gamma_2^{(i)} = \Gamma_{L2} e^{-2\gamma l_2^{(i)}}, \tag{5.166}$$

$$\Gamma_3^{(i)} = \Gamma_{L3} e^{-2\gamma l_3^{(i)}}, \tag{5.167}$$

for $i = 2, 3$, and

$$\Gamma_2^{(1)} = \Gamma_1^{(2)} e^{-2\gamma l_2^{(1)}}, \tag{5.168}$$

$$\Gamma_3^{(1)} = \Gamma_1^{(3)} e^{-2\gamma l_3^{(1)}}. \tag{5.169}$$

Finally, the array input reflection coefficient is found to be

$$\Gamma_{in} = \Gamma_1^{(1)} e^{-2\gamma l_1^{(1)}}. \tag{5.170}$$

The wave amplitudes entering the radiators, see figure 5.25, are given by

$$b_{22} = b_2^{(2)} e^{-\gamma l_2^{(2)}}, \tag{5.171}$$

$$b_{32} = b_3^{(2)} e^{-\gamma l_3^{(2)}}, \tag{5.172}$$

$$b_{23} = b_2^{(3)} e^{-\gamma l_2^{(3)}}, \tag{5.173}$$

$$b_{33} = b_3^{(3)} e^{-\gamma l_3^{(3)}}, \tag{5.174}$$

where, using the known wave amplitude *ratios*

$$b_2^{(2)} = \frac{b_2^{(2)}}{a_1^{(2)}} a_1^{(2)}, \tag{5.175}$$

$$b_3^{(2)} = \frac{b_3^{(2)}}{a_1^{(2)}} a_1^{(2)}, \tag{5.176}$$

$$b_2^{(3)} = \frac{b_2^{(3)}}{a_1^{(3)}} a_1^{(3)}, \tag{5.177}$$

$$b_3^{(3)} = \frac{b_3^{(3)}}{a_1^{(3)}} a_1^{(3)}. \tag{5.178}$$

Herein:

$$a_1^{(2)} = b_2^{(1)} e^{-\gamma l_2^{(1)}}, \tag{5.179}$$

$$a_1^{(3)} = b_3^{(1)} e^{-\gamma l_3^{(1)}}, \tag{5.180}$$

where

$$b_2^{(1)} = \frac{b_2^{(1)}}{a_1^{(1)}} a_1^{(1)}, \tag{5.181}$$

$$b_3^{(1)} = \frac{b_3^{(1)}}{a_1^{(1)}} a_1^{(1)}, \tag{5.182}$$

and

$$a_1^{(1)} = a_{in} e^{-\gamma l_1^{(1)}}. \tag{5.183}$$

5.5 SIMULATION, REALISATION AND MEASUREMENT

The four-element array is constructed from two identical two-element subarrays. One such a subarray, with all the dimensions (in mm) is shown in figure 5.26.

The whole array consists of two of these subarrays, maintaining the same element distance in the complete array antenna. This element distance is chosen such that the mutual coupling between the elements is expected to be negligible.

5.5.1 Realisation

Microstrip patch antennas are meant to be photoetched from a copperclad microwave laminate. These microwave laminates are characterised by a well-known and stable dielectric constant as well as - in general - low losses. They normally come without a photoresist layer. If the availability of such a laminate with a photoresist layer is a problem, standard (FR4) PCB laminate may form an alternative, provided that the dielectric constant of approximately four and the high losses are acceptable for

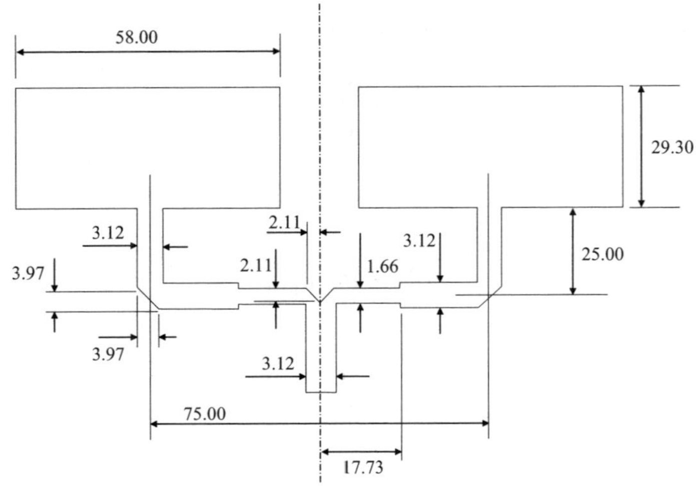

Fig. 5.26 Two-element subarray with dimensions in mm.

the chosen application. Tests with microstrip transmission lines of known physical dimensions and thus of known characteristic impedance should be undertaken to obtain the dielectric constant of the material at hand more precisely.

If photoetching facilities are not available, microstrip antennas operating at not too high frequencies may be constructed using copper tape and a knife. If the frequencies become too high, the physical tolerances will become so tight that handiwork is no longer an option. For our array antenna, operating at 2.4GHz, we may get good results using tape and a knife. The process is as follows. We start with a board of known dielectric constant. As shown in the following figures, we worked with a board of FR4, where on both sides the copper was removed completely by a photoetching process.

One side of the board is covered completely with copper tape. The other side of the board is covered with copper tape in the area where the array antenna needs to be realised. On top, we fix a sheet of paper that has the outlines of the array antenna printed on it in real size, see figure 5.27 and figure 5.28.

Next, the array antenna contour is cut out, using a ruler and a knife, see figure 5.29 and figure 5.30.

For a good electrical contact, the seams of copper tape are soldered and, finally, a connector is soldered to the input transmission line, see figure 5.31.

The final result is shown in figure 5.32.

5.5.2 Simulation and Measurement

The input reflection coefficient of the four-element array is calculated making use of the 'dominant mode' cavity model for the calculation of the radiator reflection coefficient. This reflection coefficient, Γ, follows from the radiator input

Fig. 5.27 Antenna board, underneath completely covered with copper tape, on top covered with copper tape in the area where the array antenna needs to be. On top a sheet of paper with the real-size array antenna lay-out is held in a fixed position.

Fig. 5.28 Sheet of paper with the real-size array antenna lay-out on top of the antenna board.

impedance, Z_{in}

$$\Gamma = \frac{Z_{in} - Z_0}{Z_{in} + Z_0}, \tag{5.184}$$

where Z_0 is the characteristic impedance of the microstrip transmission line connected to the microstrip patch radiator.

The input reflection coefficient, thus calculated as function of frequency, is shown in figure 5.33 together with the calculated results obtained from a general full-wave analysis method. These last results nearly coincide with measurement results that are therefore not shown in the graph.

Although we have employed a (cavity) model that is not completely suited for the specific rectangular microstrip patch radiator at hand, we do see that the radiators with corporate feed network are correctly accounted for around the desired resonance

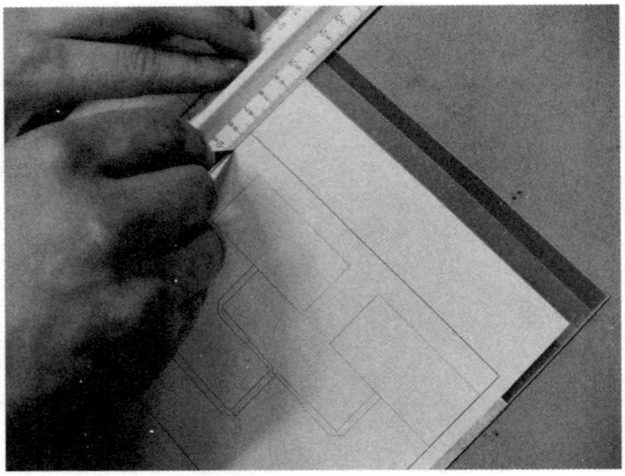

Fig. 5.29 Cutting out the array antenna contour.

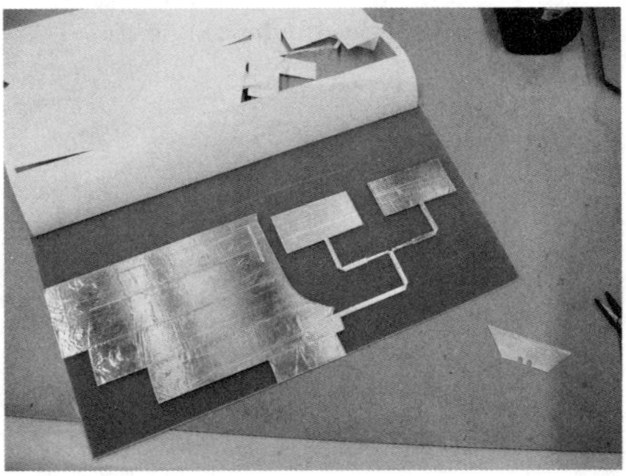

Fig. 5.30 Part of the array antenna realised.

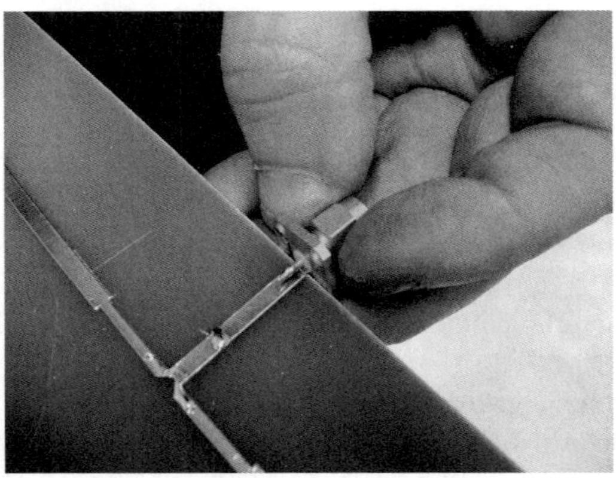

Fig. 5.31 A connector will be soldered to the input transmission line.

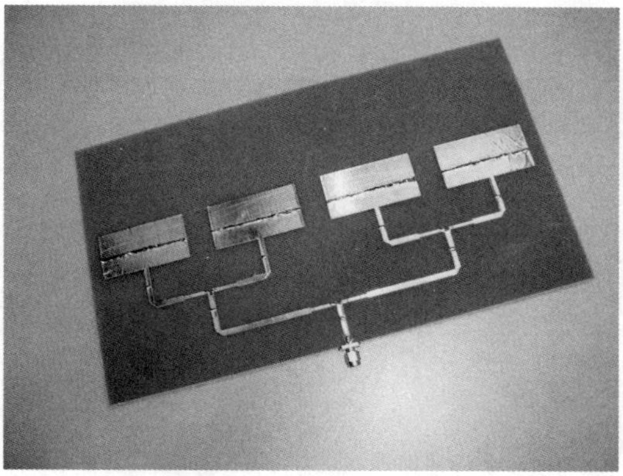

Fig. 5.32 Final result.

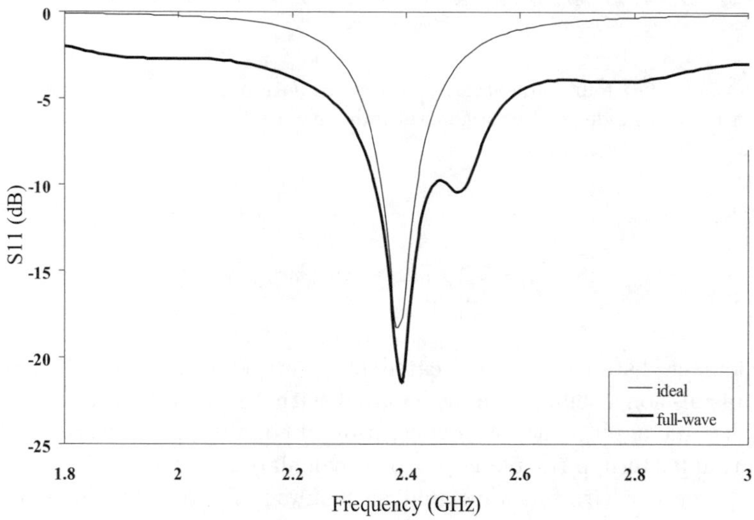

Fig. 5.33 Corporate-fed four-element array input reflection coefficient, calculated by the ideal model and by a general full wave method.

frequency.[5] This demonstrates that the employment of our fast but reduced accuracy model is of practical use in the *design* of an array antenna. A full-wave analysis method may be employed to validate and/or fine-tune an initial design based on our fast model.

The power radiation patterns at 2.4GHz are obtained by multiplying the single radiator gain pattern with the array factor (thus assuming that mutual coupling may

[5]The filtering characteristics of the quarter lambda transformers in the corporate feeding network around the desired centre frequency suppress the higher order resonances of the radiators.

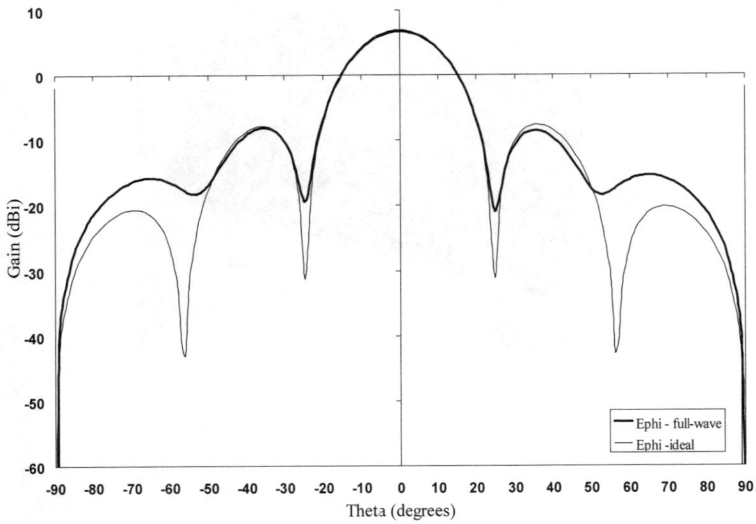

Fig. 5.34 Corporate-fed four-element array power radiation pattern in the plane $\varphi = 0°$, calculated by the ideal model and by a general full-wave method.

be neglected),

$$0.5 \sum_{n=1}^{4} e^{-jk_0(n-1)d\sin(\vartheta)}, \tag{5.185}$$

where the element distance is as indicated in figure 5.26 and the factor 0.5 is the (voltage) transmission coefficient as calculated with the theory from the preceding sections. This value agrees with our expectation of equal power division at 2.4GHz.

The radiation pattern in the plane $\varphi = 0°$ is calculated for the dominant mode only and is shown, together with full-wave analysis results in figure 5.34. The figure clearly indicates the validity of the used model and the assumption of negligible mutual coupling. The only effect of higher-order modes and mutual coupling is visible in the filling of the 'nulls'.

REFERENCES

1. K.R. Carver and J.W. Mink, Microstrip Antenna Technology, *IEEE Transactions on Antennas and Propagation*, Vol. AP-29, January 1981, pp. 2–24.

2. W.F. Richards, Y.T. Lo and D.D. Harrison, An Improved Theory for Microstrip Antennas and Applications, *IEEE Transactions on Antennas and Propagation*, Vol. AP-29, January 1981, pp. 38–46.

3. Roger F. Harrington, *Time-Harmonic Electromagnetic Fields*, John Wiley & Sons, New York, 2001.

4. J.R. James, P.S. Hall and C. Wood, *Microstrip Antennas*, Peter Peregrinus, London, 1981.

5. J.R. James and P.S. Hall, *Handbook of Microstrip Antennas*, Peter Peregrinus, London, 1989.

6. J. Huang, The Finite Ground Plate Effect on the Microstrip Antenna Radiation Patterns, *IEEE Transactions on Antennas and Propagation*, Vol. AP-31, July 1983, pp. 649–653.

7. P. Hammer, D. van Bouchaute, D. Verschraeven and A. van de Capelle, A Model for Calculating the Radiation Field of Microstrip Antennas, *IEEE Transactions on Antennas and Propagation*, Vol. AP-27, No. 2, March 1979, pp. 267–270.

8. Y.T. Lo, D. Solomon and W.F. Richards, Theory and Experiment on Microstrip Antennas, *IEEE Transactions on Antennas and Propagation*, Vol. AP-27, No. 2, March 1979, pp. 137–145.

9. Kazuhiro Hirasawa and Misao Haneishi (eds.), *Analysis, Design and Measurement of Small and Low-Profile Antennas*, Artech House, Norwood, MA, 1992.

10. David M. Pozar, *Microwave Engineering, second edition*, John Wiley & Sons, New York, 1998.

11. C.G. Montgomery, R.H. Dicke and E.M. Purcell, *Principles of Microwave Circuits*, McGraw-Hill, New York, 1948.

12. L.I. Parad and R.L. Moynihan, Split-Tee Power Divider, *IEEE Transactions on Microwave Theory and Techniques*, Vol. 13, No. 1, January 1965, pp. 91–95.

13. K.C. Gupta, Ramesh Garg and Rakesh Chadha, '*Computer Aided Design of Microwave Circuits*', Artech House, Norwood, MA, 1981.

14. Jerome L. Altman, *Microwave Circuits*, D. Van Nostrand Company, Princeton, New Jersey, 1964.

15. Kuan Min Lee and Ruey-Shi Chu, Analysis of Mutual Coupling Between a Finite Phased Array of Dipoles and its Feed Network, *IEEE Transactions on Antennas and Propagation*, Vol. 36, No. 12, December 1988, pp. 1681–1699.

6

The Linear Endfire Array Antenna

In this chapter we will treat the special situation where the main beam of a linear array antenna is directed alongside the array axis instead of perpendicular to this axis. The latter situation - treated in the previous two chapters - is known as broadside radiation, the first one we designate, for obvious reasons, *endfire* radiation. For an endfire array antenna, the radiating elements need to exhibit a radiation pattern that allows radiation into the endfire direction. Dipole radiators show a radiation pattern in the plane perpendicular to the dipole axis that satisfies this requirement. Therefore, the endfire specifics will be demonstrated in this chapter employing dipole radiators. To show that mutual coupling between radiators in array antenna - briefly mentioned in the previous chapter - is not necessarily an unwanted phenomenon, we will provide an approximate analysis of a Yagi–Uda array antenna in detail. A Yagi–Uda array antenna is a linear endfire array antenna consisting of dipole radiators, where only one element is driven and the other ones are excited by a mutual coupling mechanism.

6.1 INTRODUCTION

In the chapter on linear broadside array antennas we have seen that, for sufficiently small inter-element distances, a broadside beam is generated, provided that the array feeding network is not introducing phase shifts between the elements. We have also seen that if the inter-element distance is equal to or exceeds one wavelength, the array forms additional beams (grating lobes) at or near endfire.

The phase shifts introduced by the inter-element distances determine the number of beams, while additional phase shifts introduced by the feeding network or otherwise have the effect of dispositioning the main beam from its broadside direction. In this chapter we will look at the specific phase differences needed for creating endfire operation. The general subject of beam scanning will be treated in the next chapter on linear *phased* array antennas.

In the previous chapters we have employed the concept of *pattern multiplication*. The array antenna radiation pattern was obtained by multiplying the array factor with the element factor or element radiation pattern. The element radiation pattern was seen to have a filtering effect on the array factor. For the aperture elements considered thus far, the radiation pattern is cosine-like, being at maximum in broadside direction and at minimum in endfire direction. Obviously, such a radiating element will not allow an endfire operation of the linear array antenna. We have also seen that a dipole radiator has a radiation pattern that is rotationally symmetric around the dipole axis and at maximum for directions perpendicular to the dipole axis centre. Therefore, we will demonstrate endfire radiation of a linear array antenna by employing dipole elements.

To create endfire operation of a linear array antenna we need to establish a phase difference between the elements for fixed inter-element distances. These phase differences may be introduced by the linear array feeding network. Either a corporate or parallel feeding network or a series feeding network will introduce mechanical difficulties (e.g. the need for different layers, see for example [1]) and the feeding network may take part in, or even disturb the radiation.

We may considerably simplify the array antenna architecture by only feeding one dipole, having the other (short-circuited) dipole elements in the linear array acting as parasites. The desired inter-element phase shifts are accomplished then by virtue of mutual coupling. For (thin) dipole radiators, the mutual coupling can be calculated with a practical level of accuracy with relative ease,[1] thus enabling the analysis of a Yagi–Uda endfire array antenna.

Before we start with the analysis of the Yagi–Uda array antenna, we will first have to establish the theoretical description of a linear endfire array antenna, for the moment not bothering about feeding networks or how to accomplish additional phase shifts between radiating elements. Then we need to talk about mutual coupling in general and finally describe the mutual interaction between two dipoles of different lengths.

6.2 PHASE DIFFERENCES

We start with considering a linear array antenna consisting of K elements, equally interspaced a distance d. The direction of a wave is described by the angle ϑ between rays and array normal, see figure 6.1.

We describe the array antenna in the receive situation.

The array factor, $S_a(\vartheta)$, for this linear array antenna is given by

$$S_a(\vartheta) = \sum_{i=1}^{K} e^{j(K-i)kd\sin(\vartheta)}, \tag{6.1}$$

where $k = \frac{2\pi}{\lambda}$ is the wave number with λ being the wavelength.

[1]In general, for arrays of thick or non-dipole radiators, this is not true.

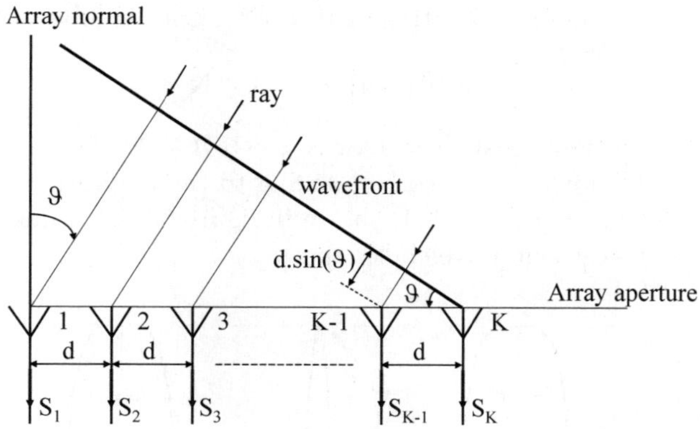

Fig. 6.1 Linear array of K elements.

If we now allow a linear phase taper to be existent over the array, meaning that every element of the array antenna adds a progressive phase α, the radiation pattern of the array antenna, $S(\vartheta)$, becomes

$$S(\vartheta) = f(\vartheta) \sum_{i=1}^{K} a_i e^{j(K-i)kd\sin(\vartheta)}, \tag{6.2}$$

where

$$a_i = e^{j(K-i)\alpha}, \tag{6.3}$$

and $f(\vartheta)$ is the element pattern.

The linear array radiation pattern may be written as

$$S(\vartheta) = f(\vartheta) \sum_{i=1}^{K} e^{j(K-i)\psi}, \tag{6.4}$$

where

$$\psi = kd\sin(\vartheta) + \alpha. \tag{6.5}$$

The maximum of the radiation pattern is obtained for $\psi = 0$, so for endfire operation at $\vartheta = 90°$, the additional phase difference, α, between two adjacent elements should be

$$\alpha = -kd\sin(90°) = -kd. \tag{6.6}$$

For endfire operation at $\vartheta = -90°$, the phase difference should be

$$\alpha = -kd\sin(-90°) = kd. \tag{6.7}$$

As we have already discussed, grating lobes occur whenever the argument ψ is an integer multiple of 2π,

$$\psi = n2\pi, \; n \in \mathbf{N}. \tag{6.8}$$

So, when $\alpha = -kd$ (endfire operation at $\vartheta = 90°$), grating lobes occur for

$$\frac{d}{\lambda}\left[\sin(\vartheta) - 1\right] = n, \ n \in \mathbf{N}. \tag{6.9}$$

For $d = \lambda$, four solutions exist: $\vartheta = 0$ for $n = -1$, $\vartheta = 90°$ for $n = 0$, $\vartheta = 180°$ for $n = -1$ and $\vartheta = -90°$ for $n = -2$, corresponding to three additional (grating) lobes next to the main beam ($n = 0$). This is shown in figure 6.2 for a linear array of eight isotropic elements, placed one wavelength apart.

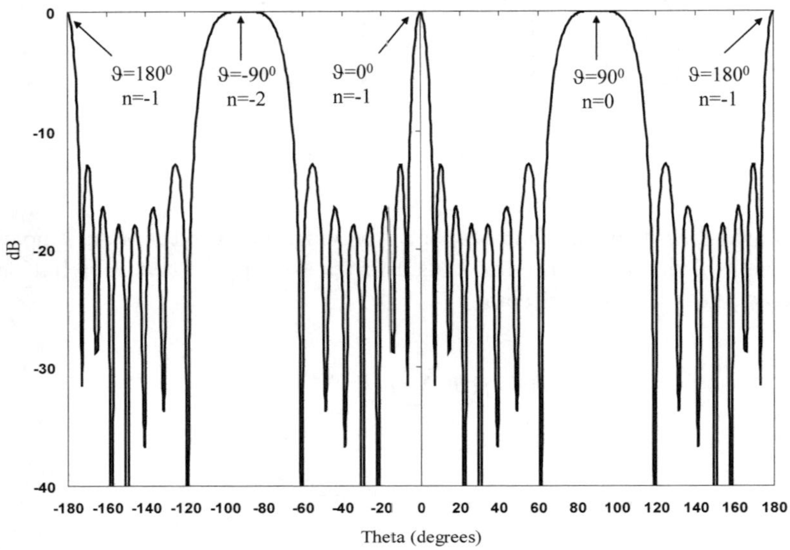

Fig. 6.2 Linear array of eight isotropic elements, placed one wavelength apart, phased for endfire operation at $\vartheta = 90°$, normalised to radiation maximum.

If we reduce the element spacing to a half wavelength, two solutions exist: $\vartheta = 90°$ for $n = 0$ and $\vartheta = -90°$ for $n = -1$, see figure 6.3 for a linear array of eight isotropic elements, placed half a wavelength apart.

So, for a broadside linear array antenna we could place the elements as far apart from each other as one wavelength, before grating lobes started to occur. For a endfire linear array antenna it appears that the maximum allowable element distance preventing the occurrence of grating lobes has reduced to half a wavelength. If we reduce the element spacing below half a wavelength, the grating lobe at $\vartheta = -90°$ in figure 6.3 will disappear.

The theoretical explanation of this phenomenon will be given in the next chapter where we will discuss beam scanning in a linear array antenna.

We will conclude the discussion on grating lobes in endfire array antennas, by showing that for a linear array, having an inter-element spacing smaller than half a wavelength, indeed only one endfire beam does exist. Figure 6.4 shows the normalised radiation pattern of a linear array consisting of eight isotropic elements, spaced four-tenths of a wavelength apart.

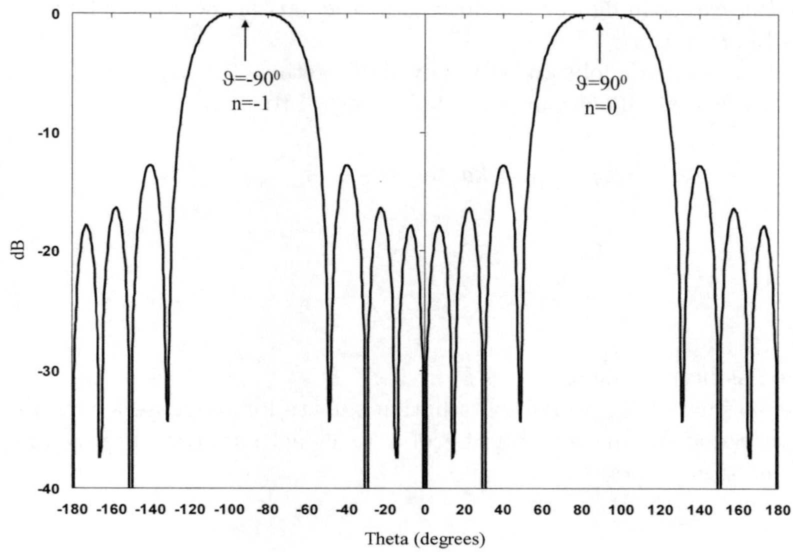

Fig. 6.3 Linear array of eight isotropic elements, placed half a wavelength apart, phased for endfire operation at $\vartheta = 90^\circ$, normalised to radiation maximum.

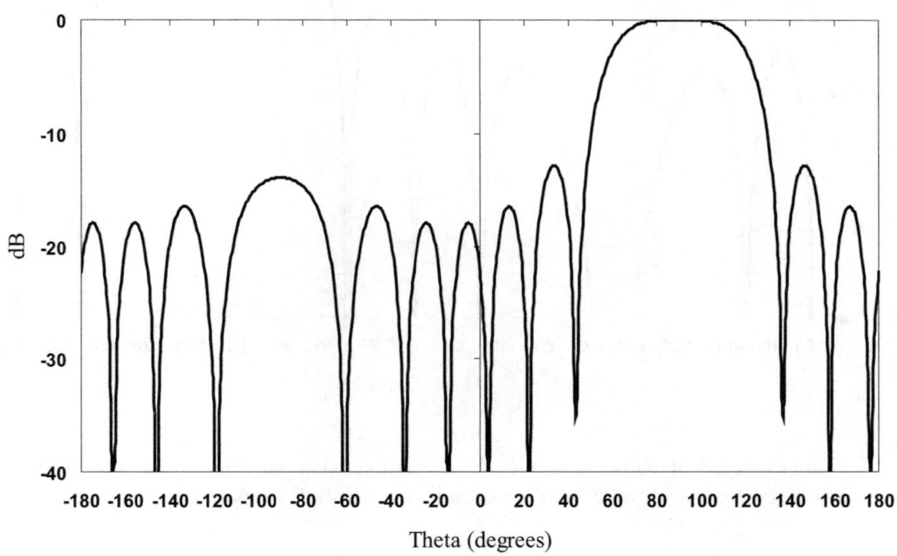

Fig. 6.4 Linear array of eight isotropic elements, placed four-tenths of a wavelength apart, phased for endfire operation at $\vartheta = 90^\circ$, normalised to radiation maximum.

In fact, by choosing an element distance smaller than half a wavelength, we have moved the maximum of the grating lobe out of the *visible* into the *invisible region.*[2]

[2]The visible region is defined by angles ϑ for which $|\sin(\vartheta)| \leq 1$. In this situation, that corresponds to angles $-180^\circ \leq \vartheta \leq 180^\circ$.

The *back lobe* we see in figure 6.4 is formed by the part of the grating lobe still present in the visible region.

If we want the back lobe to be removed altogether, the argument, ψ, for endfire operation at $\vartheta = 90°$ should, at $\vartheta = -90°$ be equal to $-\pi$,

$$\psi|_{\vartheta=-90°} = kd\left[\sin(\vartheta) - 1\right]|_{\vartheta=-90°} = -\pi, \tag{6.10}$$

so

$$d = \frac{\lambda}{4}, \tag{6.11}$$

where use has been made of $k = \frac{2\pi}{\lambda}$.

Figure 6.5 shows the normalised radiation pattern for an eight-element linear array of isotropic radiators, spaced a quarter of a wavelength apart and phased for $\vartheta = 90°$ endfire operation.

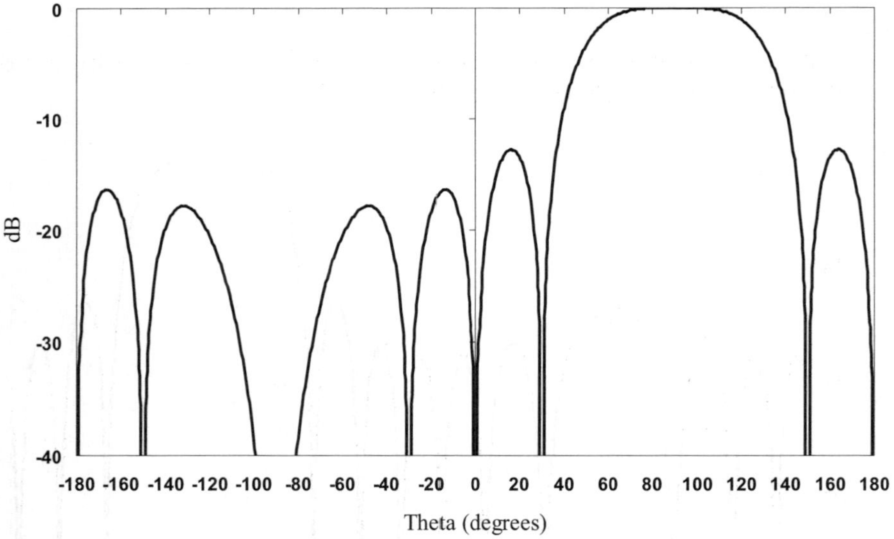

Fig. 6.5 Linear array of eight isotropic elements, placed a quarter of a wavelength apart, phased for endfire operation at $\vartheta = 90°$, normalised to radiation maximum.

The figure shows that for this element distance, the grating lobe effects - normally visible in a backfire lobe - have indeed completely disappeared.

For the linear endfire array antennas treated thus far, the linear phase taper over the array elements ($\alpha = \mp kd$) compensates exactly the phase delay of waves in the desired endfire direction. These endfire arrays are called *ordinary endfire arrays*. This naming implies also that 'extraordinary' endfire arrays must exist. In the next section we will briefly discuss the *Hansen–Woodyard endfire array*.

6.3 HANSEN–WOODYARD ENDFIRE ARRAY ANTENNA

For long array antennas (i.e. much longer than one wavelength) of closely spaced elements, a higher directivity (smaller main beam) may be obtained than for the ordinary endfire array antenna. This may be accomplished by having an additional phase difference, α', applied to adjacent array elements that is slightly larger than the phase delay a wave experiences travelling in the endfire direction

$$\alpha' = \mp(kd + \delta). \tag{6.12}$$

The minus sign applies for a radiation maximum at $\vartheta = 90°$ and the plus sign applies for a radiation maximum at $\vartheta = -90°$.

In 1938, Hansen and Woodyard [2] proposed for δ in the above equation [3, 4]

$$\delta = \frac{2.94}{K-1} \approx \frac{2.94}{K} \approx \frac{\pi}{K}. \tag{6.13}$$

The linear array antenna radiation pattern may be written then as

$$S(\vartheta) = f(\vartheta) \sum_{i=1}^{K} e^{j(K-i)\psi'}, \tag{6.14}$$

where

$$\psi' = kd \sin(\vartheta) \mp \left(kd + \frac{\pi}{K}\right), \tag{6.15}$$

in which the minus sign applies for endfire operation at $\vartheta = 90°$ and the plus sign applies for endfire operation at $\vartheta = -90°$.

Figure 6.6 shows the normalised radiation patterns of a linear array of eight isotropic radiators, phased for ordinary and for Hansen–Woodyard endfire operation at $\vartheta = 90°$ for an element spacing of half a wavelength.

We do see a smaller beam in the desired endfire direction, but at the same time we observe back lobe levels that are in excess of the main beam level. Obviously, the additional phase difference, δ, has yielded a sharper main lobe, but this additional phase shift has also caused the rise of the back lobes. The only way to decrease the back lobe level to values less than that of the main beam lies in decreasing the element spacing and thus accomplishes bringing the argument ψ' closer to $-\pi$ for $\vartheta = -90°$ [4, 5].[3]

To demonstrate this, in figure 6.7 the normalised radiation patterns are shown of a linear array of eight isotropic radiators, phased for ordinary and for Hansen–Woodyard endfire operation at $\vartheta = 90°$ for an element spacing of four-tenths of a wavelength.

We do indeed observe a lowering of the back lobe level.

If we want to achieve a maximum suppression of the back lobe - for the array antenna phased for Hansen–Woodyard endfire operation at $\vartheta = 90°$ - the following

[3]For the linear array antenna operated at $-90°$ endfire, the argument ψ' should be brought closer to $+\pi$ for $\vartheta = 90°$.

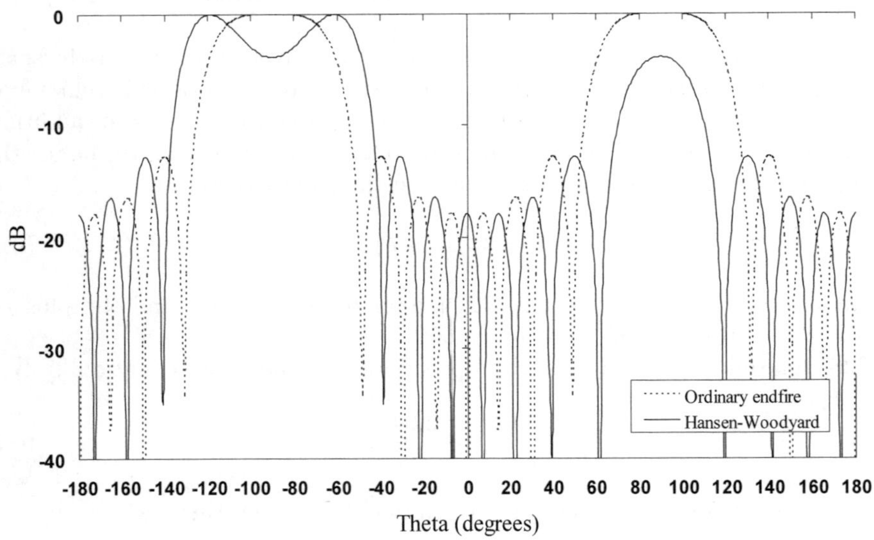

Fig. 6.6 Linear array of eight isotropic elements, placed half a wavelength apart, phased for ordinary and for Hansen–Woodyard endfire operation at $\vartheta = 90°$, normalised to radiation maximum.

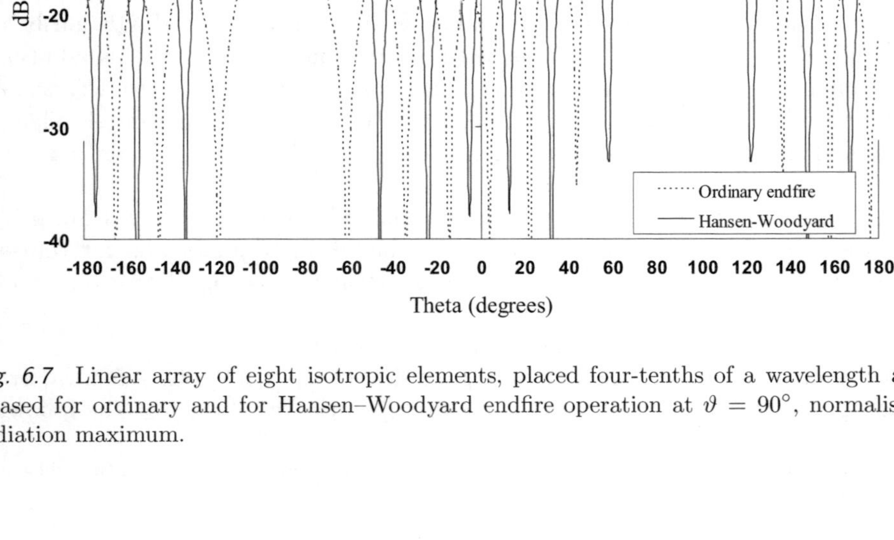

Fig. 6.7 Linear array of eight isotropic elements, placed four-tenths of a wavelength apart, phased for ordinary and for Hansen–Woodyard endfire operation at $\vartheta = 90°$, normalised to radiation maximum.

condition should apply

$$\psi'|_{\vartheta=-90^\circ} = \left[kd\sin(\vartheta) - \left(kd + \frac{\pi}{K}\right)\right]\Big|_{\vartheta=-90^\circ} = -\pi, \tag{6.16}$$

which leads to

$$d = \frac{K-1}{K}\frac{\lambda}{4} \approx \frac{\lambda}{4}. \tag{6.17}$$

Figure 6.8 shows the normalised radiation patterns of a linear array of eight isotropic radiators, phased for ordinary and for Hansen–Woodyard endfire operation at $\vartheta = 90^\circ$ for an element spacing of a quarter of a wavelength.

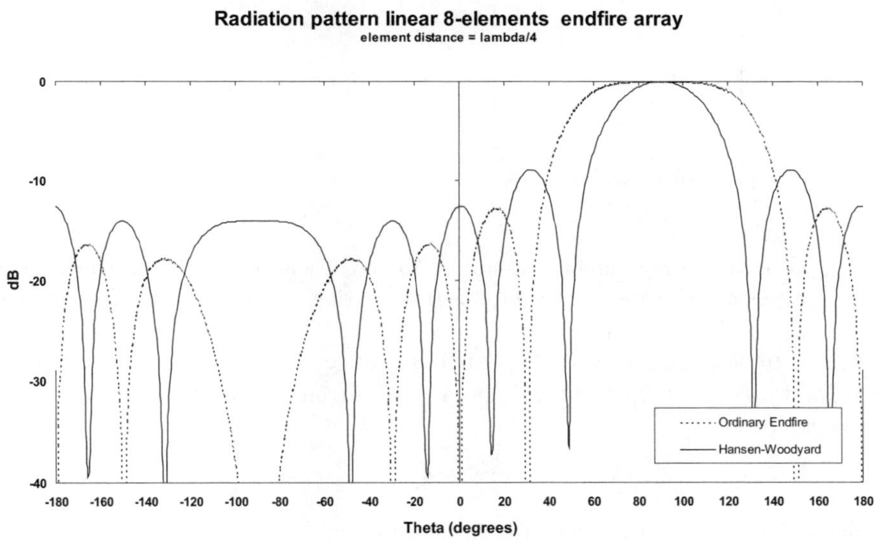

Fig. 6.8 Linear array of eight isotropic elements, placed a quarter of a wavelength apart, phased for ordinary and for Hansen–Woodyard endfire operation at $\vartheta = 90^\circ$, normalised to radiation maximum.

So, by satisfying the Hansen–Woodyard criterion ($\alpha = \mp\left(kd + \frac{\pi}{K}\right)$) in a linear endfire array antenna, a higher directivity (smaller main beam) may be accomplished. But since there is no such thing as a free lunch, the price to be paid is an increased side lobe and back lobe level, as figure 6.8 clearly demonstrates.

6.4 MUTUAL COUPLING

At the beginning of this chapter we said that we were intending to accomplish the desired phase shift between the linear endfire array antenna elements by virtue of *mutual coupling*. To that end we will first discuss mutual coupling between array antenna elements in general terms, before we go into the details of designing an endfire dipole array antenna.

Consider two antenna elements, 1 and 2 [5]. A voltage generator, V_1, is attached to antenna-element 1 that acts as transmitting element, while the current I_2, is measured at the terminals of antenna-element 2, that acts as receiving element. The set-up is shown in figure 6.9a. If we apply a voltage generator V_2 to antenna-element 2, we will measure a current I_1 at the terminals of antenna-element 1, see figure 6.9b.

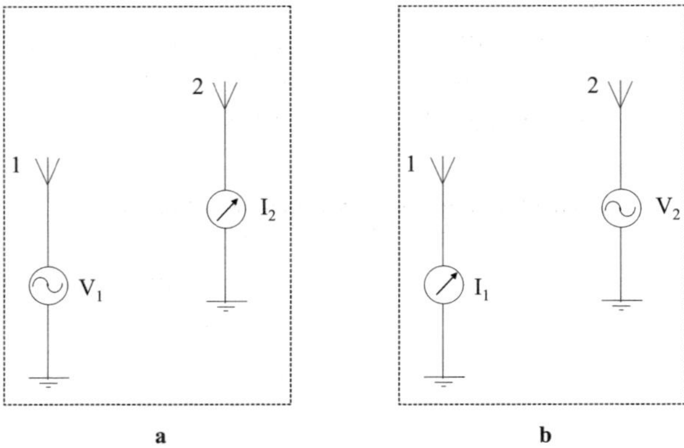

Fig. 6.9 Two-element array antenna set-up. a. Antenna-element 1 is transmitting, antenna-element 2 is receiving. b. Antenna-element 2 is transmitting, antenna-element 1 is receiving.

The ratio of voltage to current is an impedance. For the situation depicted in figure 6.9a, we call the ratio of voltage V_1 to current I_2 a *transfer impedance* or *mutual impedance*:

$$\frac{V_1}{I_2} = Z_{12}. \tag{6.18}$$

Similarly, for the situation depicted in figure 6.9b:

$$\frac{V_2}{I_1} = Z_{21}, \tag{6.19}$$

and by virtue of antenna reciprocity[4]

$$Z_{21} = Z_{12}. \tag{6.20}$$

If we now consider the situation where we only have antenna 1 present, excited by voltage generator V_1 and we measure the current I_1 at its terminals, we may define the input impedance as

$$Z_{11} = \frac{V_1}{I_1}. \tag{6.21}$$

Bringing antenna 2 in the neighbourhood of antenna 1, will mean that - through radiation[5] - a current I_2 is induced on this antenna by antenna 1. This current I_2 will

[4]If $V_1 = V_2$ in figure 6.9, then $I_2 = I_1$.

[5]Next to direct space radiation, mutual coupling may also occur through scattering from nearby objects and by reflections through and radiation from the array antenna feed network.

cause radiation from antenna 2 and thus will influence the current on antenna 1. This phenomenon is called *mutual coupling.*

The total voltage at antenna 1 may then be written as

$$V_1 = Z_{11}I_1 + Z_{12}I_2, \tag{6.22}$$

where Z_{11} and Z_{12} are as defined before

$$Z_{11} = \left.\frac{V_1}{I_1}\right|_{I_2=0}, \tag{6.23}$$

and

$$Z_{12} = \left.\frac{V_1}{I_2}\right|_{I_1=0}. \tag{6.24}$$

Similarly

$$V_2 = Z_{21}I_1 + Z_{22}I_2, \tag{6.25}$$

where

$$Z_{21} = \left.\frac{V_2}{I_1}\right|_{I_2=0}, \tag{6.26}$$

$$Z_{22} = \left.\frac{V_2}{I_2}\right|_{I_1=0}. \tag{6.27}$$

This discussion of a two-element array antenna may be generalised to a K-element array antenna for which

$$\begin{aligned} V_1 &= Z_{11}I_1 + Z_{12}I_2 + \cdots + Z_{1K}I_K \\ V_2 &= Z_{21}I_1 + Z_{22}I_2 + \cdots + Z_{2K}I_K \\ &\vdots \\ V_K &= Z_{K1}I_1 + Z_{K2}I_2 + \cdots + Z_{KK}I_K \end{aligned}, \tag{6.28}$$

where

$$Z_{mn} = \left.\frac{V_m}{I_n}\right|_{I_i=0, i\neq n}. \tag{6.29}$$

The input impedance, Z_m, of the mth element in the array, including all mutual coupling - also known as the *active impedance* - is then

$$Z_m = \frac{V_m}{I_m} = Z_{m1}\frac{I_1}{I_m} + Z_{m2}\frac{I_2}{I_m} + \cdots + Z_{mm} + \cdots + Z_{mK}\frac{I_K}{I_m}. \tag{6.30}$$

Since current is the origin of electromagnetic radiation, the mutual coupling will not only affect the input impedances of the elements in the array, but also their radiation patterns. The mutual coupling effects will in general change with element position, angle of radiation and frequency and will depend on the type of array element under consideration. Thus, in general, the calculation of mutual coupling effects will be difficult.[6] However, for some practical situations, approximations may be made that make an analytical treatment of coupling effects feasible. One such a practical situation occurs for arrays of thin dipole radiators, e.g. Yagi–Uda endfire array antennas.

6.5 YAGI–UDA ARRAY ANTENNA

The Yagi–Uda array antenna consists of a number of dipole elements, of which only one is driven. The other dipole elements act as parasitic radiators, the radiation stemming from currents induced by mutual coupling. The array is an endfire array. Usually there is one element next to the driver at one side and multiple elements next to the driver at the other side. The one element is generally longer than the driver and acts as reflector, the elements on the other side of the driver usually are shorter than the driver and act as directors, see figure 6.10.

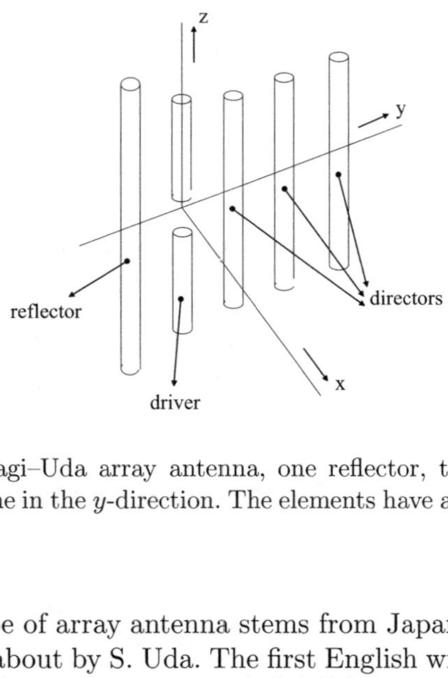

Fig. 6.10 Five-element Yagi–Uda array antenna, one reflector, three directors. The endfire operation is in the x, y-plane in the y-direction. The elements have an omnidirectional radiation pattern in the x, y-plane.

The origin of this type of array antenna stems from Japan at the end of the 1920s and was first published about by S. Uda. The first English written publication on this array antenna was by H. Yagi, a co-worker of Uda [4] and for long the antenna has been known as the Yagi-antenna. Recently the antenna has been called the *Yagi–Uda Array Antenna* to give credit to both inventors.

[6] In the chapter on array antenna measurement we will treat the concept of the *active* or *scan element pattern*, i.e. the radiation pattern of one driven element in its array environment where all other elements are terminated into matched loads. For large array antennas, an average active element pattern may be factored out of the array antenna radiation pattern, making a pattern multiplication between active element pattern and array factor without mutual coupling possible. All mutual coupling effects are accounted for then in the average active element pattern.

6.5.1 Mutual Impedance

To calculate the mutual impedances between the elements in a Yagi–Uda array antenna, we need to be able to calculate the mutual impedance between two, nonstaggered (meaning that the imaginary line through the centres of the dipoles is perpendicular to both dipoles), unequal length dipole elements as shown in figure 6.11.

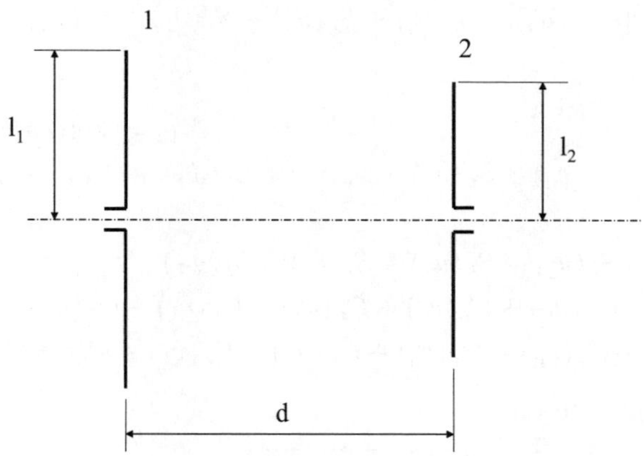

Fig. 6.11 Two, unstaggered, unequal length dipole elements.

In this figure, d is the distance between the two dipole elements and l_1 and l_2 are the half-lengths of, respectively, dipole 1 and dipole 2.

Analytical expressions for the mutual impedance, Z_{12}, for this situation are derived by King [6], employing the *induced emf* method (for an explanation of the induced emf method, see, for example [4]). The results are obtained under the assumption of infinitely thin dipoles supporting sinusoidal currents, and are shown to be valid for wire radii at least up to $10^{-5}\lambda$ [4]. λ is the used wavelength.

The currents on both dipole elements are

$$I_i = I_{i0} \begin{cases} \sin\left[k\left(z_i + l_i\right)\right] & \text{for } -l_i \leq z_i \leq 0 \\ \sin\left[k\left(l_i - z_i\right)\right] & \text{for } 0 \leq z_i \leq l_i \end{cases} \tag{6.31}$$

for $i = 1, 2$, where z_i is the local coordinate along the dipole i and l_i is the half-length of dipole i. I_{i0} is the amplitude of the current on dipole i, and k is the wave number, $k = \frac{2\pi}{\lambda}$.

The mutual impedance is then found to be [6]

$$Z_{12} = R_{12} + jX_{12}, \tag{6.32}$$

where

$$\begin{aligned} R_{12} = 30 \{ & \cos [k(l_1 + l_2)] \,. \\ & [C_i(u_0) + C_i(v_0) - C_i(u_1) - C_i(v_1) - C_i(w_1) - C_i(y_1) + 2C_i(kd)] + \\ & \cos [k(l_1 - l_2)] \,. \\ & [C_i(u_0') + C_i(v_0') - C_i(u_1) - C_i(v_1) - C_i(w_1) - C_i(y_1) + 2C_i(kd)] + \\ & \sin [k(l_1 + l_2)] \,. [-S_i(u_0) + S_i(v_0) + S_i(u_1) - S_i(v_1) - S_i(w_1) + S_i(y_1)] + \\ & \sin [k(l_1 - l_2)] \,. [-S_i(u_0') + S_i(v_0') + S_i(u_1) - S_i(v_1) + S_i(w_1) - S_i(y_1)] \}, \end{aligned} \tag{6.33}$$

and

$$\begin{aligned} X_{12} = 30 \{ & \cos [k(l_1 + l_2)] \,. \\ & [-S_i(u_0) - S_i(v_0) + S_i(u_1) + S_i(v_1) + S_i(w_1) + S_i(y_1) - 2S_i(kd)] + \\ & \cos [k(l_1 - l_2)] \,. \\ & [-S_i(u_0') - S_i(v_0') + Si(u_1) + S_i(v_1) + S_i(w_1) + S_i(y_1) - 2S_i(kd)] + \\ & \sin [k(l_1 + l_2)] \,. [-C_i(u_0) + C_i(v_0) + C_i(u_1) - C_i(v_1) - C_i(w_1) + C_i(y_1)] + \\ & \sin [k(l_1 - l_2)] \,. [-C_i(u_0') + C_i(v_0') + C_i(u_1) - C_i(v_1) + C_i(w_1) - C_i(y_1)] \}. \end{aligned} \tag{6.34}$$

In the above equations

$$u_0 = k \left[\sqrt{d^2 + (l_1 + l_2)^2} - (l_1 + l_2) \right], \tag{6.35}$$

$$v_0 = k \left[\sqrt{d^2 + (l_1 + l_2)^2} + (l_1 + l_2) \right], \tag{6.36}$$

$$u_0' = k \left[\sqrt{d^2 + (l_1 - l_2)^2} - (l_1 - l_2) \right], \tag{6.37}$$

$$v_0' = k \left[\sqrt{d^2 + (l_1 - l_2)^2} + (l_1 - l_2) \right], \tag{6.38}$$

$$u_1 = k \left[\sqrt{d^2 + l_1^2} - l_1 \right], \tag{6.39}$$

$$v_1 = k \left[\sqrt{d^2 + l_1^2} + l_1 \right], \tag{6.40}$$

$$w_1 = k \left[\sqrt{d^2 + l_2^2} + l_2 \right], \tag{6.41}$$

$$y_1 = k \left[\sqrt{d^2 + l_2^2} - l_2 \right]. \tag{6.42}$$

The *sine integral* $S_i(x)$ is defined as

$$S_i(x) = \int_0^x \frac{\sin(\rho)}{\rho} d\rho, \tag{6.43}$$

and the *cosine integral* is defined as

$$C_i(x) = -\int_x^\infty \frac{\cos(\rho)}{\rho} d\rho = \int_\infty^x \frac{\cos(\rho)}{\rho} d\rho. \tag{6.44}$$

Sine and cosine integrals are well tabulated, see for example [7], and subroutines are readily available for most programming languages, [8].

The real and imaginary part of the *self-impedance* of element i, R_i and X_i respectively, are taken from [1] for dipole lengths $2l_i$ and wire radii a satisfying $1.3 \leq kl \leq 1.7$ and $0.0016 \leq \frac{a}{\lambda} \leq 0.01$.

$$R_i = \sum_{m=0}^{4} \sum_{n=0}^{4} a_{mn} (kl_i)^m \left(\frac{a}{\lambda}\right)^{-n}, \tag{6.45}$$

$$X_i = \sum_{m=0}^{4} \sum_{n=0}^{4} b_{mn} (kl_i)^m \left(\frac{a}{\lambda}\right)^{-n}, \tag{6.46}$$

where, [1][7]

$$\begin{pmatrix} a_{00} & a_{01} & a_{02} & a_{03} & a_{04} \\ a_{10} & a_{11} & a_{12} & a_{13} & a_{14} \\ a_{20} & a_{21} & a_{22} & a_{23} & a_{24} \\ a_{30} & a_{31} & a_{32} & a_{33} & a_{34} \\ a_{40} & a_{41} & a_{42} & a_{43} & a_{44} \end{pmatrix} =$$

$$\begin{pmatrix} 0.484315 \cdot 10^4 & -0.475502 \cdot 10^2 & 0.237406 \\ -0.137680 \cdot 10^5 & 0.134736 \cdot 10^3 & -0.671429 \\ 0.147939 \cdot 10^5 & -0.144272 \cdot 10^3 & 0.717576 \\ -0.716807 \cdot 10^4 & 0.699523 \cdot 10^2 & -0.346901 \\ 0.134304 \cdot 10^4 & -0.130631 \cdot 10^2 & 0.644693 \cdot 10^{-1} \end{pmatrix}$$

$$\begin{pmatrix} -0.517831 \cdot 10^{-3} & 0.387725 \cdot 10^{-6} \\ 0.146303 \cdot 10^{-2} & -0.109502 \cdot 10^{-5} \\ -0.156154 \cdot 10^{-2} & 0.116801 \cdot 10^{-5} \\ 0.753029 \cdot 10^{-3} & -0.562448 \cdot 10^{-6} \\ -0.139347 \cdot 10^{-3} & 0.103804 \cdot 10^{-6} \end{pmatrix}, \tag{6.47}$$

[7] The '-n' powers in equations (6.45) and (6.46) have been introduced by the author to make these equations work with the coefficients as stated in Elliot's book [1] and as reproduced in equations (6.47) and (6.48). The original equations in [1] have '+n' powers, but for the correct equations the coefficients are erroneous. The correct coefficients are tabulated in the revised version of Elliot's book [12]. The altered equations with the original coefficients give results comparable to those obtained with the correct equations and correct coefficients.

and

$$\begin{pmatrix} b_{00} & b_{01} & b_{02} & b_{03} & b_{04} \\ b_{10} & b_{11} & b_{12} & b_{13} & b_{14} \\ b_{20} & b_{21} & b_{22} & b_{23} & b_{24} \\ b_{30} & b_{31} & b_{32} & b_{33} & b_{34} \\ b_{40} & b_{41} & b_{42} & b_{43} & b_{44} \end{pmatrix} =$$

$$\begin{pmatrix} -0.644754 \cdot 10^4 & 0.767385 \cdot 10^2 & -0.360563 \\ 0.189983 \cdot 10^5 & -0.237542 \cdot 10^3 & 0.110722 \cdot 10^1 \\ -0.209803 \cdot 10^5 & 0.267034 \cdot 10^3 & -0.124666 \cdot 10^1 \\ 0.102804 \cdot 10^5 & -0.131818 \cdot 10^3 & 0.619050 \\ -0.188863 \cdot 10^4 & 0.245077 \cdot 10^2 & -0.115922 \end{pmatrix}$$

$$\begin{pmatrix} 0.709234 \cdot 10^{-3} & -0.488904 \cdot 10^{-6} \\ -0.218661 \cdot 10^{-2} & 0.151517 \cdot 10^{-5} \\ 0.247753 \cdot 10^{-2} & -0.172674 \cdot 10^{-5} \\ -0.123956 \cdot 10^{-2} & 0.869298 \cdot 10^{-6} \\ 0.233996 \cdot 10^{-3} & -0.165155 \cdot 10^{-6} \end{pmatrix}. \tag{6.48}$$

We now assume a Yagi–Uda linear endfire array antenna consisting of K elements. The driven element is designated number 1, the reflector is number 2[8] and the $K-2$ directors are numbered in increasing order, moving away from the driven element, see figure 6.12.

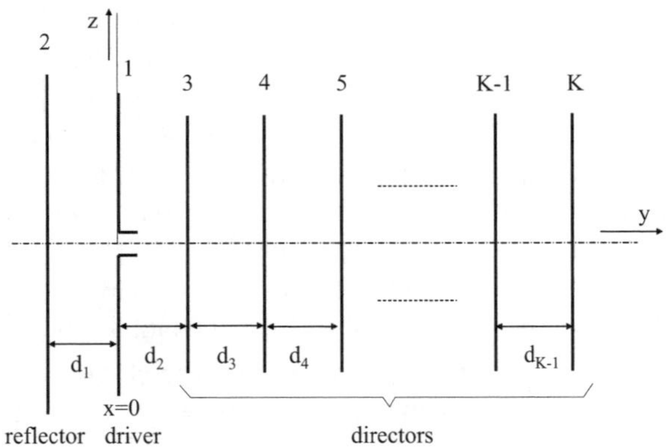

Fig. 6.12 Element designation for a K element Yagi–Uda linear endfire array antenna.

As shown in figure 6.12, all dipole elements, except for the driven one, are short-circuited. Therefore, the antenna may be described using equation (6.28), with all

[8]The reflector element is longer than the driven element, the directors in general are shorter. Because of the thus created reflective properties of the second element, adding additional reflector elements in general does not alter the antenna characteristics and in practice more than two reflecting elements are hardly ever encountered. In our analysis we restrict ourselves to having just one reflector element.

voltages, except V_1, equal to zero

$$\begin{aligned}
V_1 &= Z_{11}I_1 + Z_{12}I_2 + \cdots + Z_{1i}I_i + \cdots + Z_{1,K-1}I_{K-1} + Z_{1K}I_K \\
0 &= Z_{21}I_1 + Z_{22}I_2 + \cdots + Z_{2i}I_i + \cdots + Z_{2,K-1}I_{K-1} + Z_{2K}I_K \\
&\vdots \\
0 &= Z_{i1}I_1 + Z_{i2}I_2 + \cdots + Z_{ii}I_i + \cdots + Z_{i,K-1}I_{K-1} + Z_{iK}I_K \\
&\vdots \\
0 &= Z_{K1}I_1 + Z_{K2}I_2 + \cdots + Z_{Ki}I_i + \cdots + Z_{K,K-1}I_{K-1} + Z_{KK}I_K.
\end{aligned} \tag{6.49}$$

The last $K-1$ equations may be rewritten into

$$\begin{pmatrix} -Z_{21} \\ -Z_{31} \\ \vdots \\ -Z_{i1} \\ \vdots \\ -Z_{K-1,1} \\ -Z_{K1} \end{pmatrix} = \begin{pmatrix} Z_{22} & Z_{23} & \dots & Z_{2i} & \dots & Z_{2K} \\ Z_{32} & Z_{33} & \dots & Z_{3i} & \dots & Z_{3K} \\ \vdots & \vdots & \vdots & \vdots & \vdots & \vdots \\ Z_{i2} & Z_{i3} & \dots & Z_{ii} & \dots & Z_{iK} \\ \vdots & \vdots & \vdots & \vdots & \vdots & \vdots \\ Z_{K-1,2} & Z_{K-1,3} & \dots & Z_{K-1,i} & \dots & Z_{K-1,K} \\ Z_{K2} & Z_{K3} & \dots & Z_{Ki} & \dots & Z_{KK} \end{pmatrix} \begin{pmatrix} \frac{I_2}{I_1} \\ \frac{I_3}{I_1} \\ \vdots \\ \frac{I_i}{I_1} \\ \vdots \\ \frac{I_{K-1}}{I_1} \\ \frac{I_K}{I_1} \end{pmatrix}. \tag{6.50}$$

The system can be solved for the unknown current ratios by taking the inverse of the impedance matrix

$$[I] = [Z]^{-1}[V], \tag{6.51}$$

where

$$[I] = \left[\frac{I_2}{I_1}, \frac{I_3}{I_1}, \dots, \frac{I_K}{I_1}\right]^T, \tag{6.52}$$

$$[V] = [-Z_{21}, -Z_{31}, \dots, -Z_{K1}]^T, \tag{6.53}$$

and

$$[Z] = \begin{bmatrix} Z_{22} & Z_{23} & \dots & Z_{2i} & \dots & Z_{2K} \\ Z_{32} & Z_{33} & \dots & Z_{3i} & \dots & Z_{3K} \\ \vdots & \vdots & \vdots & \vdots & \vdots & \vdots \\ Z_{i2} & Z_{i3} & \dots & Z_{ii} & \dots & Z_{iK} \\ \vdots & \vdots & \vdots & \vdots & \vdots & \vdots \\ Z_{K-1,2} & Z_{K-1,3} & \dots & Z_{K-1,i} & \dots & Z_{K-1,K} \\ Z_{K2} & Z_{K3} & \dots & Z_{Ki} & \dots & Z_{KK} \end{bmatrix}, \tag{6.54}$$

where $^{-1}$ means taking he inverse and ^{T} means taking the transpose.

Once the system has been solved for the unknown induced currents, $I_2, I_3, \ldots, I_K$, in terms of the current I_1, the input impedance, Z_{in}, of the Yagi–Uda antenna may be calculated as

$$Z_{in} = \frac{V_1}{I_1} = Z_{11} + Z_{12}\frac{I_2}{I_1} + Z_{13}\frac{I_3}{I_1} + \cdots + Z_{1i}\frac{I_i}{I_1} + \cdots + Z_{1,K-1}\frac{I_{K-1}}{I_1} + Z_{1K}\frac{I_K}{I_1}. \quad (6.55)$$

When the impressed and induced currents are known, the radiated fields can be computed. I_1 is the impressed current, $I_2, I_3, \ldots, I_K$ are the induced currents.

6.5.2 Radiation

The far field of one finite length dipole element only has $\mathbf{E}_\vartheta$ and $\mathbf{H}_\varphi$ components[9] and - since the Maxwell equations that relate electric and magnetic fields are linear - the same applies to a Yagi–Uda antenna, see figure 6.13.

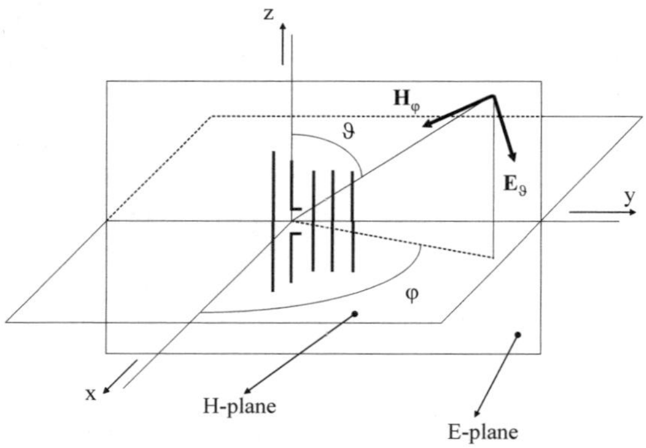

Fig. 6.13 Yagi–Uda antenna orientation.

The electric field component generated by dipole element i, E_{ϑ_i}, is calculated as [4]

$$E_{\vartheta_i} = -j\omega A_{\vartheta_i}, \quad (6.56)$$

where

$$A_{\vartheta_i} = -\frac{\mu_0 e^{-jkr}}{4\pi r}\sin(\vartheta)\int_{-l_i}^{l_i} I_i e^{jk(x_i\sin(\vartheta)\cos(\varphi)+y_i\sin(\vartheta)\sin(\varphi)+z_i\cos(\vartheta))}dz_i =$$

$$-\frac{\mu_0 e^{-jkr}}{4\pi r}\sin(\vartheta)e^{jk(x_i\sin(\vartheta)\cos(\varphi)+y_i\sin(\vartheta)\sin(\varphi))}\int_{-l_i}^{l_i} I_i e^{jkz_i\cos(\vartheta)}dz_i. \quad (6.57)$$

Herein, l_i is the half-length of element i.

With the current distribution I_i as stated in equation (6.31) and for the situation as depicted in figures 6.12 and 6.13 ($x_i = 0\ \forall i$) the electric field component is calculated

[9]These components are perpendicular to one another and are interrelated through $H_\varphi = \frac{E_\vartheta}{\eta}$, where $E_\vartheta = |\mathbf{E}_\vartheta|$, $H_\varphi = |\mathbf{H}_\varphi|$ and $\eta = \sqrt{\frac{\mu_0}{\varepsilon_0}} \approx 120\pi\Omega$ and is called the *free space characteristic impedance.*

as

$$E_{\vartheta_i} = j60 I_{i0} \frac{e^{-jkr}}{r} e^{jky_i \sin(\vartheta)\sin(\varphi)} \frac{\cos[kl_i \cos(\vartheta)] - \cos[kl_i]}{\sin(\vartheta)}, \tag{6.58}$$

where use has been made of

$$\frac{2\omega\mu_0}{4\pi k} = \frac{\eta}{2\pi} \approx \frac{120\pi}{2\pi} = 60. \tag{6.59}$$

The electric field component for the complete K-element Yagi–Uda array antenna may then be written as

$$E_\vartheta = j\frac{60}{\sin(\vartheta)} \frac{e^{-jkr}}{r} \sum_{i=1}^{K} I_{i0} e^{jky_i \sin(\vartheta)\sin(\varphi)} \left(\cos[kl_i \cos(\vartheta)] - \cos[kl_i]\right). \tag{6.60}$$

The antenna will be analysed in the E-plane and H-plane.

The E-plane is the plane containing the electric field vector and the direction of maximum radiation. In figure 6.13, the E-plane is the yz-plane ($\varphi = 90°$). The H-plane is the plane containing the magnetic field vector and the direction of maximum radiation. In figure 6.13, the H-plane is the xy-plane ($\vartheta = 90°$).

The electric fields in the E-plane and H-plane are given by, respectively

E-plane:

$$E_\vartheta = j\frac{60}{\sin(\vartheta)} \frac{e^{-jkr}}{r} \sum_{i=1}^{K} I_{i0} e^{jky_i \sin(\vartheta)} \left(\cos[kl_i \cos(\vartheta)] - \cos[kl_i]\right), \tag{6.61}$$

H-plane:

$$E_\vartheta = j60 \frac{e^{-jkr}}{r} \sum_{i=1}^{K} I_{i0} e^{jky_i \sin(\varphi)} \left(1 - \cos[kl_i]\right). \tag{6.62}$$

The field radiation pattern is proportional to $|rE_\vartheta|$. The normalised (field) radiation pattern $F(\vartheta)$ is given by

$$F(\vartheta) = \frac{|rE_\vartheta|}{|rE_{\vartheta_{max}}|}, \tag{6.63}$$

where $E_{\vartheta_{max}}$ is obtained from substitution of $\vartheta = 90°$ in equation (6.61) or substitution of $\varphi = 90°$ in equation (6.62).

$$E_{\vartheta_{max}} = j60 \frac{e^{-jkr}}{r} \sum_{i=1}^{K} I_{i0} e^{jky_i} \left(1 - \cos[kl_i]\right). \tag{6.64}$$

6.5.3 Antenna Design

The main advantage of the analysis described in the preceding paragraphs lies in its simplicity, allowing for the creation of a computer program that can analyse a multitude of configurations in a short time and thus optimise designs given some user-defined constraints.

To understand this, we will try a straightforward approach in the design of a Yagi–Uda array antenna, using the theory developed.

We will look at the most basic configuration, that consists of a driven dipole element and a single parasitic dipole element, longer than the driven element, that will act as reflector, see figure 6.14.

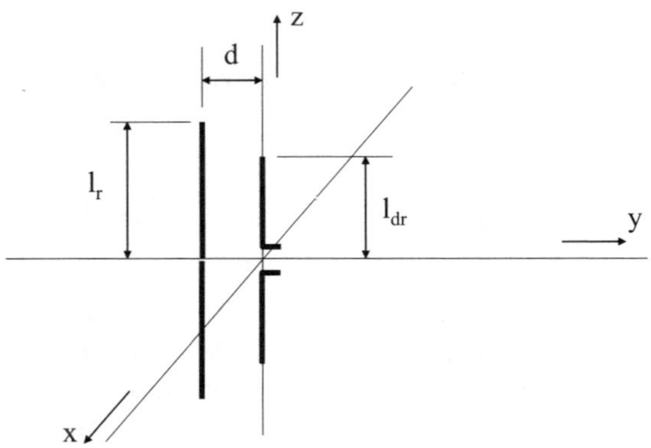

Fig. 6.14 Basic two-element Yagi–Uda array antenna configuration.

We choose the length of the driven dipole to be $2l_{dr} = 0.475\lambda$ and the length of the reflector $2l_r = 0.500\lambda$ and try to find the element distance d that ensures endfire operation along the positive y-axis.

We know that to ensure endfire operation in the desired direction, the phasing of the elements should be such that a phase difference $\alpha = -kd$ exists, see equations (6.5) and (6.6). For element distances, d, ranging between 0.025λ and 0.50λ, we have used the theory of the preceding paragraphs to calculate the phase of $\frac{I_2}{I_1}$, where I_1 is the excitation current of the driven element and I_2 is the induced current in the reflector. The results are shown in figure 6.15. In the same figure also $-kd$ as function of element distance d is shown.

We see that nowhere in the element distance range do we find a phasing that applies to the 90° endfire condition (see also [1]).

As a next try we do not pose the constraint of endfire operation at 90°, but we pose the constraint of a minimum at −90°, thus ensuring endfire operation at +90° [1]. Equations (6.5) and (6.6) learn that the phase of $\frac{I_2}{I_1}$ needs to be equal then to $-\pi + kd$. The curve of $-\pi + kd$ versus d is also shown in figure 6.15. Again there is no crossing with the phase curve of $\frac{I_2}{I_1}$.

As a last try, we look at the possibility to create endfire operation at −90°. To this end the phase of $\frac{I_2}{I_1}$ should be equal to $+kd$. In figure 6.15 we see that the phase curve of $\frac{I_2}{I_1}$ crosses the kd-curve at $d \approx 0.3\lambda$.

As a verification of the thus found element spacing, the H-plane radiation patterns of the two-element Yagi–Uda antenna are shown for different element spacings in figure 6.16.

The figure shows that the found element spacing $d \approx 0.30\lambda$ does *not* lead to an 'optimum' endfire result in the desired direction! The figure reveals an optimum

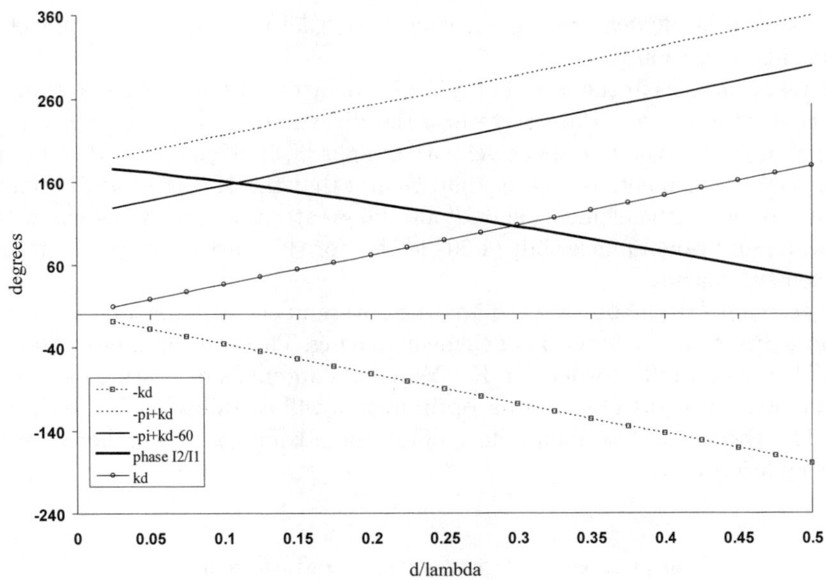

Fig. 6.15 Element phasing for the basic two-element Yagi–Uda array antenna configuration.

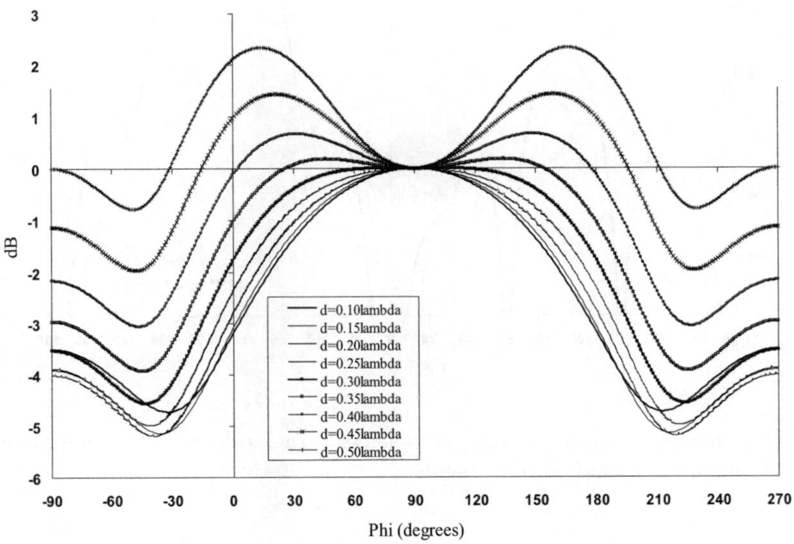

Fig. 6.16 H-plane radiation patterns for the basic two-element Yagi–Uda array antenna for different element spacings.

element distance $d \approx 0.15\lambda$. We also see that if the element distance increases, the phasing between the elements changes in such a way that the endfire operation changes into broadside operation.

The explanation lies in the fact that a Yagi–Uda array antenna is a travelling-wave antenna that supports a surface wave into the direction of endfire that has a phase velocity below that of the free space velocity of light [3, 9]. This means that the phase delay between the elements is greater than kd and that therefore the additional phase difference α between the elements should also be greater than kd. As figures 6.15 and 6.16 show, an additional phase shift of 60° leads - for this particular situation - to the desired element spacing.

As suggested in [1], the best way to find the optimum element spacing is to compute the antenna directivity as function of element spacing. The same applies to finding the optimum driver and reflector length. For Yagi–Uda antennas consisting of more than three elements, some sort of computer optimisation will certainly facilitate the design process. The theory as described before offers the advantage of simplicity, resulting in fast calculation times.

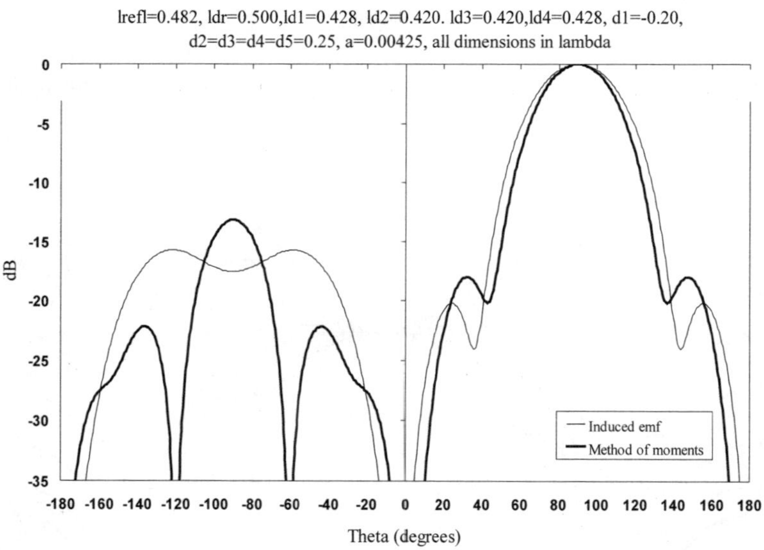

Fig. 6.17 E-plane power patterns for a six-element Yagi–Uda array antenna analysed by induced-emf method and method of moments.

This simplicity can also be a drawback, especially in the design of larger Yagi–Uda antennas. The assumption of sinusoidal currents on the array dipole elements is too crude and will lead to gross inaccuracies in the radiation pattern away from the main beam, especially in the back lobe region. It should be noted therefore that the theory was developed as an educational tool and care should be taken in employing this theory in a real Yagi–Uda antenna design tool. To that purpose, a point-matching

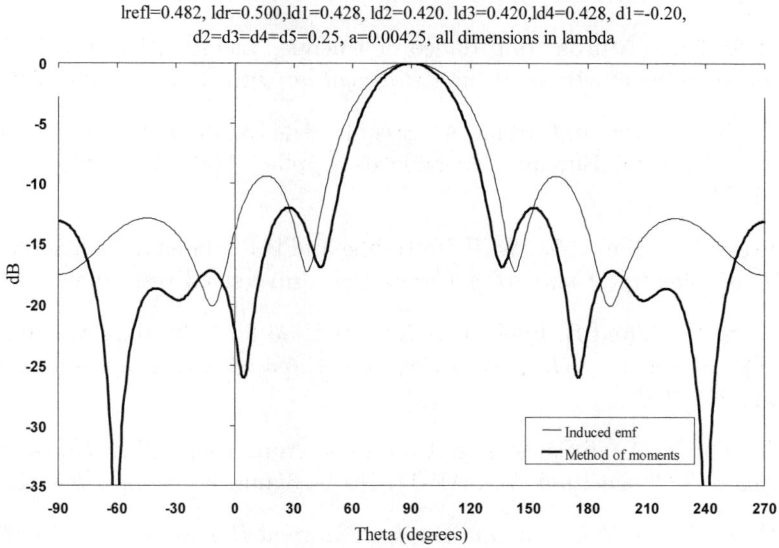

Fig. 6.18 H-plane power patterns for a six-element Yagi–Uda array antenna analysed by induced-emf method and method of moments.

version of the method of moments for the analysis of Yagi–Uda antennas [10] that is added as a Fortran routine to [4] may be employed.

As a demonstration, a six-element Yagi–Uda antenna is analysed, both by the induced-emf method and the method of moments. The results for the E-plane power pattern and the H-plane power pattern are shown in, respectively, figures 6.17 and 6.18. The dimensions of the antenna are taken from [11].

The fact that the E-plane radiation patterns seem to be more alike than the H-plane radiation patterns is due to the element-filtering characteristics in the E-plane. In the H-plane the element pattern is a constant.

REFERENCES

1. Robert S. Elliot, *Antenna Theory and Design*, Prentice-Hall, Englewood Cliffs, New Jersey, 1981.

2. W.W. Hansen and J.R. Woodyard, A New Principle in Directional Antenna Design, *Proceedings IRE*, Vol. 26, March 1938, pp. 333–345.

3. Warren L. Stutzman and Gary A. Thiele, *Antenna Theory and Design, second edition*, John Wiley & Sons, New York, 1998.

4. Constantine A. Balanis, *Antenna Theory and Design, second edition*, John Wiley & Sons, New York, 1997.

5. John D. Kraus, *Antennas*, McGraw-Hill, New York, 1950.

6. Howard E. King, Mutual Impedance of Unequal Length Antennas in Echelon, *IRE Transactions on Antennas and Propagation*, July 1957, pp. 306–313.

7. Milton Abramowitz and Irene A. Stegun (eds.), *Handbook of Mathematical Functions*, National Bureau of Standards Applied Mat., Government Printing Office, 1964.

8. W.H. Press, S.A. Teukolsky, W.T. Vetterling and B.P. Flanery, *Numerical Recipes, The Art of Scientific Computing*, Cambridge University Press, 1986.

9. H.W. Ehrenspeck and H. Poehler, A New Method for Obtaining Maximum Gain from Yagi Antennas, *IRE Transactions on Antennas and Propagation*, October 1959, pp. 379–386.

10. Gary A. Thiele, Analysis of Yagi–Uda Type Antennas, *IEEE Transactions on Antennas and Propagation*, Vol. AP-17, No. 1, January 1969, pp. 24–31.

11. Peter P. Viezbicke, Yagi Antenna Design, *National Bureau of Standard Technical Note 688*, December 1976.

12. Robert S. Elliot, *Antenna Theory and Design, revised edition*, John Wiley & Sons, New York, 2003.

7

The Linear Phased Array Antenna

As we already have seen in the chapter on endfire array antennas, a well-chosen phase taper over the elements of a linear array antenna may make the beam point in endfire direction. Choosing another phase taper may point the beam at another angle between broadside and endfire. Being able to adjust the phase taper will make it possible to *scan* the beam between these two extreme pointing angles. Obviously, it will be very difficult to realise this in a parasitic array antenna, and other means than mutual coupling are needed to accomplish the desired phase taper. Before we discuss means of realising phase taper, we will first establish the theoretical background of phase-steered or *phased* array antennas and discuss the peculiarities of the scanned beam. Objections exist about the term *phased array antenna* for a *scanned beam array antenna*, based on the fact that a non-scanned array antenna is in fact a phased array antenna, since the operation relies on relative phases between the elements. Notwithstanding this argument, we still choose for the term *phased* in connection with *beam-steered*, thereby following the historical development.

7.1 LINEAR PHASE TAPER

As in the previous chapters we start with considering a linear array antenna consisting of K elements, equally interspaced a distance d. The direction of a wave is described by the angle ϑ between rays and array normal. The difference with the previous situations (broadside linear array antenna and endfire linear array antenna) is that now, in the (corporate) feed network, we add a microwave two-port between every antenna element and its branch of the feed network, see figure 7.1.

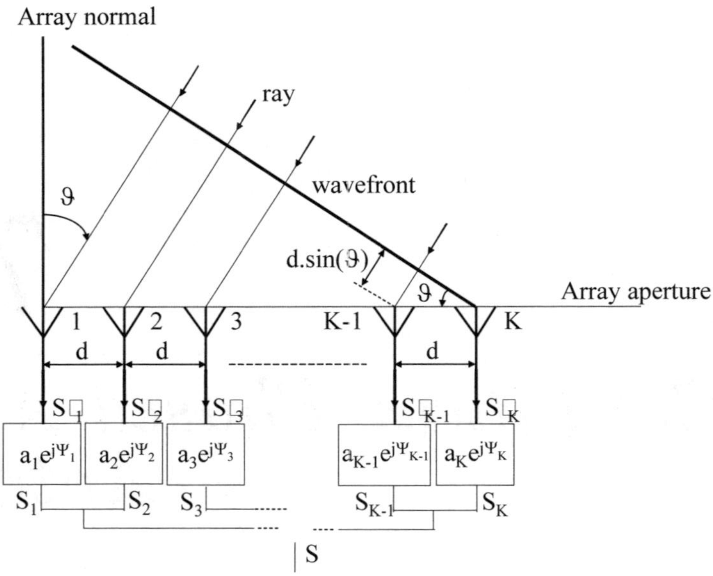

Fig. 7.1 A linear *phased* array of K radiators, equidistantly positioned along a straight line, where a plane wave is incident under an angle ϑ with respect to the array normal.

The transfer function, $H_i(\vartheta)$, of the microwave two-port i, $i = 1, 2, \ldots, K$, is given by

$$H_i(\vartheta) = \frac{S_i(\vartheta)}{S_i'(\vartheta)} = a_i e^{j\psi_i}. \tag{7.1}$$

How to realise such a two-port is left for the moment and will be treated further on in this chapter. It suffices to say now that the two-port will allow us to change the amplitude of every received signal $S_i'(\vartheta)$ and - what is more important for the moment - it will allow us to change the *phase* of the received signal. The two-ports open up the opportunity to operate a *phased array antenna.*

Since $S_i(\vartheta) = S_i'(\vartheta)H_i(\vartheta)$ and we already know from the theory of the linear broadside array antenna that $S_i'(\vartheta) = S_e(\vartheta)e^{jk_0(K-i)d\sin(\vartheta)}$, where $k_0 = \frac{2\pi}{\lambda}$ is the free space wave number and $S_e(\vartheta)$ is the element radiation pattern, we may now write for the array radiation pattern, see also figure 7.1.

$$S(\vartheta) = \sum_{i=1}^{K} S_i(\vartheta) = S_e(\vartheta) \sum_{i=1}^{K} a_i e^{j[k_0(K-i)d\sin(\vartheta)+\psi_i]}. \tag{7.2}$$

In this equation we implicitly have assumed that mutual coupling effects between the array antenna elements are negligible, allowing for a common element radiation pattern that is taken out of the summation.

All the coefficients a_i form an amplitude taper. In order not to obscure the *phased* array antenna discussion, we assume a uniform, normalised amplitude distribution:

$$a_i = 1 \text{ for } i = 1, 2, \ldots, K. \tag{7.3}$$

Now, we only have to consider the array factor, $S_a(\vartheta)$, that is given by

$$S_a(\vartheta) = \sum_{i=1}^{K} e^{j[k_0(K-i)d\sin(\vartheta)+\psi_i]}. \tag{7.4}$$

If, next, we choose a linear phase taper that is equal to

$$\psi_i = -k_0(K-i)d\sin(\vartheta_0) \quad \text{for } i = 1, 2, \ldots, K, \tag{7.5}$$

where $-90° \leq \vartheta_0 \leq 90°$, the array factor may be written as

$$S_a(\vartheta) = \sum_{i=1}^{K} e^{jk_0(K-i)d[\sin(\vartheta)-\sin(\vartheta_0)]}. \tag{7.6}$$

For $\vartheta_0 = 0$, the phase taper is zero or non-existent and we encounter the linear broadside array antenna situation. The maximum of the array factor was encountered in that situation for $\vartheta = 0$, or - more precisely - for $\sin(\vartheta) = 0$.

For the linear phased array antenna situation we now find the array factor maximum for

$$\sin(\vartheta) - \sin(\vartheta_0) = 0, \tag{7.7}$$

or, provided that $-90° \leq \vartheta, \vartheta_0 \leq 90°$, for $\vartheta = \vartheta_0$.

So, by choosing a desired beam-pointing direction ϑ_0 and subsequently phasing the linear array antenna elements according to $\psi_i = -k_0(K-i)d\sin(\vartheta_0)$, the array factor will have its maximum at the desired angle $\vartheta = \vartheta_0$.

Example 1 Consider a linear array antenna consisting of eight elements. The element voltage radiation pattern is given by

$$S_e(\vartheta) = \cos(\vartheta). \tag{7.8}$$

Given this hypothetical radiation pattern for an aperture element in an infinite ground plane, calculate and show the element power radiation pattern, the array factor power pattern and the power radiation pattern of the total array as function of the angle ϑ relative to the array normal (broadside) for the following element distances d:

1. $d = \frac{\lambda_0}{4}$;
2. $d = \frac{\lambda_0}{2}$;
3. $d = \lambda_0$;
4. $d = \frac{5\lambda_0}{4}$,

for a phasing aimed at a beam pointing to $\vartheta_0 = 30°$.

Using equations (7.2), (7.6) and (7.8) results in the radiation power patterns shown in figures 7.2, 7.3, 7.4 and 7.5 for, respectively, $d = \frac{\lambda_0}{4}$, $d = \frac{\lambda_0}{2}$, $d = \lambda_0$ and $d = \frac{5\lambda_0}{4}$. The element power pattern is calculated as $20\log(|S_e(\vartheta)|)$, the normalised array factor

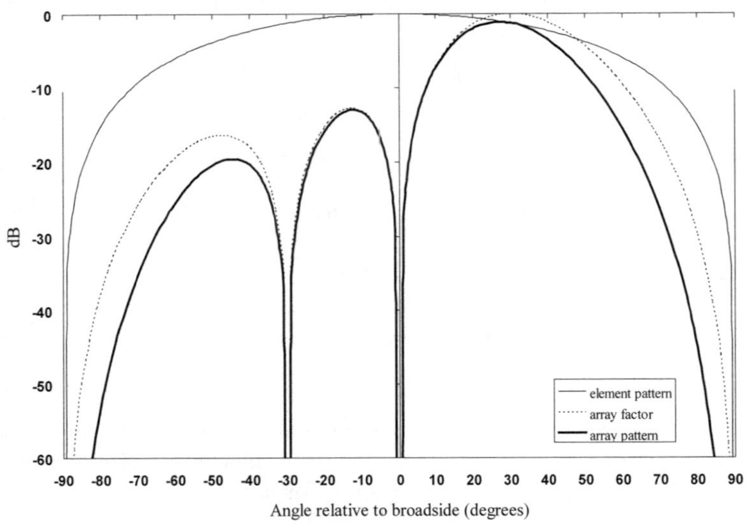

Fig. 7.2 Power radiation patterns of the element factor, the array factor and the total array of a linear 8-element phased array antenna with element distance $d = \frac{\lambda_0}{4}$, phased for beam pointing at $\vartheta_0 = 30°$.

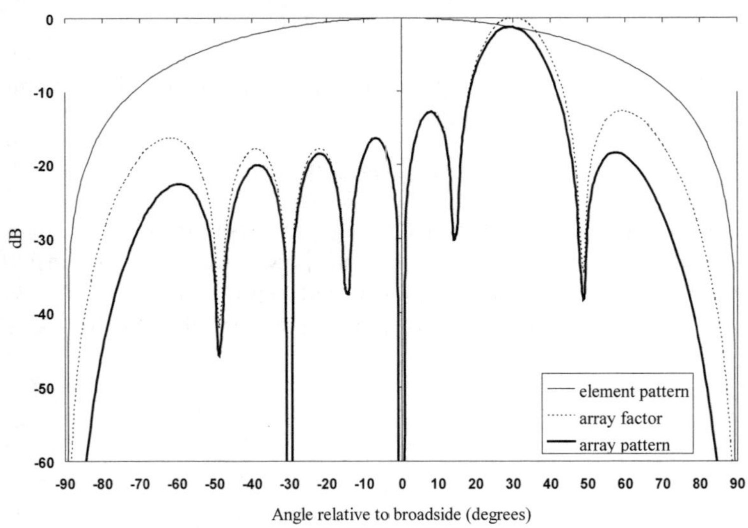

Fig. 7.3 Power radiation patterns of the element factor, the array factor and the total array of a linear 8-element phased array antenna with element distance $d = \frac{\lambda_0}{2}$, phased for beam pointing at $\vartheta_0 = 30°$.

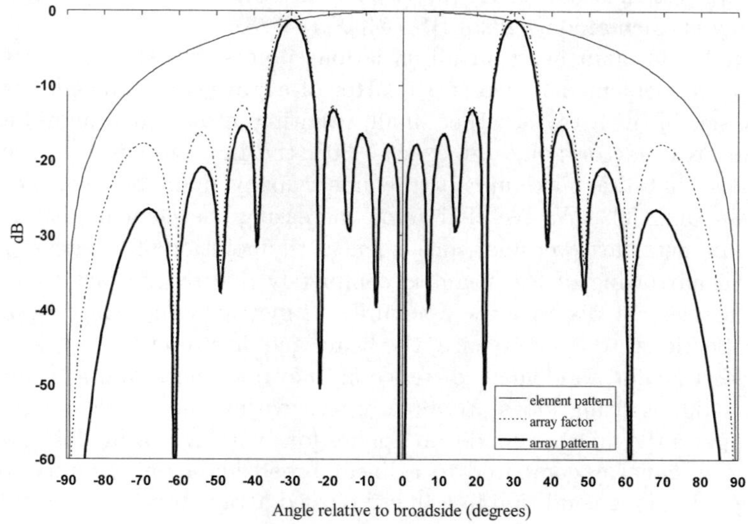

Fig. 7.4 Power radiation patterns of the element factor, the array factor and the total array of a linear 8-element phased array antenna with element distance $d = \lambda_0$, phased for beam pointing at $\vartheta_0 = 30°$.

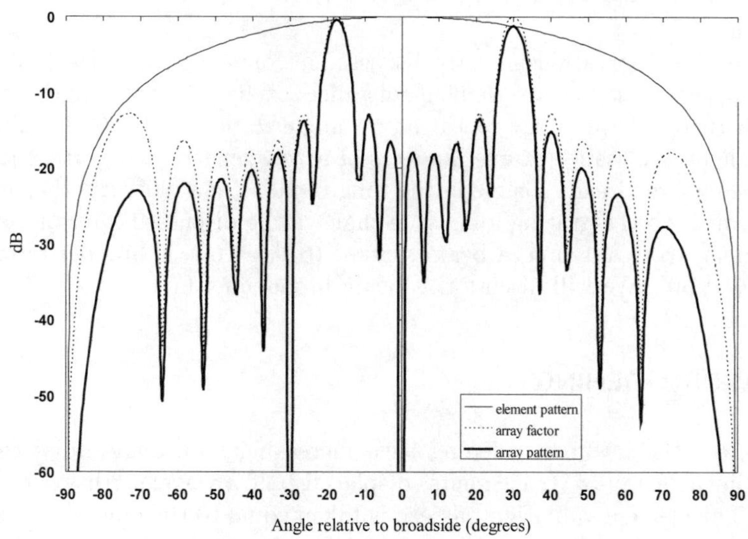

Fig. 7.5 Power radiation patterns of the element factor, the array factor and the total array of a linear 8-element phased array antenna with element distance $d = \frac{5\lambda_0}{4}$, phased for beam pointing at $\vartheta_0 = 30°$.

power pattern is calculated as $20\log(|S_a(\vartheta)|/8)$ and the normalised power pattern of the total array is calculated as $20\log(|S_e(\vartheta)||S_a(\vartheta)|/8)$.

As expected, the main beam in all situations points into the desired $\vartheta_0 = 30°$ direction. For an element distance of a quarter of a wavelength though (figure 7.2), the aperture size of the array is rather small, resulting into a broad main beam and - although the array factor points at exactly 30°, the filtering effect of the element pattern means that the maximum of the linear array main beam is at an angle ϑ slightly less than 30°. We see that upon increasing the element distance of the array, the beam gets narrower and, consequently, the pointing of the beam gets more accurate. The narrowing of the beam is completely due to the fact that with the increase of the element distance the aperture size increases and, since beamwidth is inversely proportional to aperture size, the beam gets narrower.

We also see that for an element distance of one wavelength and a beam directed towards $\vartheta_0 = 30°$, grating lobes are already well within the visible range ($-90° \leq \vartheta \leq 90°$). Apparently, another condition applies for avoiding grating lobes in a linear *phased* array antenna as compared to a linear broadside array antenna. Before we discuss the grating lobe condition in a linear phased array antenna, we will first have a look at another example.

Example 2 Consider the same linear array antenna and perform the same calculations but now for a beam pointing to $\vartheta_0 = 60°$.

Using equations (7.2), (7.6) and (7.8) results in the normalised radiation power patterns shown in figures 7.6, 7.7, 7.8 and 7.9 for, respectively, $d = \frac{\lambda_0}{4}$, $d = \frac{\lambda_0}{2}$, $d = \lambda_0$ and $d = \frac{5\lambda_0}{4}$.

Apart from the observations already discussed in connection with the first example, we now also see that for an element distance of half a wavelength, already a considerable rising of the array factor at an angle ϑ close to $-90°$ is taking place and for an element distance of five quarters of a wavelength, two grating lobes next to the main lobe are already visible in the range $-90° \leq \vartheta \leq 90°$. Another important observation, next to the grating lobe issue that will be discussed later on, is that for the same linear array antenna, a beam steered to $\vartheta_0 = 60°$ is broader than a beam steered to $\vartheta_0 = 30°$. We will discuss this beam broadening first.

7.2 BEAM BROADENING

To demonstrate the beam broadening with increasing scan angle more clearly, we will take a linear array of 32 elements, displaced half a wavelength with respect to each other. The element radiation pattern is taken equal to the one for the 8-element linear array antennas used in examples 1 and 2. The normalised array radiation power patterns for beams scanned to $\vartheta_0 = 0°$, $\vartheta_0 = 15°$, $\vartheta_0 = 30°$, $\vartheta_0 = 45°$, $\vartheta_0 = 60°$ and $\vartheta_0 = 75°$, are shown in figure 7.10.

The figure clearly shows the beam broadening with scan angle. It also clearly shows the influence of the element pattern and demonstrates that only for broad beams a pointing error results from the filtering effect of the element pattern.

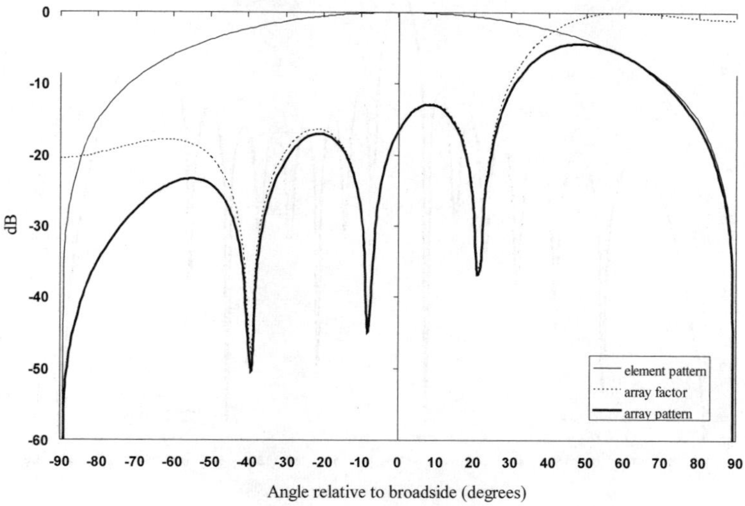

Fig. 7.6 Power radiation patterns of the element factor, the array factor and the total array of a linear 8-element phased array antenna with element distance $d = \frac{\lambda_0}{4}$, phased for beam pointing at $\vartheta_0 = 60°$.

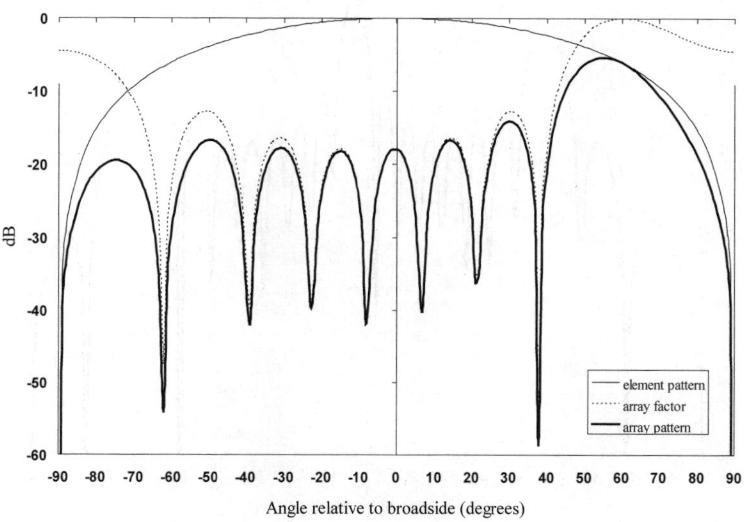

Fig. 7.7 Power radiation patterns of the element factor, the array factor and the total array of a linear 8-element phased array antenna with element distance $d = \frac{\lambda_0}{2}$, phased for beam pointing at $\vartheta_0 = 60°$.

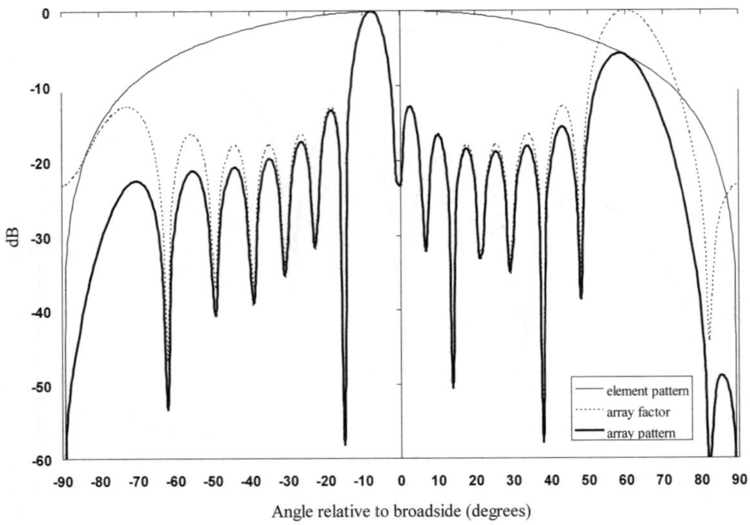

Fig. 7.8 Power radiation patterns of the element factor, the array factor and the total array of a linear 8-element phased array antenna with element distance $d = \lambda_0$, phased for beam pointing at $\vartheta_0 = 60°$.

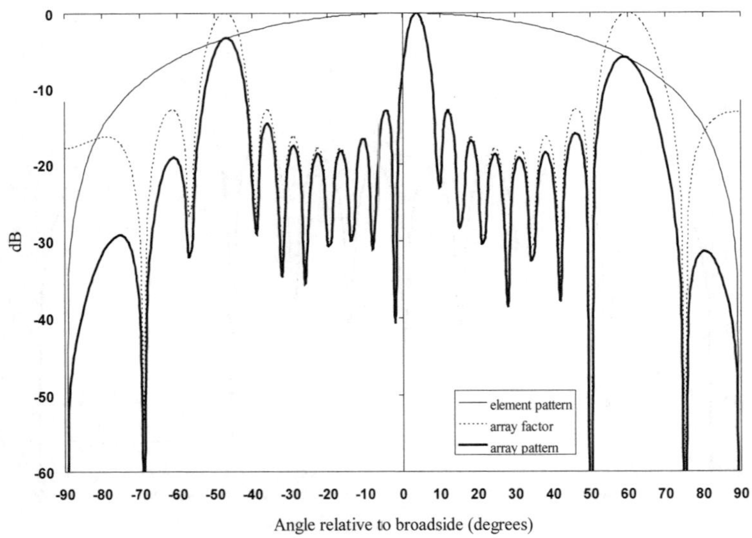

Fig. 7.9 Power radiation patterns of the element factor, the array factor and the total array of a linear 8-element phased array antenna with element distance $d = \frac{5\lambda_0}{4}$, phased for beam pointing at $\vartheta_0 = 60°$.

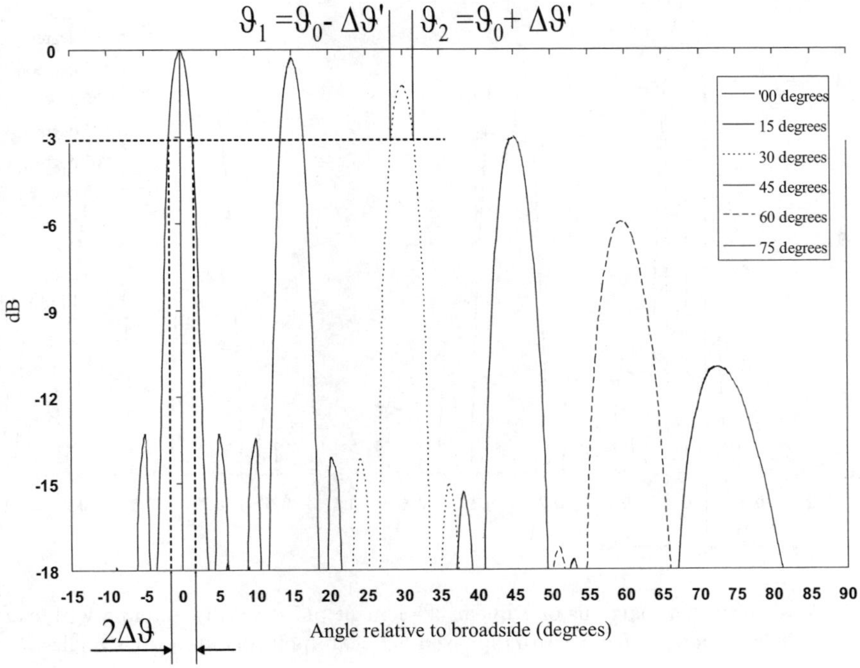

Fig. 7.10 Power radiation patterns of a linear 32-element phased array antenna with element distance $d = \frac{\lambda_0}{2}$, phased for beam pointing at various angles ϑ_0.

To explain the beam broadening with increasing scan angle we will take a closer look at the array factor, equation (7.6)

$$S_a(\vartheta) = \sum_{i=1}^{K} e^{jk_0(K-i)d[\sin(\vartheta)-\sin(\vartheta_0)]} = \sum_{i=1}^{K} e^{jk_0(K-i)d[u-u_0]}, \tag{7.9}$$

where $u = \sin(\vartheta)$ and $u_0 = \sin(\vartheta_0)$.

If we plot the array radiation pattern, for the various beam-pointing directions, as function of u instead of as function of ϑ [1], we see that the beamwidth is invariant with u, figure 7.11.

For the beam directed at broadside ($u_0 = \sin(0) = 0$), the 3dB-beamwidth is - in ϑ-coordinates - $2\Delta\vartheta$, see figure 7.10.

In u-coordinates, the beamwidth is $2\Delta u$, see figure 7.11.

The half-beamwidths $\Delta\vartheta$ and Δu are related through

$$\Delta u = \sin(\Delta\vartheta), \tag{7.10}$$

so

$$\Delta\vartheta = \arcsin(\Delta u). \tag{7.11}$$

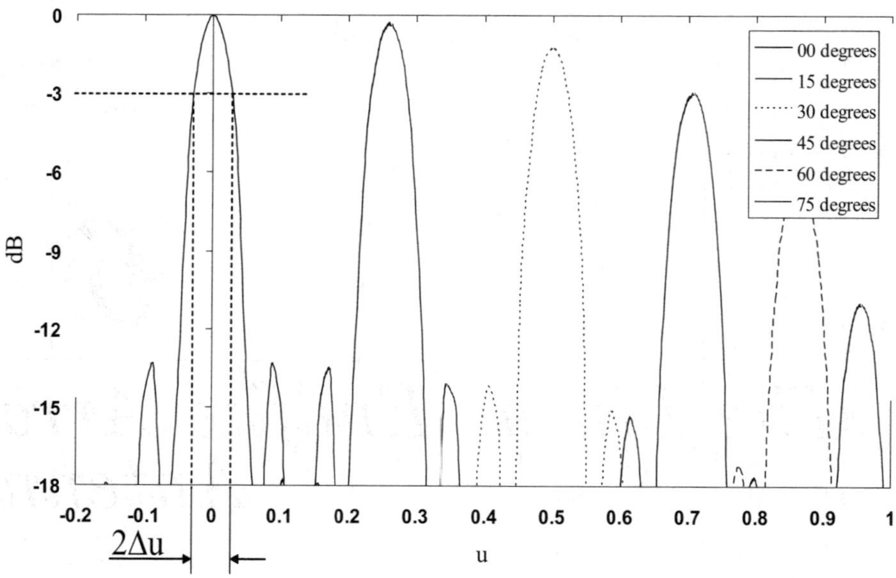

Fig. 7.11 Power radiation patterns of a linear 32-element phased array antenna with element distance $d = \frac{\lambda_0}{2}$ as function of $u = \sin(\vartheta)$, phased for beam pointing at various angles ϑ_0.

Now, we scan the beam to $\vartheta_0 \neq 0$ or, along the u-axis, $u_0 = \sin(\vartheta_0)$. The halfpower beamwidth - which is invariant in u-coordinates - is now

$$2\Delta u = \sin(\vartheta_2) - \sin(\vartheta_1), \tag{7.12}$$

where

$$\vartheta_1 = \vartheta_0 - \Delta\vartheta', \tag{7.13}$$

$$\vartheta_2 = \vartheta_0 + \Delta\vartheta', \tag{7.14}$$

and $2\Delta\vartheta'$ is the beamwidth for the beam scanned to ϑ_0, see figure 7.10.

Substitution of equations (7.13) and (7.14) into equation (7.12) leads to

$$\Delta u = \cos(\vartheta_0)\sin(\Delta\vartheta'), \tag{7.15}$$

so that we find for the halfpower beamwidth

$$2\Delta\vartheta' = 2\arcsin\left(\frac{\Delta u}{\cos(\vartheta_0)}\right). \tag{7.16}$$

For narrow beam antennas, we may approximate the arcsin-function by its argument, so

$$2\Delta\vartheta' \approx \frac{2\Delta u}{\cos(\vartheta_0)}. \tag{7.17}$$

For narrow beam antennas $2\Delta u$ is also approximately equal to the beamwidth measured in ϑ-coordinates, see equation (7.11), so that, finally

$$2\Delta\vartheta' \approx \frac{2\Delta\vartheta}{\cos(\vartheta_0)}. \tag{7.18}$$

So, the beamwidth of a scanned, narrow beam, linear array antenna, broadens approximately inversely to the cosine of the scan angle.

7.3 GRATING LOBES AND VISIBLE SPACE

When we take a closer look at the array factor again, equation (7.9), we see that whenever

$$k_0 d\left[u - u_0\right] = n2\pi, \tag{7.19}$$

where n is an integer number, the array factor repeats itself. The main beam is identified by $n = 0$, all other integer values of n identify grating lobes. The first grating lobe ($n = 1$) satisfies

$$\frac{d}{\lambda} = \frac{1}{(u - u_0)}, \tag{7.20}$$

where use has been made of $k_0 = \frac{2\pi}{\lambda}$.

We have seen in the previous section, that for $-90° \leq \vartheta \leq 90°$, the u-coordinate varies between -1 and $+1$. So when a grating lobe (maximum) just appears at $\vartheta = \pm 90°$, the corresponding u-value is $u = \pm 1$. The above equation for the first grating lobe then tells us that grating lobes may be just or just not avoided if the element distance relative to the wavelength satisfies

$$\frac{d}{\lambda} \leq \frac{1}{1 + |u_{0_{max}}|}, \tag{7.21}$$

where $u_{0_{max}} = \sin(\vartheta_{0_{max}})$ and $\vartheta_{0_{max}}$ is the maximum scan angle. The equality sign applies to the 'just' or 'just not' situation. For the equality sign, the maximum of the grating lobe is present at $\vartheta = \pm 90°$. As for a non-scanning array antenna, for a scanning array antenna the grating lobe condition should be applied more restrictively if also the slope of the first grating lobe should be suppressed.

For $\vartheta_{0_{max}} = 0$ (no scanning at all), we find $\frac{d}{\lambda} \leq 1$. This is the situation we already encountered for the broadside linear array antenna. If, on the other hand, we do not want a grating lobe maximum being present in visible space for all possible scan angles, we should take $\vartheta_{0_{max}} = \pm 90°$, which will lead to

$$\frac{d}{\lambda} \leq \frac{1}{2}. \tag{7.22}$$

This explains the grating lobes we have seen for a scanning antenna in examples 1 and 2 for element distances smaller than one wavelength.

When looking at the radiation patterns as a function of the angle ϑ relative to broadside, we see - upon enlarging the element distance - grating lobes appear at

$\vartheta = \pm 90°$. Upon further increasing the element distance, or scanning the beam further to endfire, these grating lobes move from endfire direction closer to broadside direction. The range $-90° \leq \vartheta \leq 90°$, corresponding to $|u| = |\sin(\vartheta)| \leq 1$, is called the *visible region*. Grating lobes enter the visible region coming from the *invisible region*. The invisible region, for which $|\sin(\vartheta)| > 1$ corresponds to complex angles ϑ, which are difficult to display graphically. Now, the usefulness of the parameter u becomes evident, since we can easily display the radiation pattern as function of u for $-\infty < u < \infty$, thus displaying visible region, $|u| \leq 1$, and invisible region, $|u| > 1$, in one graph.

As an example - demonstrating the usefulness of the parameter u - we will take a linear array antenna of 32 elements, equally displaced three-quarters of a wavelength with respect to one another and scan the main beam successively in the directions $\vartheta_0 = 0°$, $\vartheta_0 = 30°$ and $\vartheta_0 = 60°$. Figures 7.12 to 7.14 show the array factor as function of u and as function of ϑ and demonstrate the grating lobe moving in from invisible space to visible space.

In these figures, we have indicated the main beam with a white spot and the grating lobes with black spots at the top of the graphs. The visible region in the u-domain, $-1 \leq u \leq 1$, is indicated by two dashed lines and transfers to the shown ϑ-domain for $-90° \leq \vartheta \leq 90°$. The movement of the main beam and the grating lobes for the beam being scanned from 0° to 60° is indicated by, respectively, a white arrow and black arrows. Upon moving the main beam from 0° to 30°, figures 7.12 and 7.13, we see that the grating lobes move with the main beam. The grating lobe to the right of the main beam has moved further away from the visible region; the one on the left has entered the visible region. If the beam is scanned further, to $\vartheta_0 = 60°$, figures 7.13 and 7.14, we see that the grating lobe on the right of the main beam has moved out of our shown u-window, the grating lobe to the left of the main beam has moved further into the visible region and further to the left a new grating lobe has formed in the invisible region that is moving towards the visible region.

7.4 MEANS OF PHASE SHIFTING

So far, we have discussed the phased array antenna beam positioning by applying a phase shift to the linear array antenna elements without specifying how this phase shifting may be accomplished. In this paragraph we will briefly outline some (certainly not all) methods available for accomplishing a desired phase shift.

To this end we first start by identifying the phase of a signal. In all our calculations thus far we have assumed the signals to be time harmonic, i.e. varying according to $e^{j\omega t}$, meaning that a physical realisable signal $s(\omega)$ varies according to the real part of the complex signal $e^{j\omega t}$,

$$s(\omega) \sim \cos(\omega t), \tag{7.23}$$

where $\omega = 2\pi f$, f being the frequency of the signal. The argument of the cosine is known as the *phase*, ψ. The time t we may express as the ratio of distance, l, to velocity of the signal, c, where $c = \frac{1}{\sqrt{\varepsilon\mu}}$, so that we find for ψ

$$\psi = 2\pi f l \sqrt{\varepsilon\mu}, \tag{7.24}$$

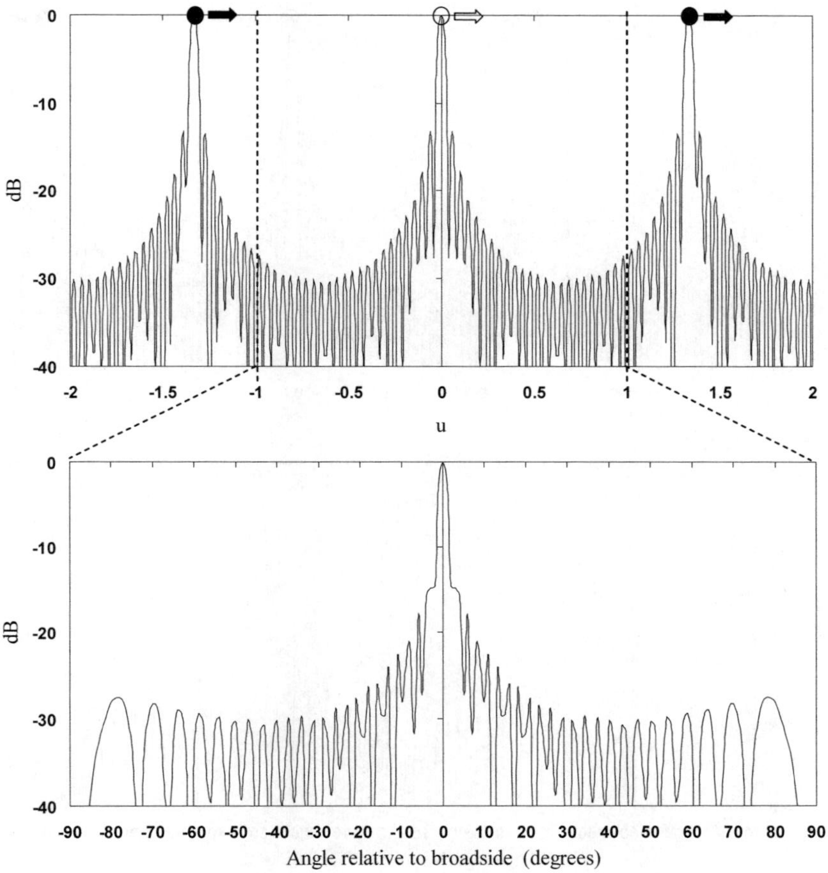

Fig. 7.12 Array factor of a linear 32-element phased array antenna with element distance $d = \frac{3\lambda_0}{4}$ as function of $u = \sin(\vartheta)$ and as function of ϑ, phased for beam pointing at $\vartheta_0 = 0°$.

where ε is the permittivity of the medium the signal is travelling through and μ is the permeability of this medium.

This equation [2] reveals all phase-shifting possibilities at a glance. The possibilities are:

- phase shifting by changing frequency;
- phase shifting by changing length;
- phase shifting by changing permittivity (dielectric constant);
- phase shifting by changing permeability.

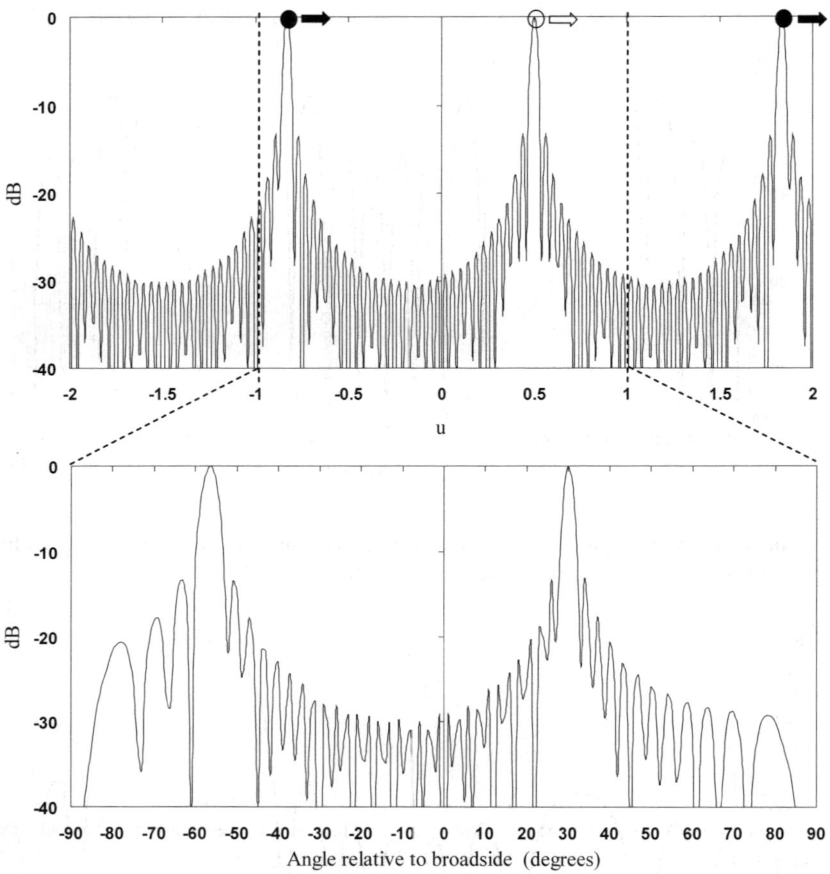

Fig. 7.13 Array factor of a linear 32-element phased array antenna with element distance $d = \frac{3\lambda_0}{4}$ as function of $u = \sin(\vartheta)$ and as function of ϑ, phased for beam pointing at $\vartheta_0 = 30°$.

7.4.1 Phase Shifting by Changing Frequency

Phase shifting by changing frequency or *frequency scanning* is accomplished by series feeding the array antenna elements, having the elements equidistantly positioned along the line and changing the frequency, see figure 7.15.

We have seen that changing the frequency makes the phase change. Another way of looking at this phase change is taking the *electrical length* into account. Using $c = \frac{1}{\sqrt{\varepsilon\mu}}$, $f = \frac{c}{\lambda}$ and $k = \frac{2\pi}{\lambda}$, substituted into equation (7.24), results in

$$\psi = kl, \tag{7.25}$$

which is known as the *electrical length*.

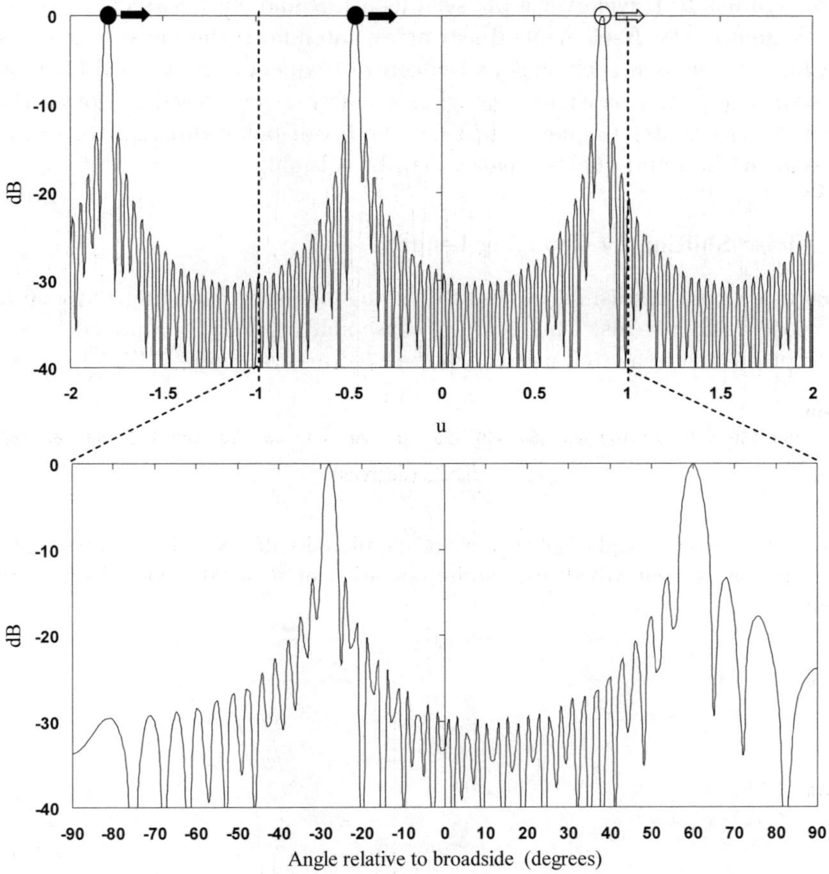

Fig. 7.14 Array factor of a linear 32-element phased array antenna with element distance $d = \frac{3\lambda_0}{4}$ as function of $u = \sin(\vartheta)$ and as function of ϑ, phased for beam pointing at $\vartheta_0 = 60°$.

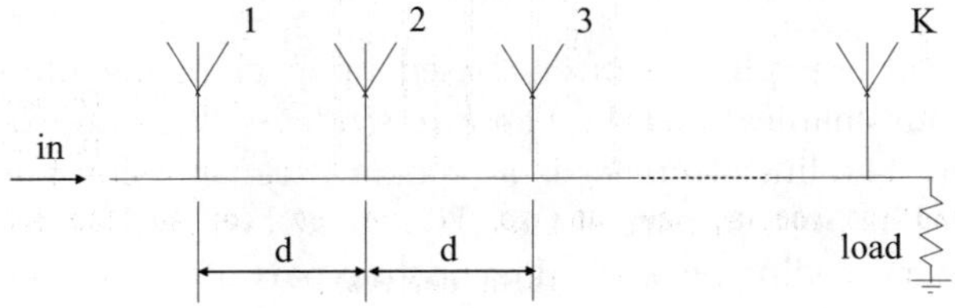

Fig. 7.15 Series-fed linear array antenna consisting of K identical elements, equidistantly displaced d with respect to one another.

This last equation now clearly shows that by changing the frequency, we create a changing linear phase taper over the array antenna elements, since the input signal in figure 7.15 has to travel over a physical length il and electrical length ikl to reach the ith element of the K-elements linear array antenna. If the physical lengths of the feeding lines are chosen such that at the centre frequency the phased array antenna beam is directed to broadsight, changing the frequency to values lower than and greater than the centre frequency will get the beam being directed to, respectively, angles smaller than and angles greater than broadsight.

7.4.2 Phase Shifting by Changing Length

Another way of accomplishing a desired phase shift is by changing physical lengths, as equation (7.23) reveals. This type of phase shifting may be applied to series-fed arrays, figure 7.16a,b, as well as to corporate-fed arrays, figure 7.16c [3].

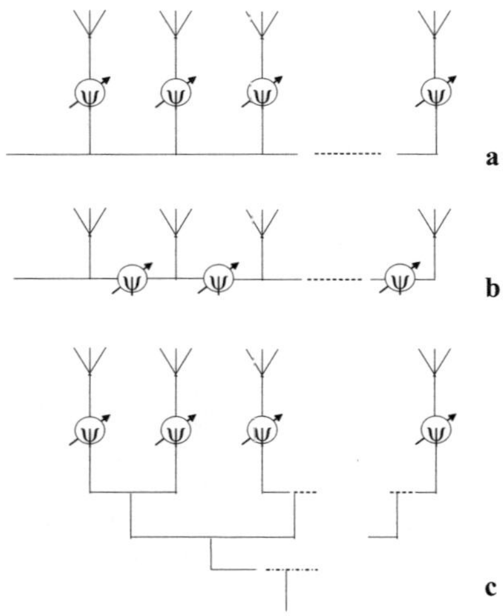

Fig. 7.16 Phase shifters in a linear phased array antenna feed network. a, b. Series-fed linear phased array antenna. c. Corporate-fed linear phased array antenna.

In the pre-digital era, phase shifters based upon changing physical length were realised by electromechanical means. The *line stretcher* [2] is an example of an early type phase shifter. The line stretcher is a (coaxial) transmission line section, bent in the form of a 'U'. The bottom part of this 'U' is attached to the two 'arms' that form part of the stationary feeding network. The bottom part of the 'U' acts as a telescoping section that may be stretched by electromechanical means, thus lengthening and shortening the transmission line section, without changing the position of the 'arms' of the 'U'.

Nowadays, different lengths of transmission line are selected digitally. A schematic view of a cascaded, four-bit, digitally switched phase shifter is shown in figure 7.17 [2, 3].

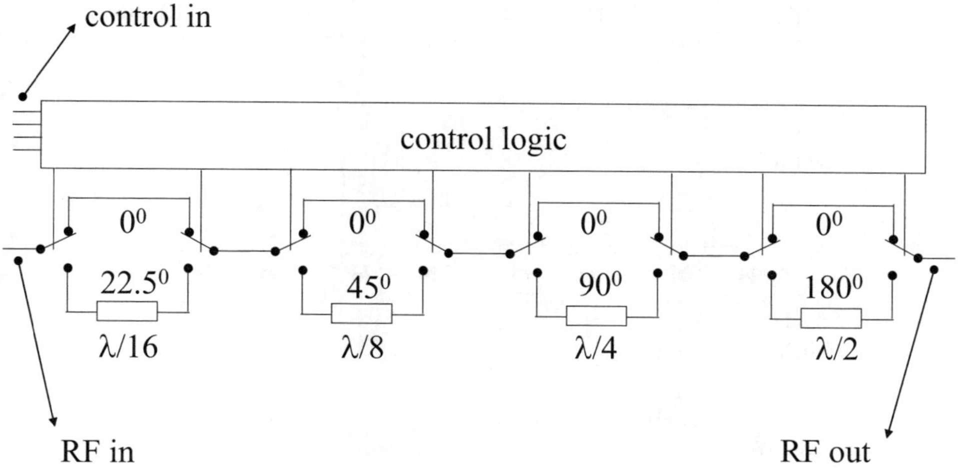

Fig. 7.17 Cascaded, four-bit, digitally switched phase shifter.

The switches in every section are used to either switch a standard length of transmission line into the network or a piece of transmission line that adds to this standard length a piece of predetermined length. These lengths are chosen such that when the cascade of standard length is taken as reference, having a phase $\psi = 0°$, 16 phases (4 bits), ranging from $\psi = 0°$ to $\psi = 337.5°$, in steps of 22.5° (least significant bit) may be selected.

PIN diodes - employed in forward and reverse bias - are often used as switching elements [2, 3]. The switched phase shifters may be realised in microstrip technology, using high dielectric constant substrate material, thus minimising physical phase shifter dimensions.

Another way of switching physical line lengths is found in the cascaded hybrid-coupled phase shifter, see figure 7.18 [2, 3].

A 3dB hybrid is a device that divides the power at input port 1, equally over output ports 2 and 3 and passes no power to output port 4, see figure 7.18. The reflections of the signals that have left ports 2 and 3 return into the hybrid and combine at output port 4, none of the power being returned to input port 1. The diode switches in every segment (bit) of the cascaded hybrid-coupled phase shifter are either returning the signals leaving ports 2 and 3 directly, or after having travelled the extra line length $\frac{\Delta l}{2}$ twice. For the four-bit phase shifter shown in figure 7.18, $\frac{\Delta l}{2} = \frac{\lambda}{32}$ for the least significant bit, and for the following three bits, respectively, $\frac{\Delta l}{2} = \frac{\lambda}{16}$, $\frac{\Delta l}{2} = \frac{\lambda}{8}$ and $\frac{\Delta l}{2} = \frac{\lambda}{4}$.

7.4.3 Phase Shifting by Changing Permittivity

As equation (7.23) shows, a phase shift may be accomplished by changing the permittivity, ε, or dielectric constant of the material a signal is propagating through.

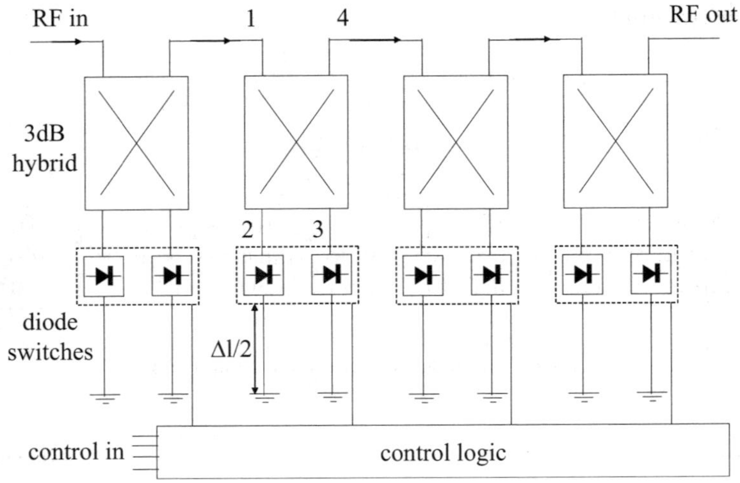

Fig. 7.18 Cascaded, four-bit, hybrid-coupled phase shifter.

One way is to use a gaseous discharge or plasma, where the dielectric constant - and thus the phase shift - is changed by changing the current through the device [2].

Another way is provided by making use of so-called *ferroelectric* materials. Ferroelectric materials are materials for which the permittivity is a function of the applied electric field over the material.

7.4.4 Phase Shifting by Changing Permeability

Equation (7.23) shows that a change in permeability, μ, works equally well in changing the phase. *Ferrimagnetic* materials, or *ferrites* are materials for which the permeability changes as function of the change in an applied magnetic field in which

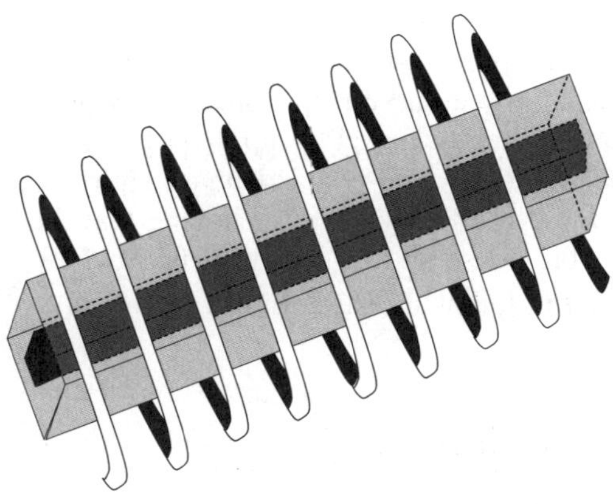

Fig. 7.19 Basic Reggia–Spencer phase shifter configuration.

the material is positioned. Ferrite-based phase shifters have been in use for a long time, especially in combination with waveguide transmission line technology.

The *Reggia–Spencer phase shifter*, in its most basic form, consists of a rod of ferrimagnetic material, centrally positioned inside a waveguide, where a solenoid is wound around the waveguide, see figure 7.19 [2, 3].

By changing the current through the solenoid, the magnetic field is changed and thereby the permeability of the ferrimagnetic rod and thus the phase of a wave going through the waveguide is changed. The phase can be changed continuously, making the Reggia–Spencer phase shifter an *analog* phase shifter.

A section (bit) of a *digital* ferrimagnetic phase shifter is shown in figure 7.20 [2, 3].

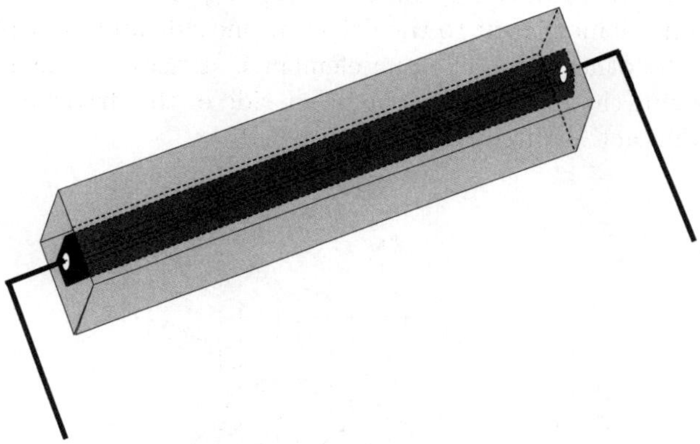

Fig. 7.20 Single section (bit) of a latched ferrite phase shifter.

The function of the solenoid is taken over by a current wire through the ferrimagnetic rod. By cascading different lengths of ferrimagnetic rods, different (discrete) phase shifts may be realised.

REFERENCES

1. N. Amitay, R.G. Pecina and C.P. Wu, Radiation Properties of Large Planar Arrays, *Monograph 5047*, Bell Telephone Laboratories, Inc., February 1965.

2. Merill I. Skolnik, *Introduction to Radar Systems, second edition*, McGraw-Hill, Auckland, 1981.

3. R.C. Johnson (ed.) *Antenna Engineering Handbook, third edition*, McGraw-Hill, New York, 1993.

8

A Frequency Scanned Slotted Waveguide Array Antenna

In this chapter we will put the theory of linear phased array antennas into practice. We will design a so-called *slotted waveguide array antenna* and provide beam scanning by changing the frequency. We also give an introductory example of beam synthesis, putting constraints on the side lobe level of the array radiation pattern for the beam scanned to broadside. The antenna will be constructed in WR90 waveguide, will have 21 radiating elements, is expected to scan from minus 15° from broadside to plus 15° from broadside and at broadside have a side lobe level of -30dB. We will construct this antenna, measure the radiation patterns for different frequencies and discuss the results.

8.1 SLOTTED WAVEGUIDE ARRAY ANTENNA

A frequency scanned array antenna may be realised by a so-called *travelling wave array antenna* [1]. A linear travelling wave array antenna may be considered as a transmission line along which radiating elements are connected. The line is excited at one end and loaded at the other end. When the excited wave travels towards the load, parts of its energy are radiated by the radiating elements until, after the last element, a small fraction of the energy remains. This last fraction of the excitation energy is dissipated into the load. By controlling the path-lengths between the radiating elements (in hardware or by changing the frequency), the beam pointing direction may be controlled. The elements should then be spaced at intervals that avoid reflections at the elements to add in phase [1]. The array is therefore *nonresonant.*

Travelling wave antennas may practically be realised as dipole arrays and microstrip patch arrays, but the most often encountered type is the *rectangular slotted waveguide array antenna*, see figure 8.1. This type of travelling wave antenna offers the greatest control over feeding and array element excitation [3].

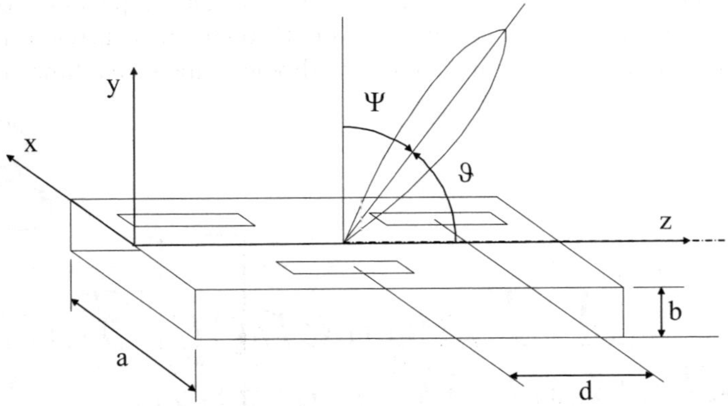

Fig. 8.1 A section of rectangular slotted waveguide array antenna. The waveguide width and height are, respectively, a and b. The angle ϑ is the angle the beam makes with the z-axis.

In this configuration, the aforementioned transmission line is the rectangular waveguide and the array radiating elements are the slots cut into the waveguide wall. Shown in figure 8.1 is the most commonly encountered type where longitudinal slots are cut into the broad wall of the waveguide.

The current flowing on the broad upper wall of the rectangular waveguide, $\mathbf{J}$ is proportional to [3]

$$\mathbf{J} \propto \cos\left(\frac{\pi x}{a}\right)\mathbf{u}_x - j\frac{ka}{\pi}\sin\left(\frac{\pi x}{a}\right)\mathbf{u}_z, \tag{8.1}$$

where $\mathbf{u}_x$ and $\mathbf{u}_z$ are unit vectors in, respectively, the x- and z-direction and $k = 2\pi/\lambda_g$ is the wavenumber in the waveguide. λ_g is the wavelength in the waveguide. For an air-filled waveguide, $\lambda_g > \lambda_0$.

The resonant length of a slot is close to $\lambda_0/2$ [3], where λ_0 is the free space wavelength.

In general, a slotted waveguide may be employed as a *resonant array* or as a *nonresonant array.*

8.1.1 Resonant Slotted Waveguide Array Antenna

A resonant slotted waveguide array antenna is employed to form a broadside ($\vartheta = 90°$) beam. For the resonant slotted waveguide to obtain a beam directed at $\vartheta = 90°$ (broadside or boresight), all slots must be excited in phase. This is accomplished by spacing the slots λ_g apart. But since the slot spacing should be less than λ_0 to avoid the creation of *grating lobes* (also known as *secondary* lobes) and since $\lambda_g > \lambda_0$, a feasible spacing will be larger than λ_0 and therefore will result in the forming of more than one lobe.

This problem is solved by placing the slots $\lambda_g/2$ apart and placing every other slot on the adjacent side of the broad wall centre line. This will add an additional phase shift of π to every other slot as can be seen from equation (8.1).

The resonant slotted waveguide is terminated into a short circuit, placed at a distance $\lambda_g/4$ or $3\lambda_g/4$ from the last slot. This ensures that the reflected wave also excites the slots in phase (a short circuit placed at another distance would result in a reflected wave that would launch a beam in a different direction than broadside).

For our purposes we do not need a fixed beam at broadside, but a beam that changes position with frequency. For that purpose, we need to employ the slotted waveguide in a nonresonant setup.

8.1.2 Nonresonant Slotted Waveguide Array Antenna

The nonresonant slotted waveguide array is very similar to the resonant slotted waveguide array. The waveguide is now terminated into a matched load that prohibits reflected waves to travel in the opposite direction and form beams in undesired directions. The energy of the excitation wave is gradually radiated by the elements (slots) encountered by the wave as it travels from source towards load. The element spacing may be larger than or smaller than but preferably *not* equal to half a wavelength (assuming every other slot placed adjacent to the broad wall centre line) to avoid reflections (from the slots) adding up in phase. The slotted waveguide therefore is *nonresonant.*

We assumed that adjacent elements are on opposite sides of the broad wall centre line, to allow for a slot spacing close to half a wavelength [6].

8.1.2.1 Radiation Pattern Applying array theory to the slotted waveguide array of figure 8.1 [1, 2], the far-field radiation pattern $F(\Psi)$ in the y, z-plane may be written as

$$F(\Psi) = f(\Psi) \sum_{n=1}^{N} a_n e^{-j\delta_n} e^{j[k_0 n d \sin(\Psi) + n\pi]}, \tag{8.2}$$

where $f(\Psi)$ is the radiation pattern of a single slot, a_n is the excitation amplitude of the nth slot, δ_n is the excitation phase of the nth slot and $k_0 = 2\pi/\lambda_0$ is the free space wavenumber. The array consists of N slots, equally spaced at a distant d, every other slot displaced on adjacent sides of the broad wall centre line, explaining the factor $n\pi$ in the exponent of the kernel of the summation. The mutual coupling between the slots is considered to be negligible, making all slot radiation patterns identical. This allows us to separate the slot radiation pattern and array factor as we have done in the above equation.

If we apply a linear phase taper over the slots, $\delta_n = n\delta$, the radiation pattern transforms to

$$F(\Psi) = f(\Psi) \sum_{n=1}^{N} a_n e^{jn\left[\frac{2\pi}{\lambda_0} d \sin(\Psi) + \pi - \delta\right]}. \tag{8.3}$$

The array pattern is at maximum for

$$\frac{2\pi}{\lambda_0} d \sin(\Psi_m) + \pi - \delta = 2\pi m, \; m = 0, \pm 1, \pm 2, \ldots, \tag{8.4}$$

where Ψ_m is the direction of the far-field maximum (beam pointing direction).

8.1.2.2 Grating Lobes The directions of the first couple of secondary or grating lobes, for a main beam in the direction Ψ_m, are found for the values $m-1$ and $m+1$, substituted in the above equation

$$\sin(\Psi_{m\pm 1}) = \sin(\Psi_m) \pm \frac{\lambda_0}{d}. \tag{8.5}$$

In general we do not want grating lobes to be present in *visible space*, i.e. for real values of Ψ, or - in other words - for values of Ψ that satisfy the condition $|\sin(\Psi)| \leq 1$.

If we allow grating lobes to be *just not* present at $\Psi = \pm\pi/2$, we find - upon substitution of these values for $\Psi_{m\pm 1}$ in the above equation - that the inter-element distance should satisfy

$$d < \frac{\lambda_0}{1 + |\sin(\Psi_m)|}. \tag{8.6}$$

For all angles $-\Psi_m < \Psi < \Psi_m$, no grating lobes will appear in visible space.

In figure 8.2, the maximum element distance (in wavelengths) versus maximum absolute beam pointing angle is shown.

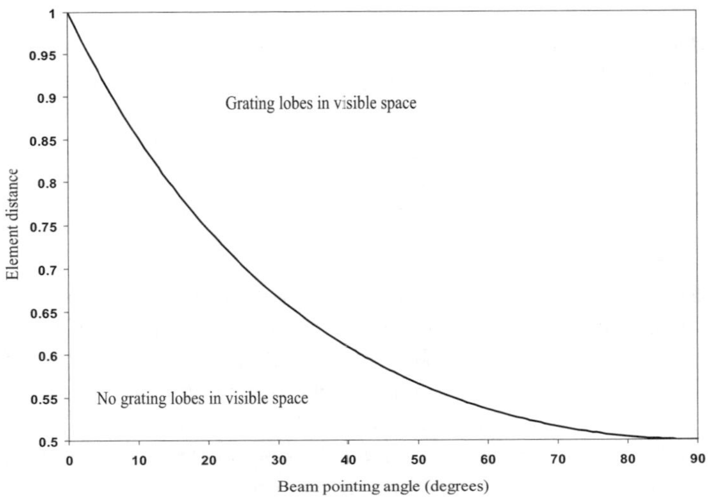

Fig. 8.2 Maximum element distance vs. beam-pointing angle for grating lobe free visible space.

We see that the maximum inter-element spacing is independent from the element phase excitation. Furthermore we see that for a fixed beam at broadside ($\Psi_m = 0$), the element distance may be as large as one free space wavelength, λ_0. If, on the other hand, we want the beam to scan over the whole hemisphere ($-90° < \Psi < 90°$) in the y, z-plane, the element distance may not exceed one half of a free space wavelength.

Of course, allowing the grating lobes to be *just not* present at $\Psi = \pm\pi/2$, still leaves us with the slope of the grating lobe close to $\Psi = \pm\pi/2$. Although this slope will be greatly suppressed by the slot radiation pattern, ($f(\Psi)$), it will not hurt to be a bit more restrictive on the element distance than suggested by equation (8.6).

In the above, we have not yet specified the integer m that was introduced in equation (8.4). We take the first maximum that occurs as the main lobe and the next one as the first grating lobe. This means that $m = 0$.

8.1.2.3 Frequency Scanning Before we look at frequency scanning in detail, we need to determine the phase, δ. This phase is, due to the slotted waveguide structure - where the elements are physically displaced a distance d - simply

$$\delta = \frac{2\pi}{\lambda_g} d, \tag{8.7}$$

where λ_g is the wavelength in the waveguide. This wavelength is related to the free space wavelength, λ_0, through [6]

$$\lambda_g = \frac{\lambda_0}{\sqrt{1 - \left(\frac{\lambda_0}{\lambda_c}\right)^2}} = \frac{\lambda_0}{\sqrt{1 - \left(\frac{\lambda_0}{2a}\right)^2}}. \tag{8.8}$$

In the above equation, λ_c is the *cut-off wavelength* of the waveguide, i.e. the maximum wavelength[1] that will propagate through the guide. This cut-off wavelength is equal to twice the width of the guide, $\lambda_c = 2a$, see figure 8.1.

We want the frequency scanning to occur relative to a chosen centre frequency, f_{cntr}. At this centre frequency, the ratio of free space wavelength to waveguide wavelength will be designated a constant, C_1. At the centre frequency we want the antenna to radiate into the broadside direction ($\Psi_m = 0$). The element distance may also be related to the centre wavelength, λ_{cntr}

$$\frac{\lambda_0}{\lambda_g} = C_1, \text{ for } \lambda_0 = \lambda_{cntr}, \tag{8.9}$$

$$d = C_2 \lambda_{cntr}. \tag{8.10}$$

Substitution of equation (8.7) into equation (8.4) and rearranging terms results in

$$\sin(\Psi_m) = \left(m - \frac{1}{2}\right)\frac{\lambda_0}{d} + \frac{\lambda_0}{\lambda_g}. \tag{8.11}$$

At broadside ($\Psi_m = 0$), substitution of equations (8.9) and (8.10) into equation (8.11) results in

$$\left(m - \frac{1}{2}\right) = -C_1 C_2. \tag{8.12}$$

[1]Related to the minimum frequency through $f_c = \frac{c_0}{\lambda_c}$, where c_0 is the speed of light in free space.

Substitution of this result into equation (8.11) and, subsequently using equations (8.10) and (8.8) leads to

$$\sin(\Psi_m) = \frac{\lambda_0}{\lambda_g} - C_1\frac{\lambda_0}{\lambda_{cntr}} =$$
$$\sqrt{1-\left(\frac{f_c}{f}\right)^2} - C_1\frac{\lambda_0}{\lambda_{cntr}}. \tag{8.13}$$

If we next relate the cut off frequency, f_c, to the centre frequency, f_{cntr}, through

$$f_c = C_3 f_{cntr}, \tag{8.14}$$

substitution of this relation into equation (8.13), evaluated at broadside, gives

$$C_1 = \sqrt{1 - C_3^2}. \tag{8.15}$$

Therefore, we finally find for the beam-pointing dependency on frequency

$$\sin(\Psi_m) = \sqrt{1 - C^2\left(\frac{f_{cntr}}{f}\right)^2} - \sqrt{1-C^2}\left(\frac{f_{cntr}}{f}\right), \tag{8.16}$$

where

$$f_c = C f_{cntr}. \tag{8.17}$$

In figures 8.3, 8.4 and 8.5 the beam position is shown as function of the frequency normalised to the centre frequency for different values of C.

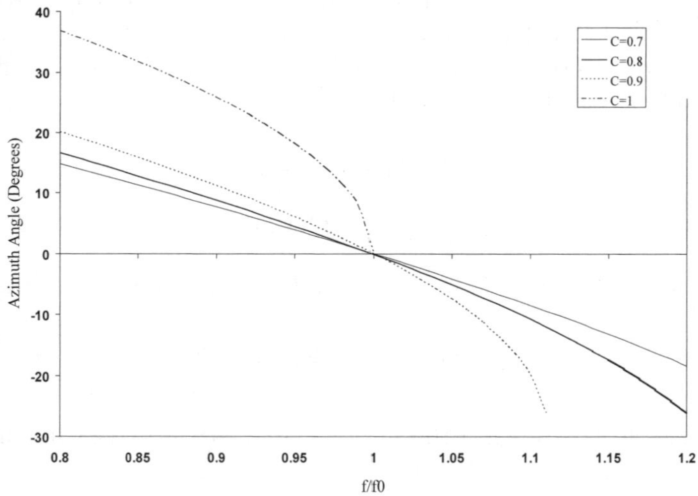

Fig. 8.3 Beam position as function of frequency for different values of C between 0.7 and 1.0.

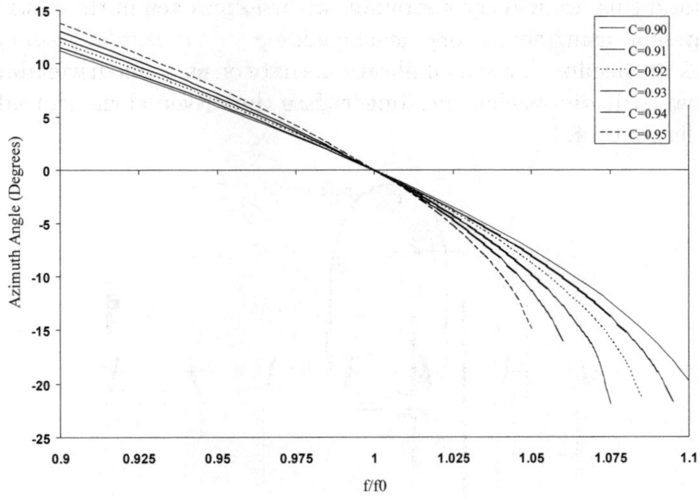

Fig. 8.4 Beam position as function of frequency for different values of C between 0.9 and 0.95.

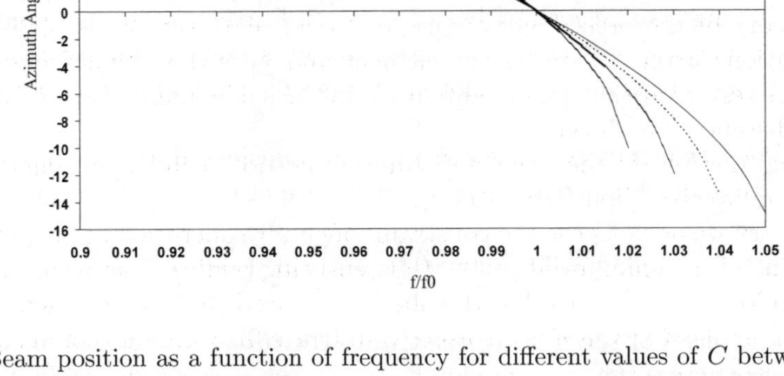

Fig. 8.5 Beam position as a function of frequency for different values of C between 0.95 and 0.98.

From these three figures we see that a linear dependency of beam-pointing angle with frequency is feasible, see especially figure 8.3.

Also clearly shown in the figures is that the angular scan with frequency is rather limited. This is due to the fixed path length dictated by the slotted waveguide configuration. If a larger scan volume is required, the waveguide in between the slots may be bent to lengthen the path, while maintaining the slot spacing, see figure 8.6 [1].

This type of feeding is usually combined with slots in the smaller waveguide side and is known as *serpentine feeding*, *snake feeding* or *sinusoidal feeding*. Since the complexity of the feeding system will increase costs and since we need a limited scan volume, we will *not* use this kind of feeding and stick to the slotted waveguide configuration of figure 8.1.

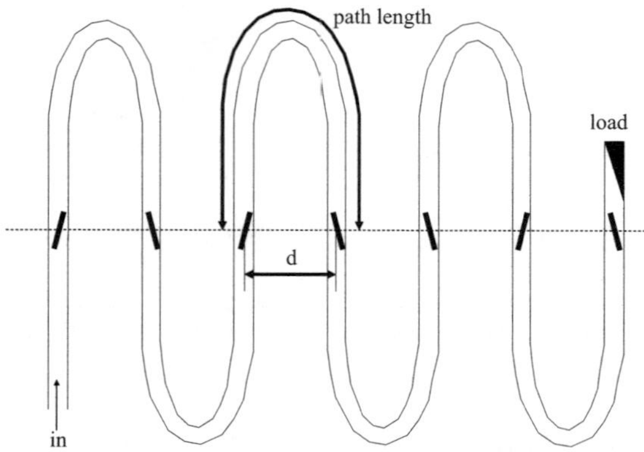

Fig. 8.6 Snake feed geometry. Slots in small side waveguide. Not drawn to scale.

8.1.2.4 Power Distribution For the determination of the power distribution we assume that at every slot power is coupled from the travelling wave to that slot and that the remaining power is transported to the next slot where a similar process is taking place and that this process takes place without the generation of reflections. This assumption is realistic, provided that the coupling per slot is small. This may be achieved by having a sufficient number of slots [1]. Furthermore, we assume that the waveguide loss may be neglected. This assumption is realistic for straight travelling wave feeds [1], as is the case in our slotted waveguide array antenna.

Normalised Slot Conductances Every slot may be considered as a shunt conductance across the waveguide. The normalised[2] conductance of the nth slot is given by [3, 5]

$$g_n = 2.09 \frac{\lambda_g}{\lambda_0} \frac{a}{b} \cos^2\left(\frac{\pi \lambda_0}{2\lambda_g}\right) \sin^2\left(\frac{\pi x_n}{a}\right), \tag{8.18}$$

where x_n is the offset of the nth slot relative to the broad wall centre line.

If the voltage across the transmission line is V, the power radiated by the nth slot is given by $|V|^2 g_n/2$. The offset of the slot therefore is a means of controlling the power distribution [3].

Given a power distribution function F_n, $n = 1, 2, \ldots, N$, the fraction of the input power that is finally dissipated into the load, L, is given by [1, 3]

$$L = \frac{P_L}{P_{in}}, \tag{8.19}$$

[2]Normalised with respect to the characteristic admittance of the rectangular waveguide TE_{10}-mode.

where P_L is the power delivered to the load and P_{in} is the power delivered to the slotted waveguide antenna at the input. The equivalent circuit and the powers and distribution coefficients involved are shown in figure 8.7.

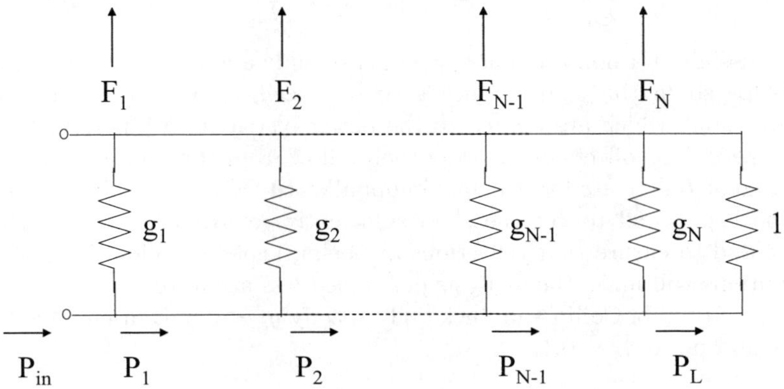

Fig. 8.7 Slotted waveguide equivalent circuit.

We start at the last slot. The voltage at this point of the equivalent transmission line is V_N. The power *available* at this point of the transmission line is then

$$P_{av} = \frac{|V_N|^2}{2} = P_N + L, \tag{8.20}$$

where P_N is the power radiated by the Nth slot, normalised to the input power P_{in}. We have seen that this radiated normalised power is

$$P_N = \frac{|V_N|^2 g_N}{2}, \tag{8.21}$$

where g_N is the normalised conductance of slot N.

We find the normalised slot conductance upon division of radiated power and available power. The radiated power is prescribed by the power function F_n, so

$$g_N = \frac{P_N}{P_{av}} = \frac{P_N}{P_N + L} = \frac{F_N}{F_N + L}. \tag{8.22}$$

For the last but one slot, we find a normalised available power $P_{av} = P_{N-1} + P_N + L$, a radiated power P_{N-1} and therefore a normalised slot conductance g_{N-1}

$$g_{N-1} = \frac{F_{N-1}}{F_{N-1} + F_N + L}. \tag{8.23}$$

Continuation of this analysis results in the expression for g_n

$$g_n = \frac{F_n}{L + \sum_{i=n}^{N} F_i}. \tag{8.24}$$

This equation may be rewritten, using $L + \sum_{i=1}^{N} F_i = 1$, which follows from the normalisation process, as

$$g_n = \frac{F_n}{L + \sum_{i=1}^{N} F_i - \left(\sum_{i=1}^{N} F_i - \sum_{i=n}^{N} F_i\right)} = \frac{F_n}{1 - \sum_{i=1}^{n-1} F_i}. \tag{8.25}$$

For the design of a nonresonant slotted waveguide antenna, the fraction of power to be dissipated in the load, L, needs to be specified as well as the normalised conductances, g_n, which are related to the power distribution function, F_n, through equation (8.25). The offsets of the slots then follow from equation (8.18).

The value of L is related to the maximum allowed value for g_n. The conductances need to be kept small to get the slots sufficiently removed from the edges of the waveguide and to ensure that reflections are being kept low. Violating either one or both conditions will make the analysis performed less accurate.

As a rule of thumb, Collin and Zucker [3], specify $g_n < 0.2$. Hansen [1] is even more restrictive and poses $g_n < 0.1$.

Taylor Distribution In general we want a prescribed function determining the amplitude factors, a_n, in equation (8.2). If we choose all factors equal, $a_n = 1$, we have a so-called *uniform amplitude weighting.* The benefit of such a weighting is that it results in the smallest possible half power beam width (HPBW). The drawback is that it provides - at the same time - the highest possible side lobe level (SLL) of -13.2dB. Other amplitude distributions, like for example *Dolph-Tschebyscheff, binomial, cosine squared,* etc. all make a trade-off between HPBW and SLL. A *Taylor* distribution offers an often used compromise between HPBW and SLL. In the following, a (very brief) outline of the *Taylor one-parameter distribution* will be presented. Details may be found in [7] and [8].

Start of the discussion of the Taylor one-parameter distribution - that ensures side lobe levels to decay with increasing angle, moving away from the main lobe - is a *continuous* line source (current) distribution [2]

$$I(z') = \begin{cases} J_0\left[j\pi B\sqrt{1-\left(\frac{2z'}{l}\right)^2}\right] & -\frac{l}{2} \le z' \le \frac{l}{2} \\ 0 & \text{elsewhere} \end{cases}, \tag{8.26}$$

where $J_0(x)$ is the Bessel function of the first kind and order zero of argument x and l is the total length of the source, that is directed along the z-coordinate of a regular Cartesian coordinate system. In the above equation, B is a constant that will be determined for the specified side lobe level.

The antenna pattern, $F(\vartheta)$, as function of the angle relative to the z-axis, ϑ, follows from the source distribution [2] and is given by

$$F(\vartheta) = \begin{cases} l\dfrac{\sinh\left[\sqrt{(\pi B)^2 - u^2}\right]}{\sqrt{(\pi B)^2 - u^2}} & u^2 < (\pi B)^2 \\ l\dfrac{\sin\left[\sqrt{u^2 - (\pi B)^2}\right]}{\sqrt{u^2 - (\pi B)^2}} & u^2 > (\pi B)^2 \end{cases}, \tag{8.27}$$

where

$$u = \pi \frac{l}{\lambda} \cos(\vartheta). \tag{8.28}$$

Maximum radiation takes place at $\vartheta = \frac{\pi}{2}$ or $u = 0$. The normalised (with respect to source length, l) main beam level, H_0, is therefore

$$H_0 = \frac{\sinh(\pi B)}{\pi B}. \tag{8.29}$$

For $u \gg \pi B$, the normalised pattern may be approximated by

$$F(\vartheta) = \frac{\sin\left[\sqrt{u^2 - (\pi B)^2}\right]}{\sqrt{u^2 - (\pi B)^2}} \approx \frac{\sin(u)}{u}, \tag{8.30}$$

and the first side lobe level, H_1, is then found to be

$$H_1 = 0.217233. \tag{8.31}$$

The maximum desired voltage SLL, R_0 is then calculated as

$$R_0 = \frac{H_0}{H_1} = 4.603 \frac{\sinh(\pi B)}{\pi B}. \tag{8.32}$$

We now have all the ingredients necessary to design a slotted waveguide array antenna.

8.2 ANTENNA DESIGN

Our goal is, by employing a standard *WR90* waveguide (dimensions: $a = 0.9'' = 22.86$mm, $b = 0.4'' = 10.16$mm [9]) with 21 slots, to design a slotted waveguide array antenna having a side lobe level of -30dB that can scan the beam from $-15°$ to $+15°$ relative to broadside.

Since it would be nice to have a linear dependency of beam-pointing angle on frequency, we choose - using figure 8.3 - $C = 0.7$ in equation (8.16). The cut-off frequency of the WR90 waveguide is $f_c = 6.562$GHz. With $C = 0.7$ substituted in equation (8.17), the centre frequency is found to be $f_{cntr} = 9.374$GHz. The frequency sweep needs to be 40%, see figure 8.3, so scanning will be from 7.499GHz to 11.249GHz and the frequency sweep will be $\Delta f = 3.75$GHz.

The slot length, l_g is taken to be half the centre wavelength, $l_g = \lambda_{cntr}/2 = c_0/(2f_{cntr})$ =1.60cm. We choose the width to be 1.5mm. The element distance, d, is taken to be, $d = 2.29$cm, well above half the largest free space wavelength possible. Thus, we prevent the slotted waveguide array antenna from becoming resonant while changing the frequency between lower and upper boundary.

Next, we find the SLL, R_0, to be

$$R_0 = 10^{\frac{30}{20}} = 31.62, \tag{8.33}$$

and, with use of equation (8.32)

$$\frac{e^{\pi B} - e^{-\pi B}}{\pi B} = 13.74, \tag{8.34}$$

where use is made of $\sinh(x) = \frac{1}{2}\left(e^{x} - e^{-x}\right)$.

Finally, we find (graphically)

$$B = 1.2763. \tag{8.35}$$

With use of equation (8.26), we now calculate the amplitude coefficients, a_n in equation (8.2), for $n = 11, 12, \ldots, 21$ (note that equation (8.26) is for a symmetric amplitude distribution, so we place the origin on element 11). The length l of the array is $l = 20d$ =0.458m.

$$a_{11} = I(z')|_{z'=0} = J_0(j4.010) = 11.39, \tag{8.36}$$

Next,

$$a_{12} = I(z')|_{z'=d} = J_0\left(j4.0095 \cdot \sqrt{1 - \left(\frac{0.0458}{0.458}\right)}\right) = J_0(j3.989) = 11.20. \tag{8.37}$$

These and the rest of amplitude coefficients are grouped in table 8.1.

From equation (8.18), we find - at the centre frequency - the slot offsets, x_n, as

$$x_n = 7.277 \arcsin\left(\sqrt{\frac{g_n}{1.241}}\right) \text{ (in mm)}, \tag{8.38}$$

where the slot conductances, g_n, follow from equation (8.25). The terms F_i in equation (8.25) can be calculated from the amplitude coefficients in table 8.1 according to

$$F_i = \frac{a_i^2}{2Nor}, \tag{8.39}$$

where Nor is a normalisation factor that is calculated from

$$\frac{\sum_{i=1}^{N} \frac{a_i^2}{2}}{Nor} + L = 1, \tag{8.40}$$

where N is the number of array elements and L is the fraction of power to be dissipated into the load. L needs to be specified by the designer.

We choose for a 5% dissipation into the load ($L = 0.05$) and then find the absolute offsets for the 21 slots as shown in table 8.2.

8.3 VALIDATION

The slotted waveguide has been constructed by milling the slots into the broad wall of a length of WR90 waveguide. Then flanges are soldered to both ends, where care is taken to have a sufficient length of waveguide between the end-slots and the flanges.

Table 8.1 Amplitude coefficients for 30dB SLL.

n	a_n
1	1.00
2	1.92
3	3.06
4	4.37
5	5.78
6	7.19
7	8.55
8	9.72
9	10.62
10	11.20
11	11.39
12	11.20
13	10.62
14	9.72
15	8.55
16	7.19
17	5.78
18	4.37
19	3.06
20	1.92
21	1.00

Onto one of these flanges a coaxial to rectangular waveguide adapter is connected and a waveguide load is connected to the other one, see figure 8.8.

The calculated array factor and measured, normalised power radiation patterns for three frequencies corresponding to beam positions near or at $-15°$, $0°$ and $+15°$, relative to broadside, are shown in, respectively, figure 8.9, figure 8.10 and figure 8.11.

The figures show that, although the measured side lobe level is less than -13.2dB as would have been expected for a uniform amplitude distribution, the expected -30dB is not realised. The calculations however do predict an SLL of -30dB.

Furthermore we see that the direction of the main beam moves with frequency as expected. However the beam direction deviates a few degrees from the predicted positions for all frequencies. This deviation is always in the same direction.

This last observation leads to the conclusion that a misalignment of the antenna under test must have been taking place prior to the antenna pattern measurements. The high side lobe level and the form of the main beam may have different sources. To start with, the anechoic chamber used for measuring the radiation patterns, prevented a separation between antenna under test and standard gain antenna that satisfied the far-field condition. Next to these possible measurement errors, the construction of the slotted waveguide array antenna can cause deviations from the calculated patterns. Figure 8.12 shows a detail of the slotted waveguide array antenna.

Table 8.2 Slot offsets in mm for a 21-element 30dB SLL slotted waveguide array antenna.

n	x_n
1	0.82
2	1.55
3	2.33
4	3.01
5	3.46
6	3.65
7	3.66
8	3.54
9	3.35
10	3.11
11	2.85
12	2.57
13	2.28
14	1.98
15	1.68
16	1.38
17	1.09
18	0.82
19	0.57
20	0.36
21	0.19

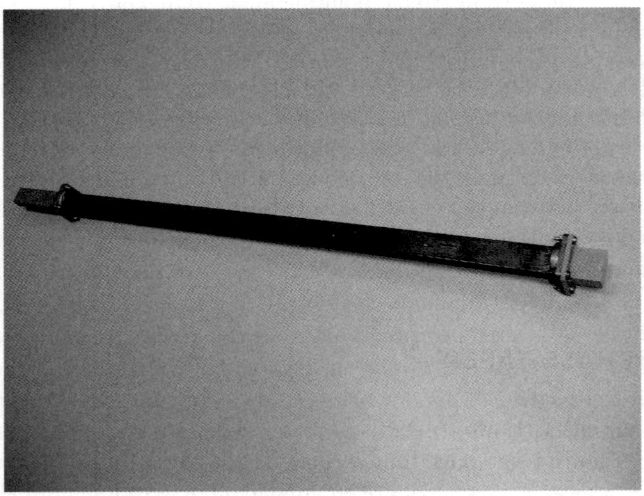

Fig. 8.8 Realised 21-slots slotted waveguide array antenna in WR90 having a coaxial to rectangular waveguide adapter connected to one side and a waveguide load to the other side.

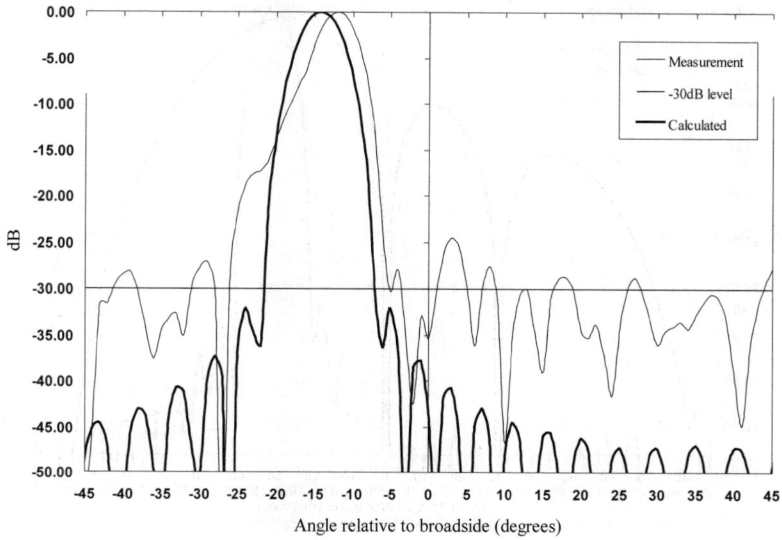

Fig. 8.9 Calculated and measured power radiation pattern for the slotted waveguide array antenna of figure 8.8 for 8.00GHz.

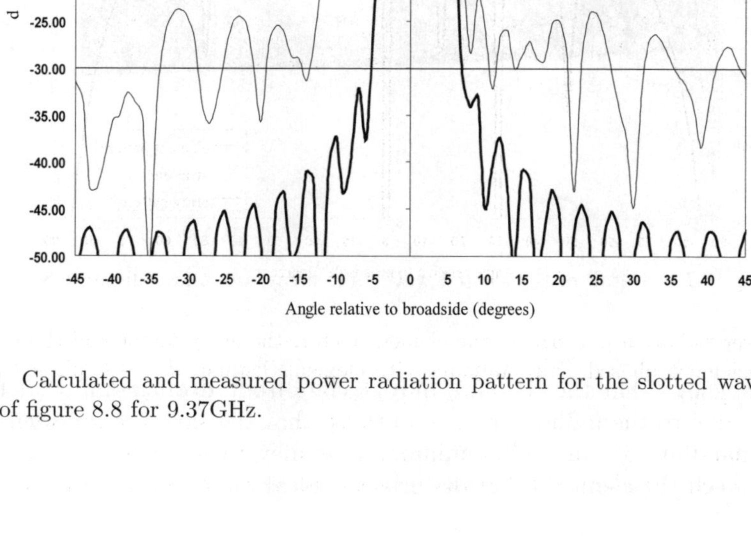

Fig. 8.10 Calculated and measured power radiation pattern for the slotted waveguide array antenna of figure 8.8 for 9.37GHz.

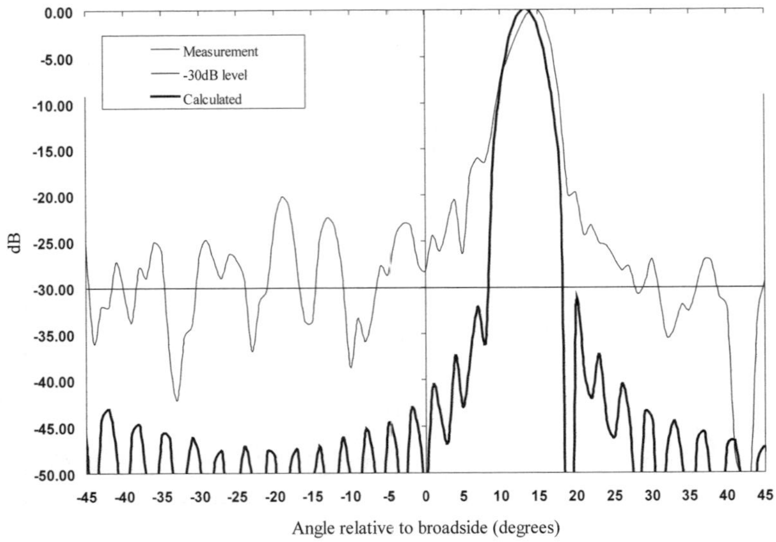

Fig. 8.11 Calculated and measured power radiation pattern for the slotted waveguide array antenna of figure 8.8 for 11.25GHz.

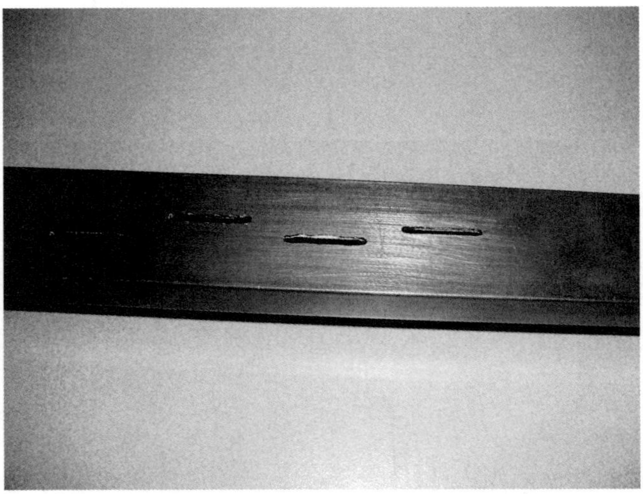

Fig. 8.12 Detail of the slotted waveguide array antenna of figure 8.8.

The figure shows that the slots not only deviate from ideal slots in that the ends are rounded, due to the milling process, but also that the slots are not identical due to a rough handling of burrs after milling. This may cause an unexpected mutual coupling between the elements that disturbs the ideal radiation pattern.

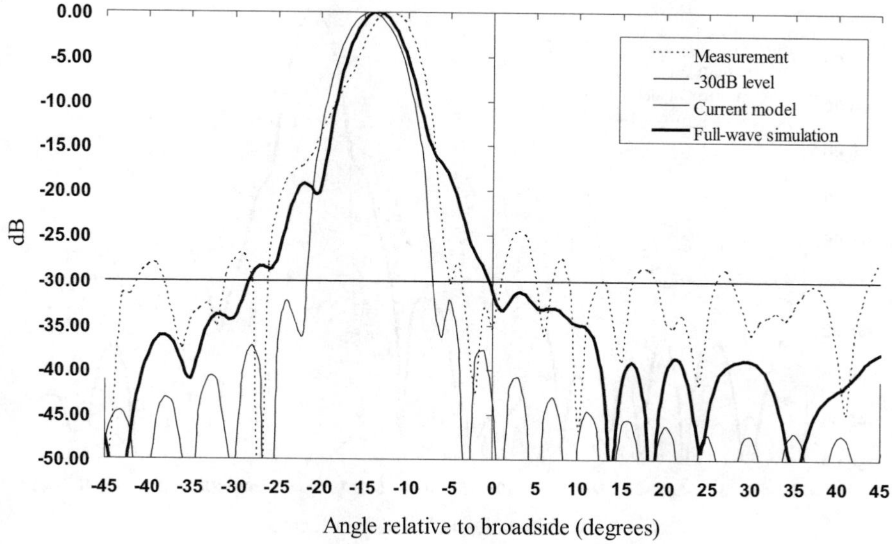

Fig. 8.13 Calculated (single mode and full wave) and measured power radiation pattern for the slotted waveguide array antenna of figure 8.8 for 8.00GHz.

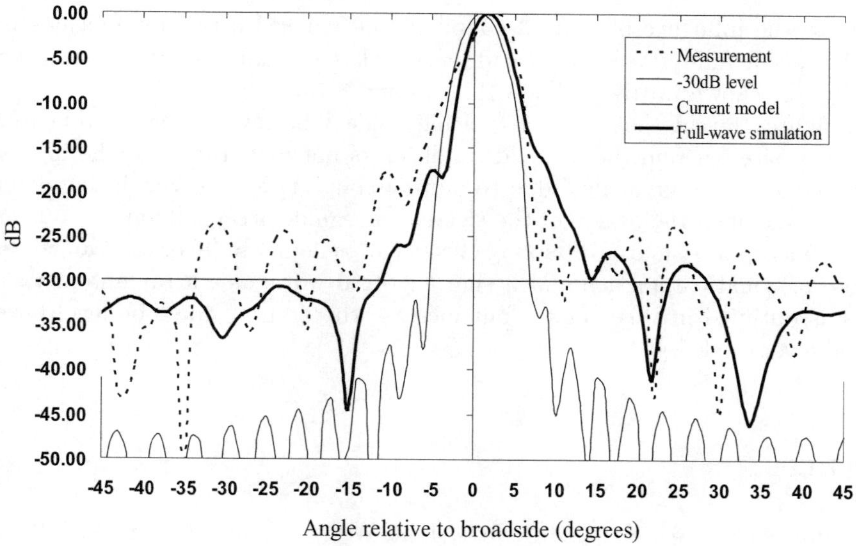

Fig. 8.14 Calculated (single mode and full wave) and measured power radiation pattern for the slotted waveguide array antenna of figure 8.8 for 9.37GHz.

Mutual coupling is not taken into consideration in the design of the slotted waveguide array antenna and even for a perfectly constructed antenna, mutual coupling may be a cause of deviations from the calculated array factors.

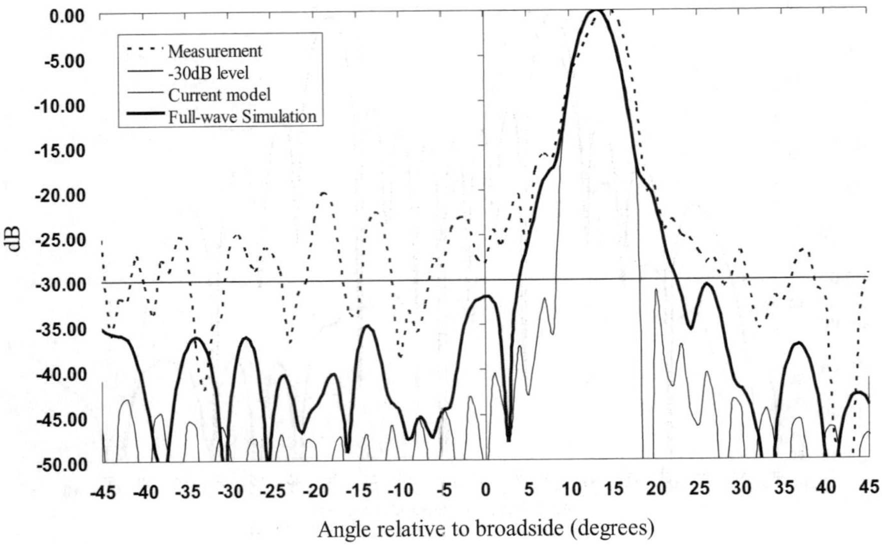

Fig. 8.15 Calculated (single mode and full wave) and measured power radiation pattern for the slotted waveguide array antenna of figure 8.8 for 11.25GHz.

To assess the influence of mutual coupling, the realised array antenna has been analysed using a full-wave analysis code that takes mutual coupling into account. The results are shown in the figures 8.13, 8.14 and 8.15.

These figures reveal that the relative high side lobe level is partly due to the measurement site wherein the far-field condition is not met. However, the high side lobe level is also for a great deal due to mutual coupling effects that have not been taken into account in the design of the slotted waveguide array antenna.

Ways to incorporate mutual coupling effects can be found in [8] or one can fall back on a more pragmatic approach and design a slotted waveguide array antenna along the lines as outlined in this chapter but increase the desired side lobe level beyond the actual level needed [4].

REFERENCES

1. R.C. Hansen, *Phased Array Antennas*, John Wiley & Sons, New York, 1998.

2. Contantine E. Balanis, *Antenna Theory, Analysis and Design, second edition*, John Wiley & Sons, New York, 1997.

3. R.E. Collin and F.J. Zucker, *Antenna Theory, Part I*, McGraw-Hill, New York, 1969.

4. Andre Dion, Nonresonant Slotted Arrays, *IRE Transactions on Antennas and Propagation*, Vol. AP 6, October 1958, pp. 360–365.

5. A.F. Stevenson, Theory of Slots in Rectangular Waveguides, *Journal of Applied Physics*, Vol. 19, January 1948, pp. 24–38.

6. R.C. Johnson (ed.), *Antenna Engineering Handbook, third edition*, McGraw-Hill, New York, 1993.

7. T.T. Taylor, Design of Line-Source Antennas for Narrow Beamwidth and Low Sidelobes, *IRE Transactions on Antennas and Propagation*, Vol. AP-3, No. 1, January 1955, pp. 16–28.

8. R.S. Elliot, Design of Line Source Antennas for Narrow Beamwidth and Asymmetric Low Sidelobes, *IEEE Transactions on Antennas and Propagation*, Vol. AP-23, No. 1, January 1975, pp. 10–107.

9. *Microwave Datamate*, Marconi Instruments Ltd, Longacres, 1987.

9

The Planar Array and Phased Array Antenna

Once the basics of linear array and linear phased array antennas are understood, the planar array and planar phased array antenna are not expected to pose any real problem. Geometrically, the planar array antenna is in fact a linear array of linear arrays; i.e. a linear array antenna where the array antenna elements are linear array antennas - along the orthogonal direction - themselves. For the phasing of the elements in a planar phased array antenna it will appear that this linear array antenna analogy still holds.

9.1 GEOMETRY

As we did for the linear array antenna, we start with the assumption that the elements of the planar array antenna are positioned on a regular lattice or grid. This is not necessary for the array or phased array antenna operation, but it will simplify the discussion and it describes the situation most often encountered in practice.

To even further simplify the discussion, we assume the lattice to be rectangular, see figure 9.1, a situation often encountered in practice. The triangular (or hexagonal) lattice, also often encountered in practice, may be seen as a superposition of two, displaced rectangular lattices, see figure 9.2.

As in the situation of the linear array antenna, we allow the elements to have directive properties (a non-isotropic radiation pattern), but - for the moment - we do not bother about physical dimensions. This leaves us the freedom to experiment (theoretically) with inter-element distances without having to restrict ourselves to situations that are constrained by the actual elements being applied.

Returning to figure 9.1, we see that $K \times L$ elements are grouped in a configuration of K rows, having an inter-element distance d_x between the rows and L columns,

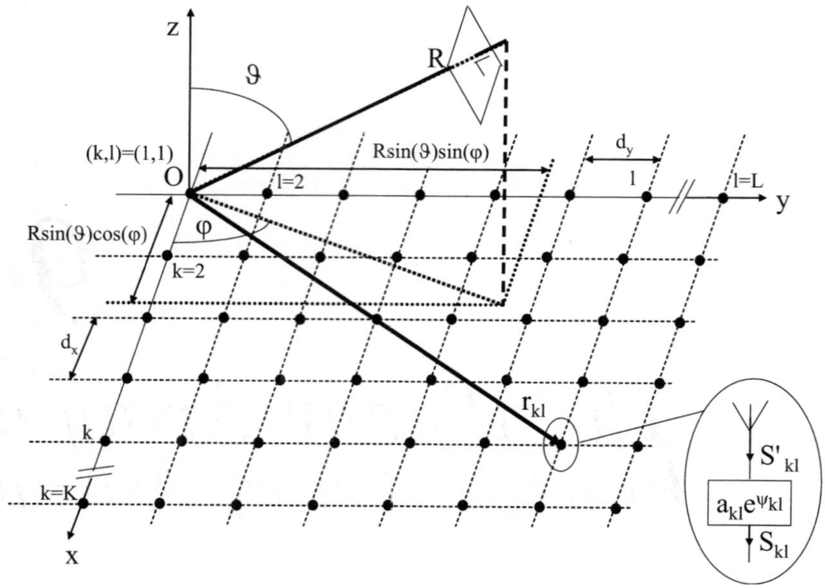

Fig. 9.1 Planar (phased) array antenna, consisting of $K \times L$ elements, positioned in a rectangular lattice in the x, y-plane of a Cartesian coordinate system.

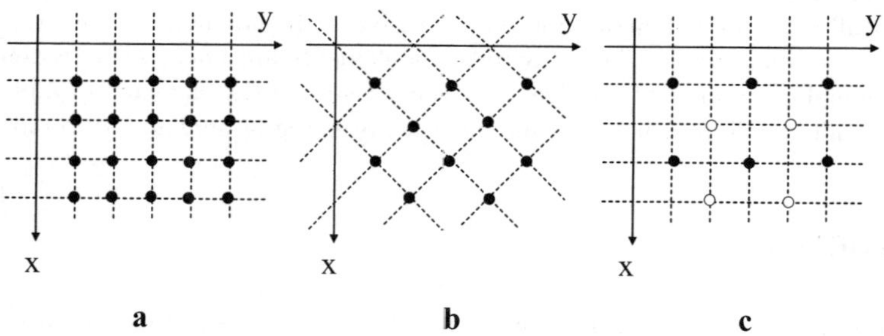

Fig. 9.2 Regular lattices. a. Rectangular lattice. b. Triangular lattice. c. Triangular lattice as a superposition of two rectangular lattices.

having an inter-element distance d_y between the columns. The planar array antenna therefore may be regarded as either a linear array antenna directed along the y-axis, having an inter-element distance d_y or a linear array antenna directed along the x-axis, having an inter-element distance d_x. In the first situation, the array antenna elements are linear arrays, directed along the x-axis, having element distance d_x, in the second situation, the array antenna elements are linear arrays, directed along the y-axis, having element distance d_y, see figure 9.3.

The position of an element (k, l) in the array, $\mathbf{r}_{kl}$, where k is a counter into the x-direction $(k = 1, 2, \ldots, K)$ and l is a counter into the y-direction $(l = 1, 2, \ldots, L)$,

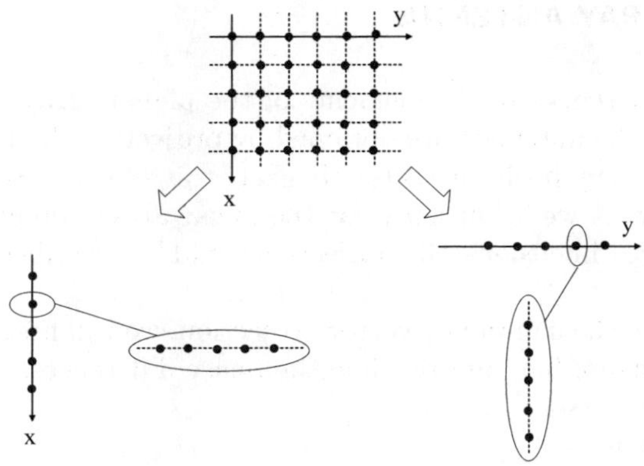

Fig. 9.3 A planar rectangular lattice array antenna may be envisaged as a linear array of linear arrays antenna.

is given by

$$\mathbf{r}_{kl} = (k-1)d_x\hat{\mathbf{u}}_x + (l-1)d_y\hat{\mathbf{u}}_y, \tag{9.1}$$

where $\mathbf{r}_{kl}$ is the vector, originating in O, see figure 9.1, pointing to element (k,l). The vectors $\hat{\mathbf{u}}_x$ and $\hat{\mathbf{u}}_y$ are unit vectors in the x- and y-direction, respectively. The element $(k,l) = (1,1)$ is positioned at O $((x,y,z) = (0,0,0))$. The planar array is assumed to be lying in the $z = 0$ plane, see figure 9.1.

The feeding of the array antenna elements may be accomplished by a variety of feeding structures, see for example [1]. For a 4×4 planar array antenna, different feeding arrangements are shown in figure 9.4.

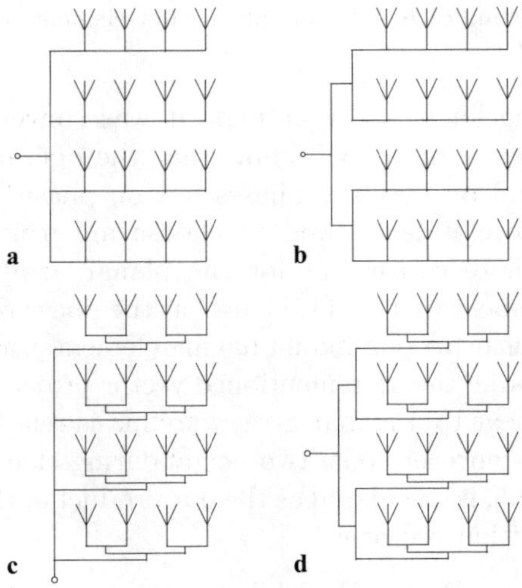

Fig. 9.4 Feeding arrangements for a 4×4 planar array antenna. a. Series-parallel feeding network. b. Series-corporate feeding network. c. Corporate-series feeding network. d. Full corporate feeding network.

9.2 PLANAR ARRAY ANTENNA

The path-length differences to the elements of the planar array antenna for a plane wave incident upon the antenna[1] are obtained by projecting the plane wave direction onto the planar element position vector. In fact, this is what we have done for the linear array antenna as well, but since for the linear array antenna the analysis was restricted to only two dimensions, the projection could be described in scalar equations directly.

Before we perform the mentioned vector projection, we will first return to the linear array antenna, see figure 9.5, and redefine the phase differences.

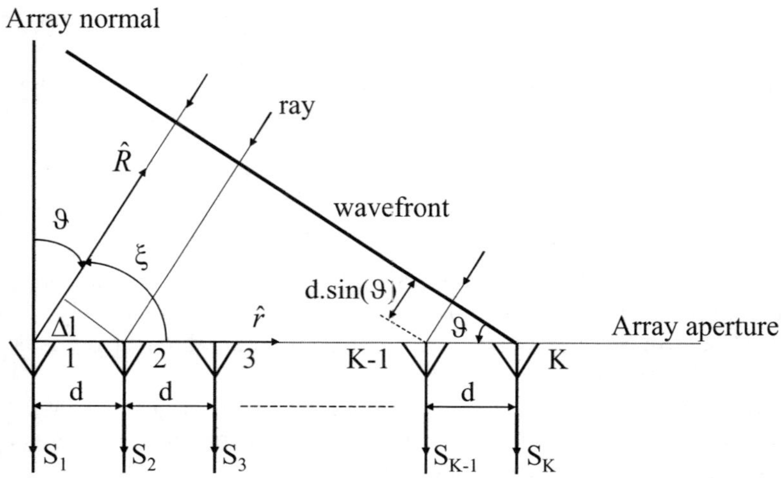

Fig. 9.5 Linear array antenna of K elements at an inter-element spacing d, receiving a plane wave from the direction $\hat{\mathbf{R}}$.

In the analysis of the linear array antenna it was convenient to take the phase reference ($\psi = 0°$) at element K. We know that the operation of the linear array antenna does not depend on absolute phases but on phase *differences* between the array elements. Therefore it is allowed to choose an arbitrary element as phase reference. Since it is more convenient for the planar array antenna as shown in figure 9.1 to take the element $(k, l) = (1, 1)$ at O as the phase reference, we will rewrite the equations for the linear array antenna having element 1 as phase reference, while at the same time explaining the aforementioned vector projection. After this exercise, the extension from a linear to a planar array antenna is relatively straightforward.

The path-length difference between two neighbouring elements in the linear array antenna, Δl, see figure 9.5, is calculated as the dot product of the plane wave direction, $\hat{\mathbf{R}}$, and the element position vector $\mathbf{r}$

$$\Delta l = \hat{\mathbf{R}} \cdot \mathbf{r} = \hat{\mathbf{R}} \cdot \hat{\mathbf{r}} d = d \cos(\xi) = d \sin(\vartheta). \tag{9.2}$$

[1]We regard the antenna as being a receive antenna. By virtue of reciprocity, the results apply equally well to the antenna in transmit mode, so the choice for transmit or receive mode depends on what is most convenient at the moment.

The phase difference, $\Delta\psi$, between two adjacent elements is thus

$$\Delta\psi = k_0 d \sin(\vartheta), \tag{9.3}$$

and the phase of element k relative to element 1, ψ_k, is given by

$$\psi_k = k_0(k-1)d\sin(\vartheta), \text{for } k = 1, 2, \ldots, K, \tag{9.4}$$

where $k_0 = \frac{2\pi}{\lambda_0}$. This result is consistent with our earlier findings, only the phase reference has been transferred from element K to element 1.

Equation (9.2) is easily translated to the planar array antenna configuration from inspection of figure 9.1. For the moment we assume that every element is not weighted ($a_{ij} = 1$) and not phased ($\psi_{ij} = 0$). For element (k, l), the difference in path length for the plane wave to this element, relative to the $(1, 1)$-element, Δl_{kl}, is

$$\begin{aligned} \Delta l_{kl} = \hat{R} \cdot \mathbf{r}_{kl} = \\ (k-1)d_x \hat{\mathbf{R}} \cdot \hat{\mathbf{u}}_x + (l-1)d_y \hat{\mathbf{R}} \cdot \hat{\mathbf{u}}_y = \\ (k-1)d_x \sin(\vartheta)\cos(\varphi) + (l-1)d_y \sin(\vartheta)\sin(\varphi). \end{aligned} \tag{9.5}$$

The phase of element (k, l) relative to the $(1, 1)$-element, ψ_{kl}, is then given by

$$\psi_{kl} = k_0(k-1)d_x \sin(\vartheta)\cos(\varphi) + k_0(l-1)d_y \sin(\vartheta)\sin(\varphi), \tag{9.6}$$

resulting in the planar array radiation pattern

$$S(\vartheta, \varphi) = S_e(\vartheta, \varphi) \sum_{k=1}^{K} \sum_{l=1}^{L} e^{j[k_0(k-1)d_x \sin(\vartheta)\cos(\varphi) + k_0(l-1)d_y \sin(\vartheta)\sin(\varphi)]}, \tag{9.7}$$

where $S_e(\vartheta, \varphi)$ is the element radiation pattern. Again, we assume that all elements are identical and that mutual coupling effects are negligible.

The planar array antenna radiation pattern may be written as the product of an element factor or element pattern and two linear array factors

$$S(\vartheta, \varphi) = S_e(\vartheta, \varphi) S_{a_1}(\vartheta, \varphi) S_{a_2}(\vartheta, \varphi), \tag{9.8}$$

where

$$S_{a_1}(\vartheta, \varphi) = \sum_{k=1}^{K} e^{jk_0(k-1)d_x \sin(\vartheta)\cos(\varphi)}, \tag{9.9}$$

is the array factor for the linear array in the x-direction and

$$S_{a_2}(\vartheta, \varphi) = \sum_{l=1}^{L} e^{jk_0(l-1)d_y \sin(\vartheta)\sin(\varphi)}, \tag{9.10}$$

is the array factor of the linear array in the y-direction.

With the planar array factor, $S_a(\vartheta, \varphi)$, thus written as the product of two linear array factors, $S_a(\vartheta, \varphi) = S_{a1}(\vartheta, \varphi) S_{a2}(\vartheta, \varphi)$, some interesting characteristics of regular grid planar arrays become visible. If, for example, we have a plane wave incident with a direction parallel to the xz-plane, i.e. for $\varphi = 0°$, the array factor S_{a_2} reduces to a constant L and the normalised array factor radiation pattern reduces to that of a K-element linear array with inter-element spacing d_x. The effect of the additional number of elements with respect to the K elements of the linear array antenna is found in the directivity multiplication factor L. For a plane wave having a direction parallel to the yz-plane, i.e. for $\varphi = 90°$, the array factor S_{a_1} reduces to a constant K and the normalised array factor radiation pattern reduces to that of an L-element linear array with inter-element spacing d_y. The directivity of this array antenna is multiplied with a factor K to account for all the elements in the planar array antenna.

By the positioning of multiple elements in a row (linear array antenna), we have achieved that - relative to a single element - the directivity has increased in the plane perpendicular to and containing the linear array antenna. In the plane perpendicular to this plane, the directivity has not changed and remains that of a single element, see figure 9.6.

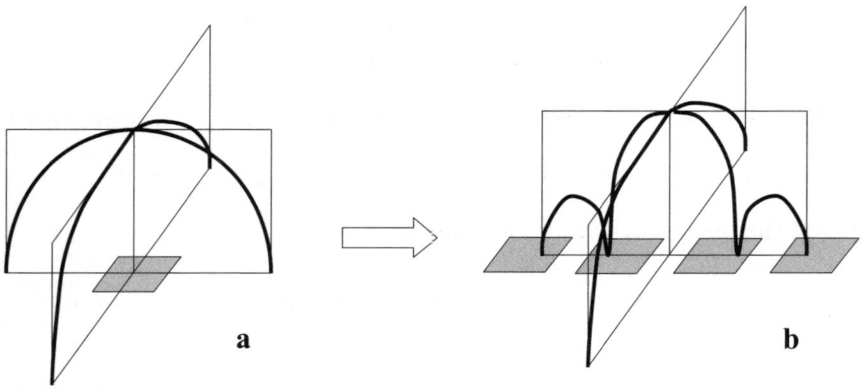

Fig. 9.6 Normalised radiation patterns of a single element and a linear array consisting of these elements in the principal planes. a. Element. b. Linear array.

By the positioning of multiple elements in a regular (rectangular) grid, we have achieved that the directivity has increased in both principal planes of the array antenna, see figure 9.7. As a matter of fact in all planes $0° \leq \varphi \leq 90°$ the directivity has increased with respect to that of a single element.

9.2.1 Radiation

We will now analyse a 64-element planar array antenna where the identical elements are arranged in a square ($d_x = d_y$) 8×8 lattice. We assume the element radiation pattern to be that of a rectangular aperture in an infinite ground plane. The voltage radiation pattern of the element is assumed to be φ-independent and is stated as

$$S_e(\vartheta) = \cos(\vartheta), \text{ for all } \varphi, \tag{9.11}$$

just as for the 8-element linear array antennas we have analysed before.

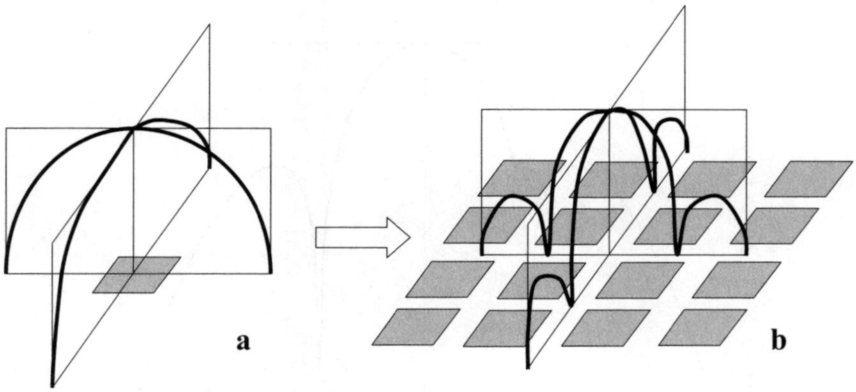

Fig. 9.7 Normalised radiation patterns of a single element and a planar array consisting of these elements in the principal planes. a. Element. b. Planar array.

The normalised radiation patterns in the principal planes ($\varphi = 0°$ and $\varphi = 90°$) for inter-element distances ranging from $\frac{\lambda}{4}$ to $\frac{5\lambda}{4}$ are identical to the ones shown earlier for the 8-element linear array antennas. Therefore we will not show them. For three planes in between these principal planes, the radiation patterns will be shown.

Figures 9.8 to 9.11 show the element radiation pattern, array factor and array radiation pattern in the plane $\varphi = 15°$ for element distances of, respectively, $\frac{\lambda}{4}$, $\frac{\lambda}{2}$, λ and $\frac{5\lambda}{4}$. The patterns are normalised to the maximum and are displayed in decibels. The element pattern is thus plotted as $20\log(|S_e(\vartheta)|)$, the array factor as $20\log\left(\frac{1}{64}|S_{a_1}(\vartheta,\varphi)|\,|S_{a_2}(\vartheta,\varphi)|\right)$ and the array pattern as

$$20\log\left(\frac{1}{64}|S_e(\vartheta)|\,|S_{a_1}(\vartheta,\varphi)|\,|S_{a_2}(\vartheta,\varphi)|\right).$$

Comparing these patterns with the ones obtained for the linear 8-element array antennas (i.e. the principal planes radiation patterns) leads to the observation that grating lobes are not present in the plane $\varphi = 15°$. Even for element distances equal to or larger than one wavelength, only a rising of the side lobes is observed. Furthermore, the distribution of the side lobe levels is less regular than for the earlier analysed linear array antenna.

Figures 9.12 to 9.15 show the element radiation pattern, array factor and array radiation pattern in the plane $\varphi = 30°$ for element distances of, respectively, $\frac{\lambda}{4}$, $\frac{\lambda}{2}$, λ and $\frac{5\lambda}{4}$.

In comparison with the radiation patterns in the plane $\varphi = 15°$, we see that the patterns are deviating further from the ones in the principal planes (linear array antenna patterns). In the plane $\varphi = 30°$ the side lobe levels are distributed less evenly than compared to the plane $\varphi = 15°$, in which plane the side lobe levels were already distributed less evenly than compared to the plane $\varphi = 0°$.

As a last example, in figures 9.16 to 9.19, we show the element radiation pattern, array factor and array radiation pattern in the plane $\varphi = 45°$ for element distances of, respectively, $\frac{\lambda}{4}$, $\frac{\lambda}{2}$, λ and $\frac{5\lambda}{4}$.

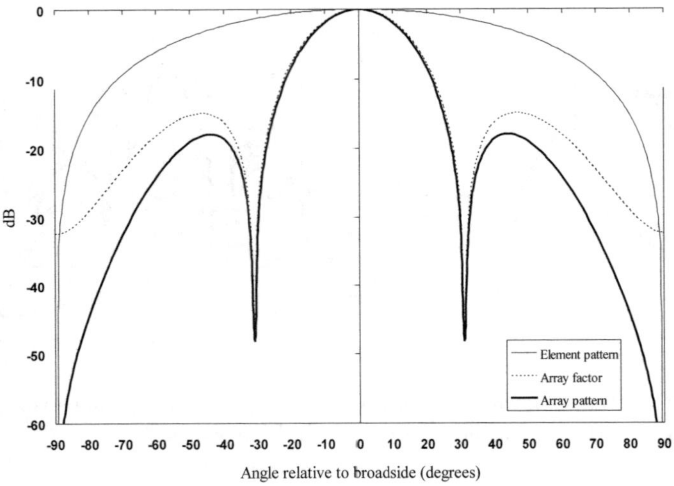

Fig. 9.8 Element radiation pattern, array factor and array radiation pattern for a square lattice 8 × 8 planar array antenna in the plane $\varphi = 15°$ for $d_x = d_y = \frac{\lambda}{4}$.

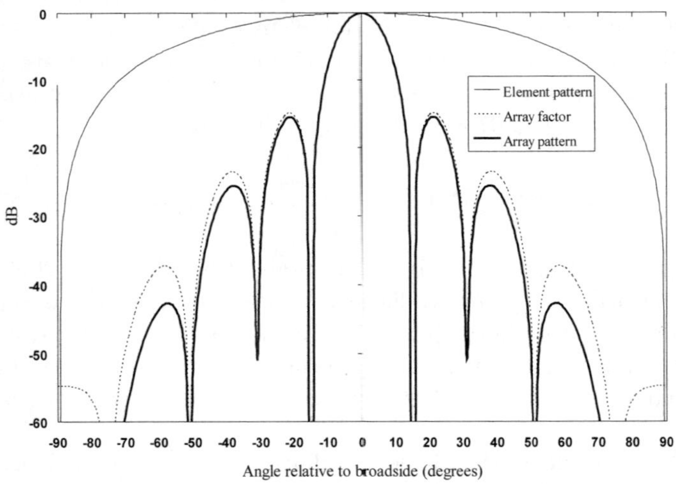

Fig. 9.9 Element radiation pattern, array factor and array radiation pattern for a square lattice 8 × 8 planar array antenna in the plane $\varphi = 15°$ for $d_x = d_y = \frac{\lambda}{2}$.

Since, in our planar, rectangular grid array antenna we have chosen the element distances in both orthogonal directions to be identical (square lattice) and have opted for an element radiation pattern that is φ-independent, the radiation patterns in the planes $\varphi = 60°$, $\varphi = 75°$ and $\varphi = 90°$ are identical to those in the planes $\varphi = 30°$, $\varphi = 15°$ and $\varphi = 0°$, respectively.

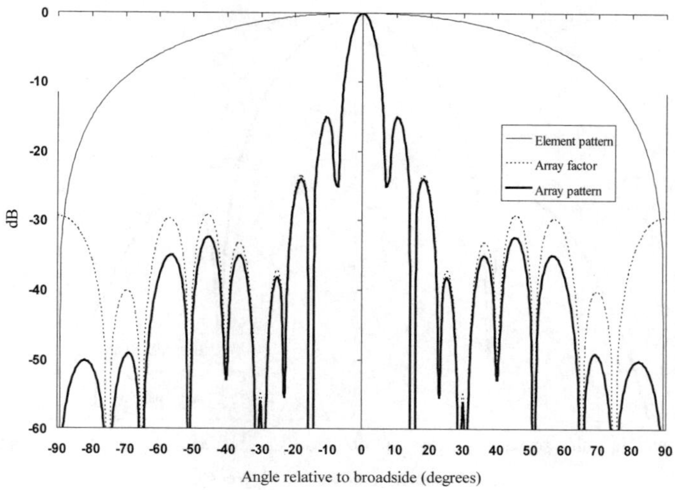

Fig. 9.10 Element radiation pattern, array factor and array radiation pattern for a square lattice 8 × 8 planar array antenna in the plane $\varphi = 15^\circ$ for $d_x = d_y = \lambda$.

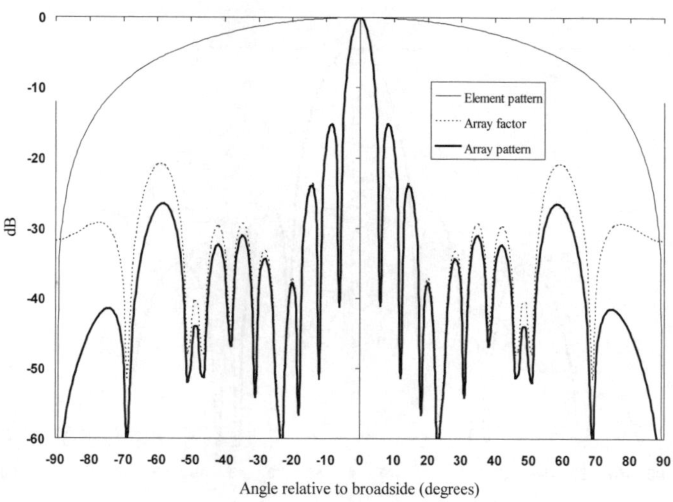

Fig. 9.11 Element radiation pattern, array factor and array radiation pattern for a square lattice 8 × 8 planar array antenna in the plane $\varphi = 15^\circ$ for $d_x = d_y = \frac{5\lambda}{4}$.

We see that in the plane that is maximally removed from the principal planes, i.e. $\varphi = 45^\circ$, we again encounter a regular side lobe level distribution and in none of the planes in between the two principal planes do we encounter grating lobes for all the element distances analysed. Only in the principal planes do grating lobes occur for element distances equal to and larger than one wavelength. The level of the first

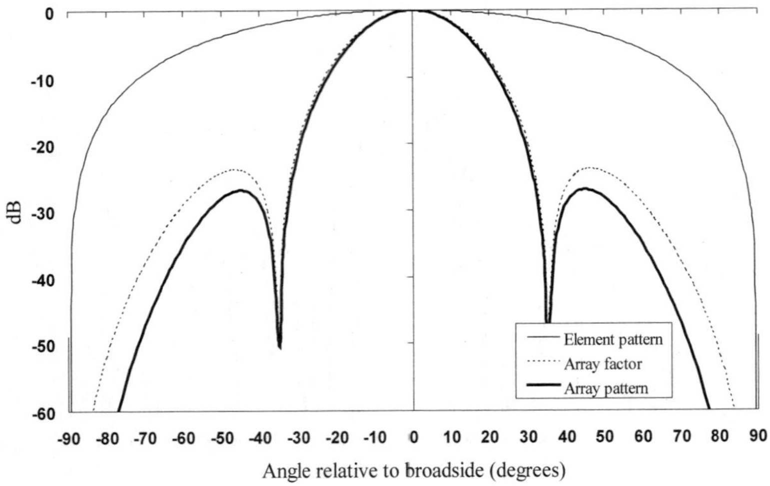

Fig. 9.12 Element radiation pattern, array factor and array radiation pattern for a square lattice 8 × 8 planar array antenna in the plane $\varphi = 30^\circ$ for $d_x = d_y = \frac{\lambda}{4}$.

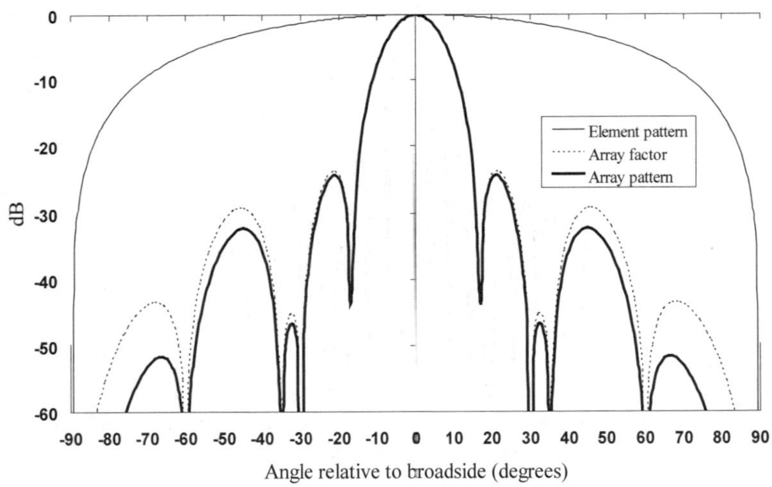

Fig. 9.13 Element radiation pattern, array factor and array radiation pattern for a square lattice 8 × 8 planar array antenna in the plane $\varphi = 30^\circ$ for $d_x = d_y = \frac{\lambda}{2}$.

side lobe in the plane $\varphi = 45^\circ$ appears to be approximately -26dB as can be seen from the array factor in the figures 9.16 to 9.19. Intuitively it can be understood that this level is not a matter of coincidence and in the remainder of this section we will demonstrate this.

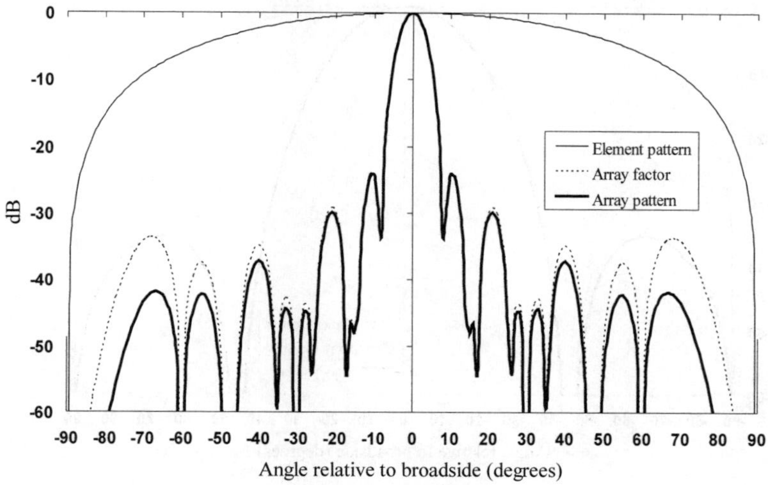

Fig. 9.14 Element radiation pattern, array factor and array radiation pattern for a square lattice 8 × 8 planar array antenna in the plane $\varphi = 30°$ for $d_x = d_y = \lambda$.

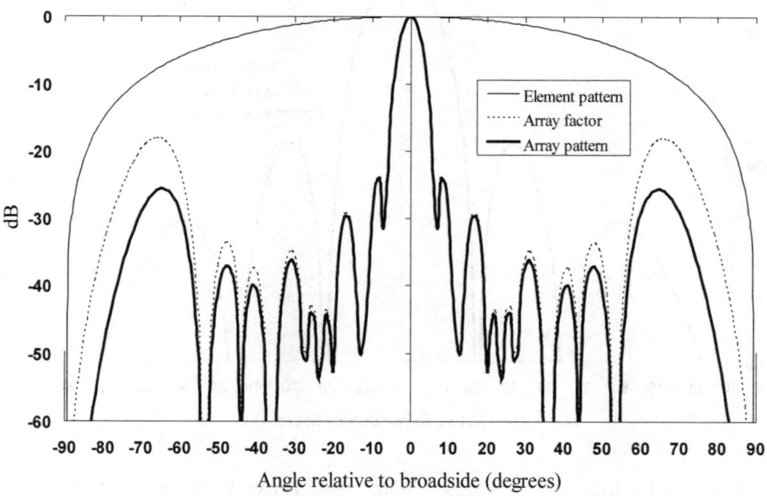

Fig. 9.15 Element radiation pattern, array factor and array radiation pattern for a square lattice 8 × 8 planar array antenna in the plane $\varphi = 30°$ for $d_x = d_y = \frac{5\lambda}{4}$.

However, we will first explain the decrease and increase in regularity of the side lobe level distribution in the radiation patterns when moving from one principal plane to the other (orthogonal) principal plane.

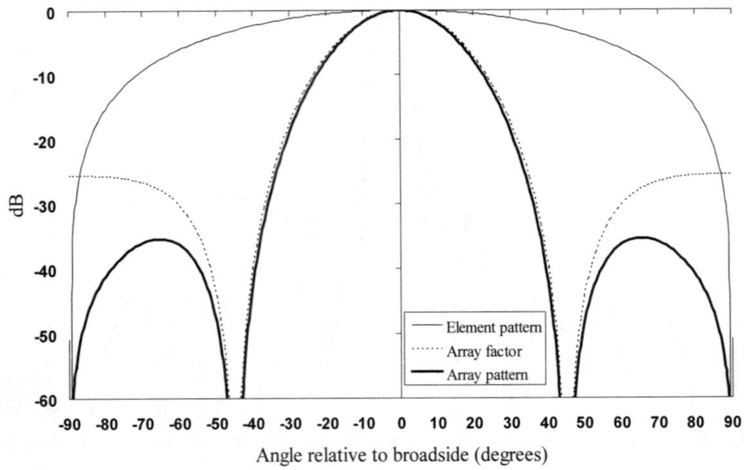

Fig. 9.16 Element radiation pattern, array factor and array radiation pattern for a square lattice 8 × 8 planar array antenna in the plane $\varphi = 45°$ for $d_x = d_y = \frac{\lambda}{4}$.

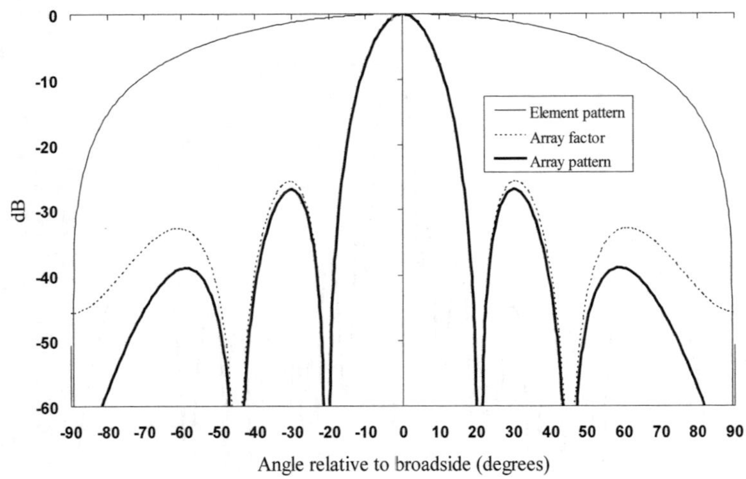

Fig. 9.17 Element radiation pattern, array factor and array radiation pattern for a square lattice 8 × 8 planar array antenna in the plane $\varphi = 45°$ for $d_x = d_y = \frac{\lambda}{2}$.

9.2.2 Side Lobe Level

Obviously, the distribution of the side lobe levels in the radiation patterns follow from the radiation pattern, as calculated by equation (9.7). We will try to explain the phenomenon not mathematically, but in a more graphical manner. We have already seen that in the principal planes ($\varphi = 0°$ and $\varphi = 90°$), the array antenna

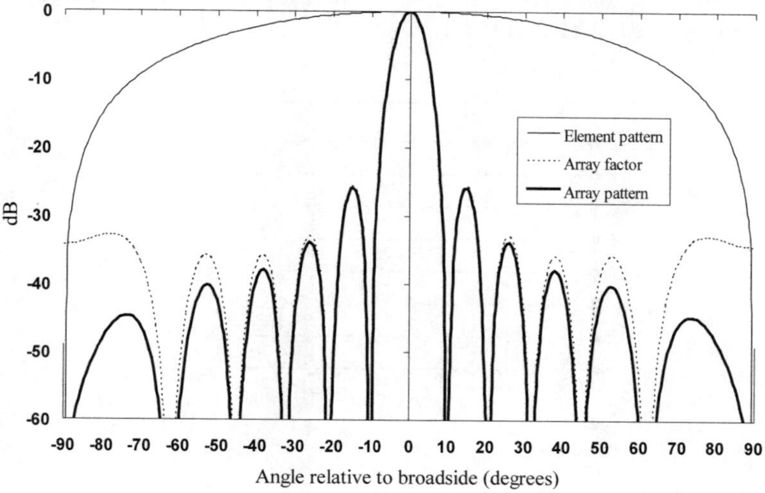

Fig. 9.18 Element radiation pattern, array factor and array radiation pattern for a square lattice 8×8 planar array antenna in the plane $\varphi = 45°$ for $d_x = d_y = \lambda$.

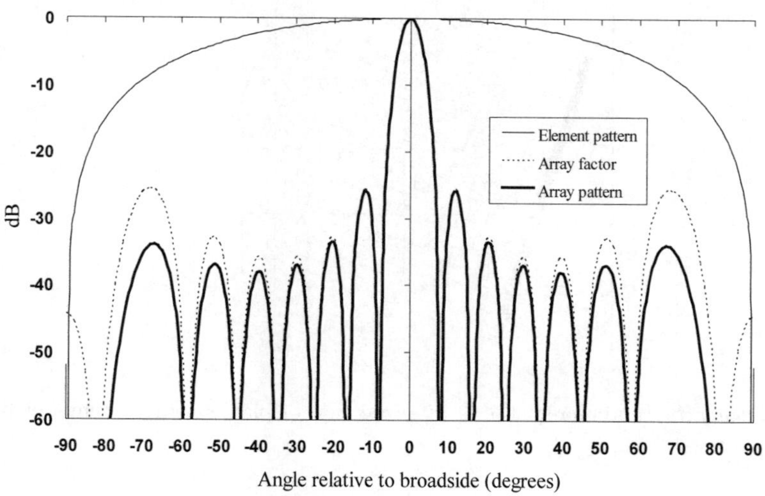

Fig. 9.19 Element radiation pattern, array factor and array radiation pattern for a square lattice 8×8 planar array antenna in the plane $\varphi = 45°$ for $d_x = d_y = \frac{5\lambda}{4}$.

radiation patterns are basically linear array antenna radiation patterns, where the only difference is in the amount of totally radiated power, not in the form of the radiation pattern. In fact, we may project all elements perpendicularly onto the φ-plane of analysis, apply an amplitude weighting factor proportional to the number of projected elements and consider - in the φ-plane of analysis - the projected elements as being elements of a linear array antenna. This process is demonstrated, for three angles φ, in figures 9.20, 9.21 and 9.22.

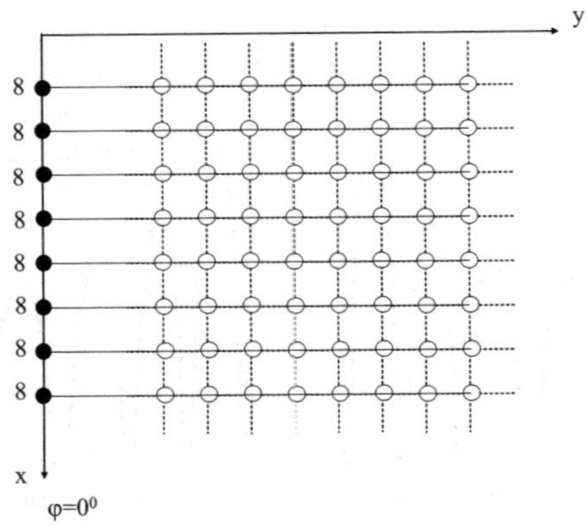

Fig. 9.20 Linear array antenna synthesis by projection of planar array antenna elements for the plane $\varphi = 0°$.

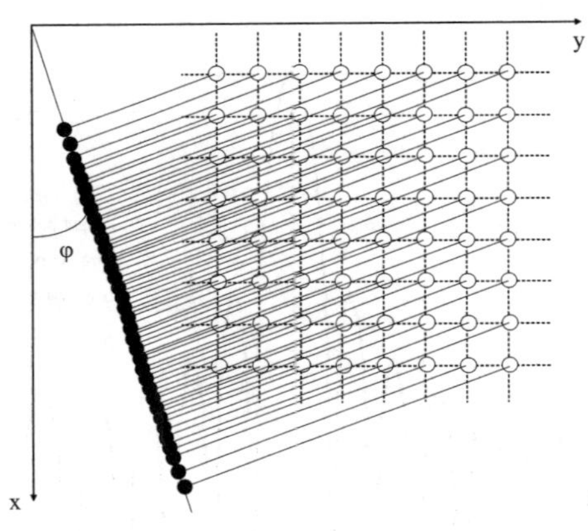

Fig. 9.21 Linear array antenna synthesis by projection of planar array antenna elements for the plane $0° < \varphi < 45°$.

In figure 9.20 we see that for the direction $\varphi = 0°$, the planar array antenna projected on the $\varphi = 0°$-plane leads to a radiation pattern identical in form to the one found for the already known 8-element linear array antenna. The only difference is in the radiated power. This effect is visualised by the (identical) weighting factors of the projected linear array antenna elements. In figure 9.21 we see what happens for directions $0° < \varphi < 45°$. We now get a projected linear array antenna, where every element is assigned a weighting factor 'one' (not shown in the figure) and where the inter-element distance is no longer identical for every pair of adjacent elements. The irregular element spacing is responsible for the observed irregular distribution of the side lobe levels. In figure 9.22, the situation is shown for the $\varphi = 45°$-plane. For this

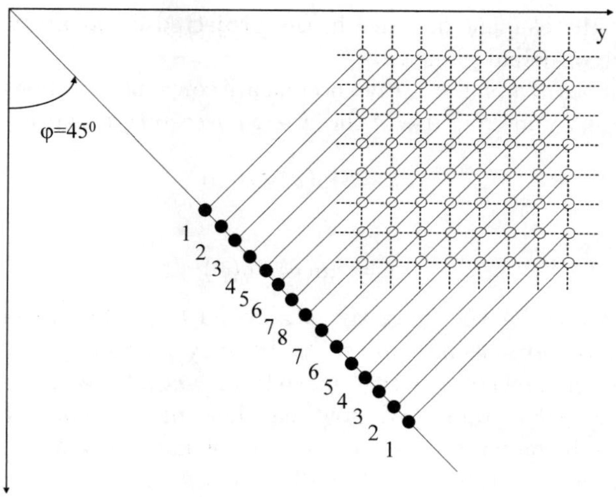

Fig. 9.22 Linear array antenna synthesis by projection of planar array antenna elements for the plane $\varphi = 45°$.

special situation, the projected linear array antenna exhibits a constant inter-element spacing (equal to $\frac{1}{2}\sqrt{2}d$), resulting in a regular side lobe level distribution. Besides, we see - for this special situation - that a *grid-inherent amplitude taper* has been created, resulting in a (first) side lobe level that is lower than that in the principal planes.

To quantify this effect, we rewrite the product of linear array factors, equation (9.8), in the way we did for the linear array antenna. We then find for the normalised array factor of a planar array antenna

$$S_a(\vartheta,\varphi) = S_{a_1}(\vartheta,\varphi)S_{a_2}(\vartheta,\varphi) \approx \frac{\sin\left(\pi\frac{Kd_x}{\lambda}\sin(\vartheta)\cos(\varphi)\right)}{\pi\frac{Kd_x}{\lambda}\sin(\vartheta)\cos(\varphi)}\frac{\sin\left(\pi\frac{Kd_y}{\lambda}\sin(\vartheta)\sin(\varphi)\right)}{\pi\frac{Kd_y}{\lambda}\sin(\vartheta)\sin(\varphi)}, \tag{9.12}$$

where we have been making use of the sine function approximation for small arguments. For $\varphi = 45°$, we find

$$S_a(\vartheta,\varphi) \approx \left[\frac{\sin\left(\pi\frac{Kd_x}{\lambda}\sin(\vartheta)\frac{1}{2}\sqrt{2}\right)}{\pi\frac{Kd_x}{\lambda}\sin(\vartheta)\frac{1}{2}\sqrt{2}}\right]^2 \tag{9.13}$$

For the first side lobe level we then find: -26.92dB.

9.2.3 Grating Lobes

Grating lobes appear whenever the inter-element distance of the projected linear array antenna is such that the argument of equation (9.8) is a multiple of 2π. From the projections shown in figures 9.20, 9.21 and 9.22, we can already deduce that if no grating lobes appear in the principal planes, all other planes φ will be free of grating lobes too, since the element distance in the projected linear array antenna is at its maximum in the principal planes.

The grating lobes of a planar array antenna are conveniently shown in the *projection coordinates* space, by making use of the *direction cosines* u and v, where

$$u = \sin(\vartheta)\cos(\varphi), \tag{9.14}$$

and

$$v = \sin(\vartheta)\sin(\varphi). \tag{9.15}$$

We have seen how for the linear array antenna ($\varphi = 0$ and thus $u = \sin(\vartheta)$ and $v = 0$) these direction cosines proved to be very convenient for showing grating lobes, especially for showing how they move from *invisible space* to *visible space* upon enlarging the element distance. We now see, that this concept applied to the linear array antenna is in fact a special case of the planar array antenna for which the direction cosines are given by equations (9.14) and (9.15).

Upon substitution of equations (9.14) and (9.15) into equation (9.7), we find that grating lobes occur when

$$\frac{d_x}{\lambda}(k-1)u + \frac{d_y}{\lambda}(l-1)v = m, \tag{9.16}$$

where m is an integer number.

This condition can only be met if, to start with,

$$u = p\frac{\lambda}{d_x}, \tag{9.17}$$

and

$$v = q\frac{\lambda}{d_y}, \tag{9.18}$$

where p an q are integer numbers.

In u, v-space, the area within the circle of radius 'one' corresponds to real angles ϑ and φ ($-90^\circ < \vartheta < 90^\circ$ and $0^\circ < \varphi < 360^\circ$)[2] and is called *visible space*. The area outside the circle, corresponding to complex angles ϑ and φ, is called *invisible space*.

In the u, v-plane, the positions of main lobe (i.e. the grating lobe for which $p = q = 0$), grating lobes, visible space and invisible space and their interrelations can be shown with great ease. For the planar array antenna we have been working with so far (8×8 elements, $d_x = d_y$), the so-called *grating lobe diagrams* are shown in figure 9.23 for four element distances. The main lobe is indicated by a white dot, the grating lobes are indicated by black dots.

The grating lobe diagrams show - in accordance with the earlier shown radiation patterns - that for an element distance equal to one wavelength, grating lobes occur in the principal planes ($u = 0$ and $v = 0$), just on the border between visible and invisible space. For an element distance greater than one wavelength ($d_x = d_y = \frac{5\lambda}{4}$),

[2]Strictly speaking, we should take $-180^\circ < \vartheta < 180^\circ$, but since we implicitly assume that we are dealing with array antenna elements that have their radiation maximum at broadside and that are embedded in a ground plane, we restrict ourselves to the forward directions.

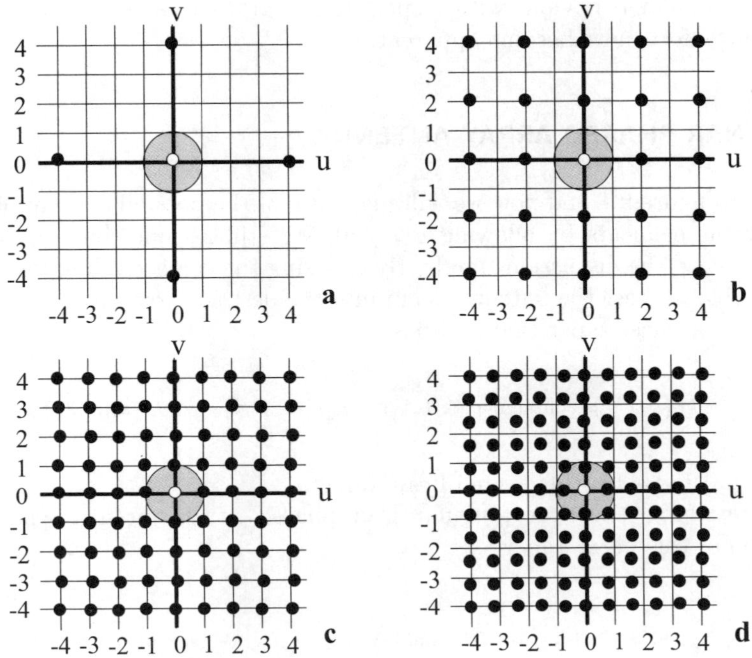

Fig. 9.23 Grating lobe diagrams for a rectangular lattice, planar array antenna. a. $d_x = d_y = \frac{\lambda}{4}$. b. $d_x = d_y = \frac{\lambda}{2}$. c. $d_x = d_y = \lambda$. d. $d_x = d_y = \frac{5\lambda}{4}$.

see figure 9.23d, the grating lobes in the principal planes have entered the visible space and in the diagonal plane ($\varphi = 45°$, or $u = v$), grating lobes are about to cross the border from invisible space to visible space.

To determine the element distances for which grating lobes move into the visible region, we may restrict ourselves to analysing the grating lobes in the principal planes. These planes, we have seen, act as 'worst case planes'. If grating lobes do not occur in visible space in the principal planes, grating lobes will not occur in visible space in any plane in between. Thus, we only have to take into account two linear array antenna array factors, the one we obtain by substituting $\varphi = 0°$ into equation (9.7) and the one we obtain by substituting $\varphi = 90°$ into equation (9.7).

Then, in complete analogy with the considerations we gave in the discussion of the linear array antenna, we find

$$\frac{d_x}{\lambda} \leq 1, \tag{9.19}$$

and

$$\frac{d_y}{\lambda} \leq 1. \tag{9.20}$$

In the discussion of the linear array antenna we have seen that at scanning the main beam, the grating lobes move with the main beam. For a planar array antenna we expect a similar behaviour with respect to the grating lobes. In the grating lobe diagram this effect must become apparent.

9.3 PLANAR PHASED ARRAY ANTENNA

We return to figure 9.1 and now we will make use of the possibility to apply a phase taper over the elements by allowing $\psi_{kl} \neq 0$. We still assume that $a_{kl} = 1$ for all elements to keep the discussion simple. By choosing the correct values for the ψ_{kl} we should be able to steer the antenna beam into the desired direction.

We choose a phase taper that satisfies

$$\psi_{kl} = -k_0(k-1)d_x \sin(\vartheta_0)\cos(\varphi_0) - k_0(l-1)d_y \sin(\vartheta_0)\sin(\varphi_0), \tag{9.21}$$

where (ϑ_0, φ_0) indicates the desired beam direction.

The array factor of a rectangular grid phased array antenna will be, upon substitution of this phase taper

$$S_a(\vartheta,\varphi) = \sum_{k=1}^{K} e^{jk_0(k-1)d_x[\sin(\vartheta)\cos(\varphi)-\sin(\vartheta_0)\cos(\varphi_0)]} e^{jk_0(l-1)d_y[\sin(\vartheta)\sin(\varphi)-\sin(\vartheta_0)\sin(\varphi_0)]}. \tag{9.22}$$

This array factor clearly demonstrates the correctness of our chosen phasing; the maximum for the array radiation will occur for $(\vartheta,\varphi) = (\vartheta_0,\varphi_0)$.

9.3.1 Radiation

As an example, we again take the 8 × 8-element, square lattice ($d_x = d_y$), array antenna for the element spacings for which we have shown the radiation patterns in figures 9.8 to 9.19. We now apply a phase taper, such that the desired beam direction is at $(\vartheta_0,\varphi_0) = (30°, 45°)$, i.e. the beam is positioned at 30° in *elevation* and 45° in *azimuth*. Figures 9.24 to 9.27 show the normalised element power radiation pattern, array factor power radiation pattern and array power radiation pattern in the plane $\varphi = 0°$ for the phased array antenna for element distances $d = d_x = d_y = \frac{\lambda}{4}$, $d = \frac{\lambda}{2}$, $d = \lambda$ and $d = \frac{5\lambda}{4}$.

The normalisation is with respect to the maximum radiation. Therefore, we see in the $\varphi = 0°$ cuts, that the array factor and array pattern amplitudes are less than 0dB. For an element distance $d_x = d_y = \frac{\lambda}{4}$, see figure 9.24, we see a distinctive lobe in the neighbourhood of $\vartheta = 30°$. Since, for this element distance, the rectangular array size is quite small, the beam will be relatively wide and the observed lobe in the $\varphi = 0°$ cut must be part of the main beam directed at $(\vartheta_0,\varphi_0) = (30°, 45°)$. For larger element distances, figures 9.25 to 9.27, the array size increases and therefore the main beam gets smaller. We do see that the most distinctive lobes in the shown cuts get smaller, move away from $\vartheta = 30°$ and decrease in amplitude. Therefore these lobes must be side lobes rather than being slices from the main lobe.

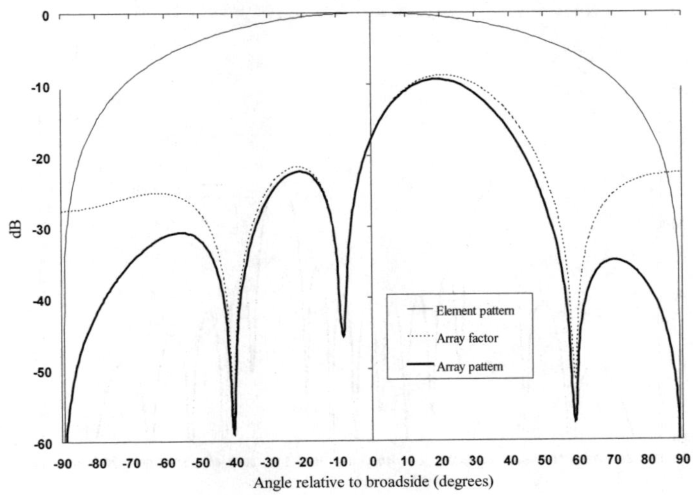

Fig. 9.24 Element radiation pattern, array factor and array radiation pattern for a square lattice 8 × 8 planar phased array antenna, phased for maximum radiation at $(\vartheta_0, \varphi_0) = (30°, 45°)$, in the plane $\varphi = 0°$ for $d_x = d_y = \frac{\lambda}{4}$.

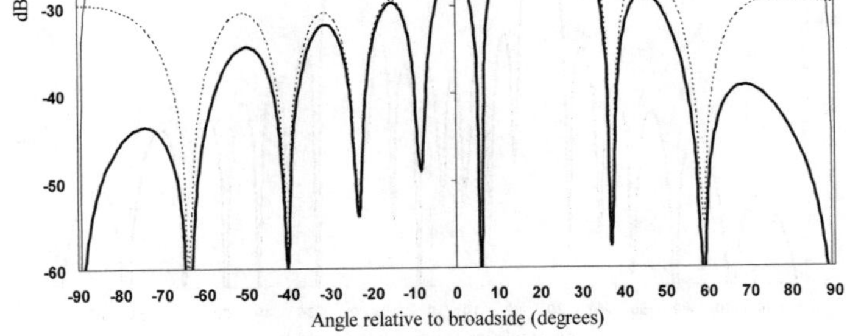

Fig. 9.25 Element radiation pattern, array factor and array radiation pattern for a square lattice 8 × 8 planar phased array antenna, phased for maximum radiation at $(\vartheta_0, \varphi_0) = (30°, 45°)$, in the plane $\varphi = 0°$ for $d_x = d_y = \frac{\lambda}{2}$.

To verify the correctness of the observations, we repeat the exercise for cuts at $\varphi = 30°$, i.e. azimuth planes closer to the azimuth direction the beam is steered to. The results are shown in figures 9.28 to 9.31.

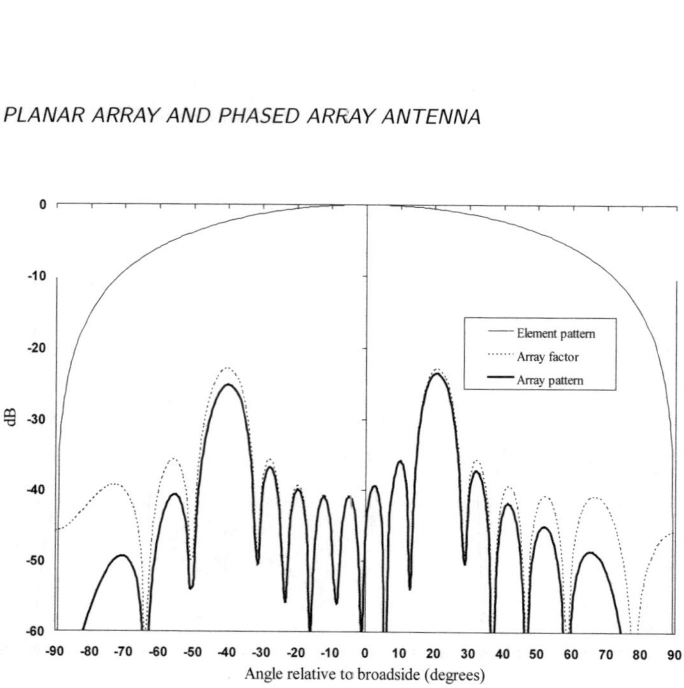

Fig. 9.26 Element radiation pattern, array factor and array radiation pattern for a square lattice 8 × 8 planar phased array antenna, phased for maximum radiation at $(\vartheta_0, \varphi_0) = (30°, 45°)$, in the plane $\varphi = 0°$ for $d_x = d_y = \lambda$.

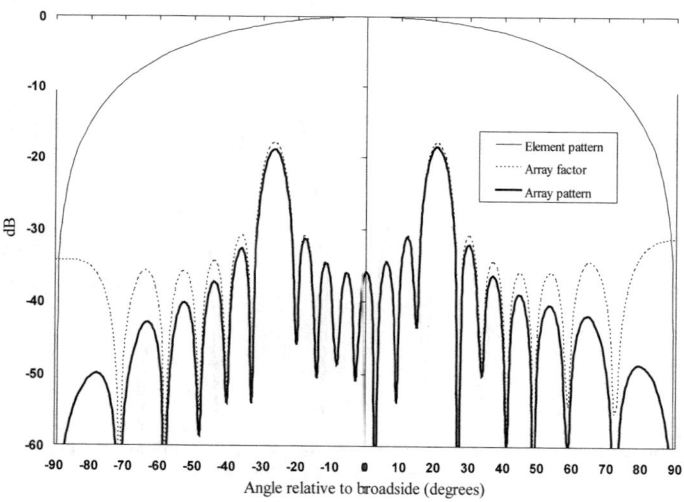

Fig. 9.27 Element radiation pattern, array factor and array radiation pattern for a square lattice 8 × 8 planar phased array antenna, phased for maximum radiation at $(\vartheta_0, \varphi_0) = (30°, 45°)$, in the plane $\varphi = 0°$ for $d_x = d_y = \frac{5\lambda}{4}$.

The cuts reveal that for element distances $d = \frac{\lambda}{4}$ and $d = \frac{\lambda}{2}$, we see slices of the main beam in the neigbourhood of $\vartheta = 30°$. In the $\varphi = 30°$ cuts for element distances

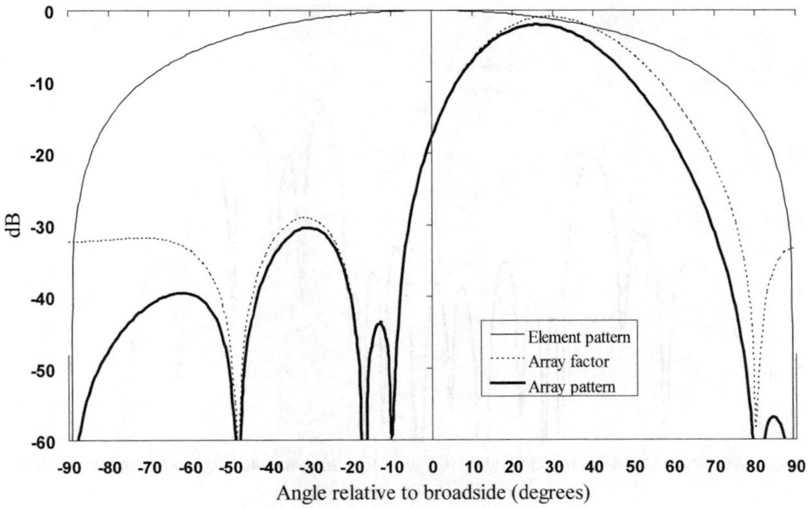

Fig. 9.28 Element radiation pattern, array factor and array radiation pattern for a square lattice 8 × 8 planar phased array antenna, phased for maximum radiation at $(\vartheta_0, \varphi_0) = (30°, 45°)$, in the plane $\varphi = 30°$ for $d_x = d_y = \frac{\lambda}{4}$.

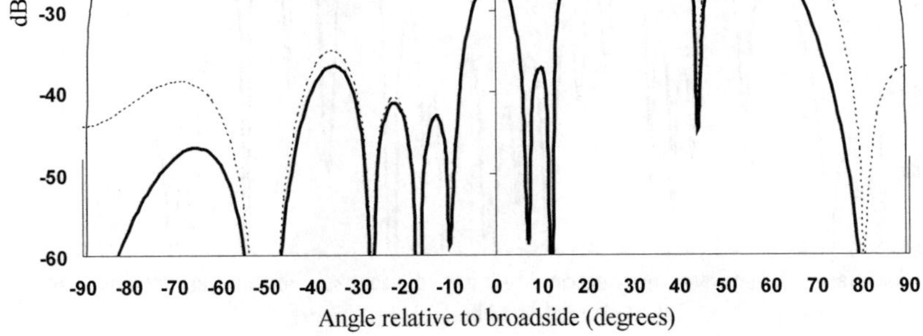

Fig. 9.29 Element radiation pattern, array factor and array radiation pattern for a square lattice 8 × 8 planar phased array antenna, phased for maximum radiation at $(\vartheta_0, \varphi_0) = (30°, 45°)$, in the plane $\varphi = 30°$ for $d_x = d_y = \frac{\lambda}{2}$.

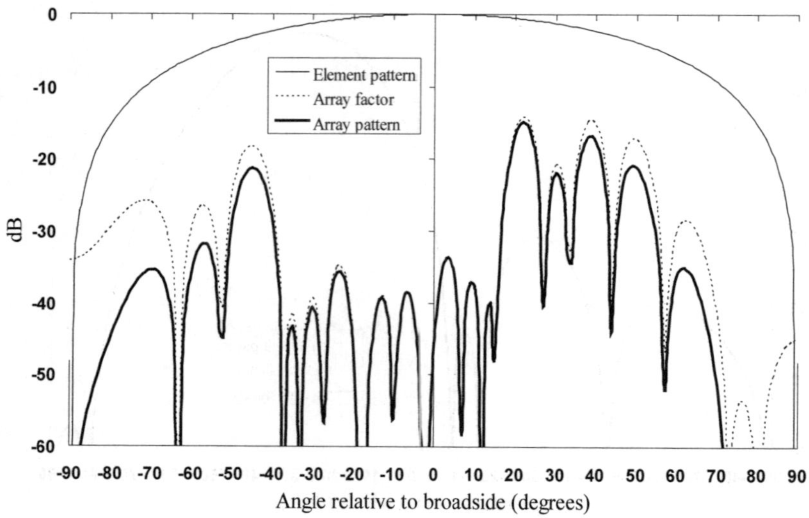

Fig. 9.30 Element radiation pattern, array factor and array radiation pattern for a square lattice 8 × 8 planar phased array antenna, phased for maximum radiation at $(\vartheta_0, \varphi_0) = (30°, 45°)$, in the plane $\varphi = 30°$ for $d_x = d_y = \lambda$.

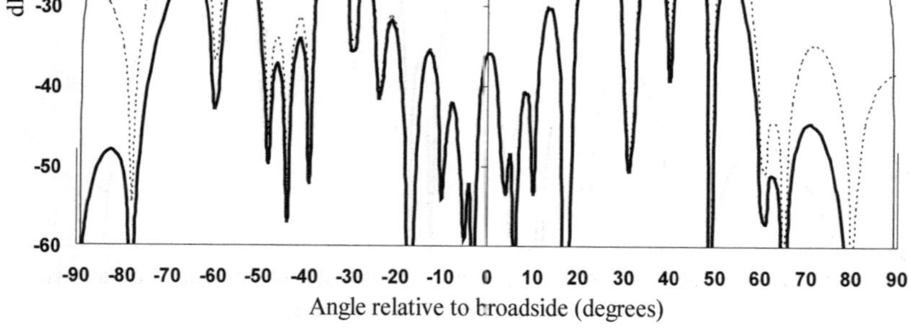

Fig. 9.31 Element radiation pattern, array factor and array radiation pattern for a square lattice 8 × 8 planar phased array antenna, phased for maximum radiation at $(\vartheta_0, \varphi_0) = (30°, 45°)$, in the plane $\varphi = 30°$ for $d_x = d_y = \frac{5\lambda}{4}$.

$d = \lambda$ and $d = \frac{5\lambda}{4}$ we are most probably looking at side lobes in the neighbourhood of $\vartheta = 30°$.

The best way to distinguish between main beam and side lobes is to look at a three-dimensional pattern instead of a couple of two-dimensional cuts. The reason that we are emphasising the use of two-dimensional cuts is twofold. First, relevant data may be read directly from these cuts, while it is difficult to read this data from a three-dimensional pattern. Second, the calculation of these cuts in general is less intensive than the calculation of a three-dimensional pattern, especially if numerical methods must be employed to account for mutual coupling effects. If the distribution of main lobe level and side lobes is important in a certain angular area, cuts in ϑ (as function of φ) and/or additional φ-cuts could be made.

Finally, we will show the patterns in the $\varphi = 45°$ plane. The results, for the different element distances, are shown in figures 9.32 to 9.35.

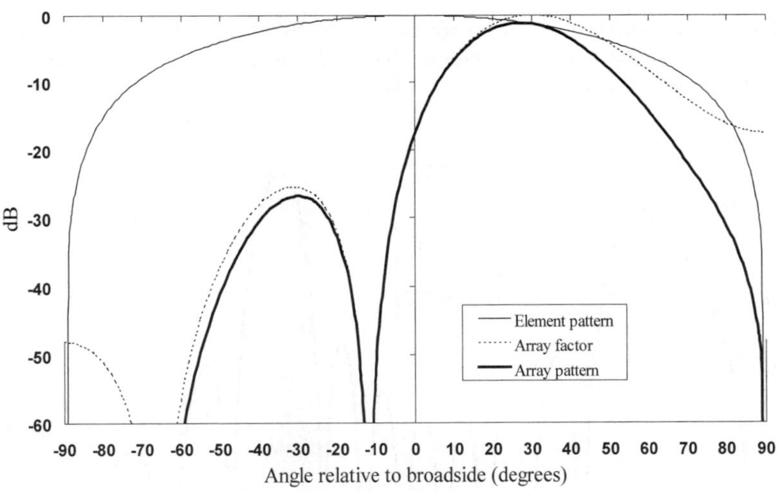

Fig. 9.32 Element radiation pattern, array factor and array radiation pattern for a square lattice 8 × 8 planar phased array antenna, phased for maximum radiation at $(\vartheta_0, \varphi_0) = (30°, 45°)$, in the plane $\varphi = 45°$ for $d_x = d_y = \frac{\lambda}{4}$.

We see that the beam is indeed scanned into the desired direction $(\vartheta_0, \varphi_0) = (30°, 45°)$.

As can be seen, especially in the diagonal plane, the grating lobes move with the main beam when this beam is scanned away from broadside ($\vartheta = 0°$). As seen in figures 9.34 and 9.35, the grating lobes, coming from invisible space, enter visible space. A better way to track the movement of the grating lobes is by plotting their positions in the grating lobe diagrams.

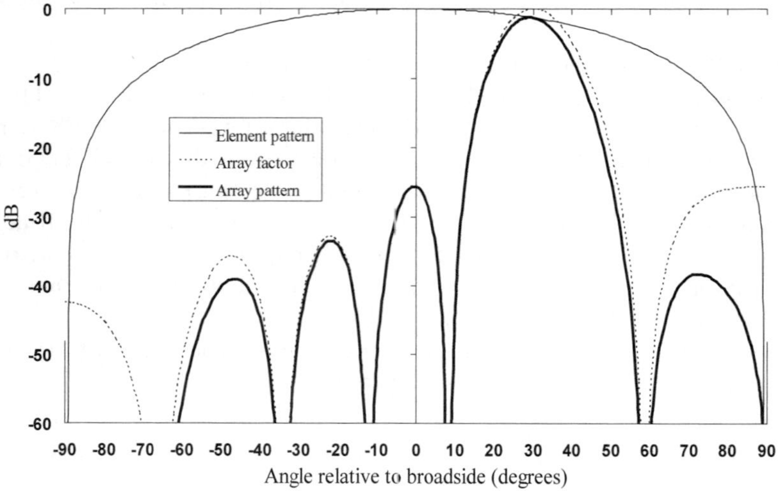

Fig. 9.33 Element radiation pattern, array factor and array radiation pattern for a square lattice 8 × 8 planar phased array antenna, phased for maximum radiation at $(\vartheta_0, \varphi_0) = (30°, 45°)$, in the plane $\varphi = 45°$ for $d_x = d_y = \frac{\lambda}{2}$.

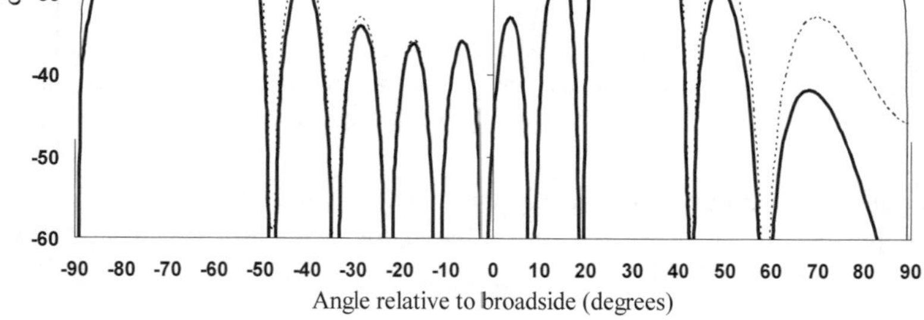

Fig. 9.34 Element radiation pattern, array factor and array radiation pattern for a square lattice 8 × 8 planar phased array antenna, phased for maximum radiation at $(\vartheta_0, \varphi_0) = (30°, 45°)$, in the plane $\varphi = 45°$ for $d_x = d_y = \lambda$.

9.3.2 Grating Lobes

Therefore we will make use of direction cosines as we did in the previous paragraph. Grating lobes are found whenever the follcwing condition applies:

$$\frac{d_x}{\lambda}(k-1)(u-u_0) + \frac{d_y}{\lambda}(l-1)(v-v_0) = m, \tag{9.23}$$

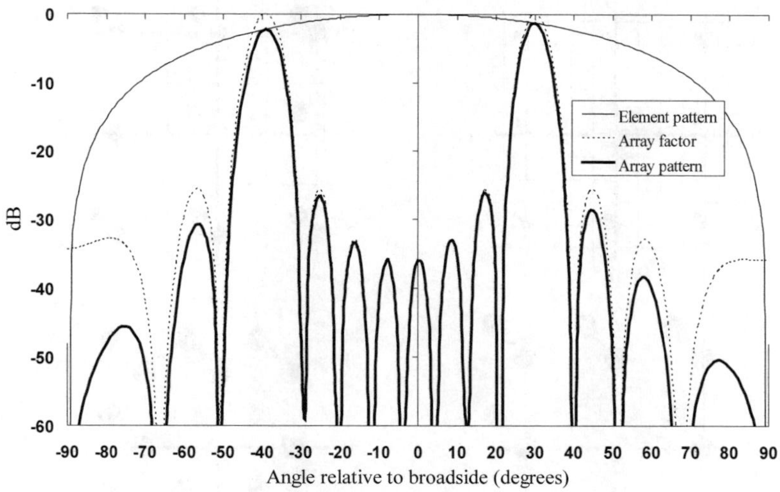

Fig. 9.35 Element radiation pattern, array factor and array radiation pattern for a square lattice 8 × 8 planar phased array antenna, phased for maximum radiation at $(\vartheta_0, \varphi_0) = (30°, 45°)$, in the plane $\varphi = 45°$ for $d_x = d_y = \frac{5\lambda}{4}$.

where m is an integer number and

$$u_0 = \sin(\vartheta_0)\cos(\varphi_0), \tag{9.24}$$

$$v_0 = \sin(\vartheta_0)\sin(\varphi_0). \tag{9.25}$$

This condition can only be met if, to start with,

$$u = p\frac{\lambda}{d_x} + u_0, \tag{9.26}$$

and

$$v = q\frac{\lambda}{d_y} + v_0, \tag{9.27}$$

where p and q are integer numbers.

If we compare these last two equations with equations (9.17) and (9.18) for the non-scanned array, we see that as a result of the applied linear phase taper, the main lobe and grating lobes are translated by $\Delta u = u_0$ and $\Delta v = v_0$.

In figure 9.36 we show for the four square grid phased array antennas, the grating lobe diagram before (non-scanned) and after scanning to $(\vartheta_0, \varphi_0) = (30°, 45°)$. The main lobe is indicated by a white dot, the grating lobes are indicated by black dots. Arrows indicate the translations due to scanning.

The angle φ is measured with respect to the positive u-axis, going into the direction of the positive v-axis. A $\varphi =$ *constant*-plane is shown in the u, v-plane as a straight line through the origin. Along this line, the coordinate is $\sin(\vartheta)$, see equations (9.24) and (9.25).

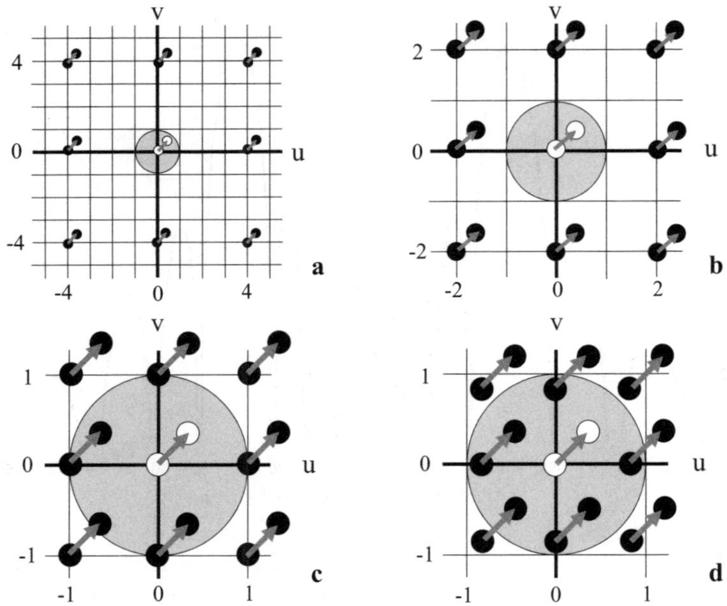

Fig. 9.36 Grating lobe diagrams for a square lattice planar phased array antenna, before and after being phased for maximum radiation at $(\vartheta_0, \varphi_0) = (30°, 45°)$. a. $d_x = d_y = \frac{\lambda}{4}$, b. $d_x = d_y = \frac{\lambda}{2}$, c. $d_x = d_y = \lambda$, d. $d_x = d_y = \frac{5\lambda}{4}$.

Figures 9.36c,d show - in the plane $\varphi = 45°$ - grating lobes entering visible space. This is in accordance with what we have seen in figures 9.34 and 9.35. The advantage of using the grating lobe diagram now also becomes evident. At a single glance we see when and *where* grating lobes will occur. The construction of a grating lobe diagram is much more efficient than plotting a great number of three-dimensional radiation patterns or an even greater number of two-dimensional cuts. After analysis of the grating lobe diagram, pattern cuts can be calculated in a more directed way, for example to analyse the side lobe behaviour in an angular region of interest.

To find the critical element distance where grating lobes (i.e. the maxima) are just entering visible space, it suffices to take into account beam steering in the principal planes only. As we have seen for the planar array antenna, in these directions grating lobes will occur first upon enlarging the element distance. If grating lobes do not occur in the principal planes, we know for sure that in all other planes grating lobes will not occur. Therefore, we will only look at the situations $\varphi = 0°$ and $\varphi = 90°$.

Analogous to the situation for the linear phased array antenna we find for the critical element distances

$$\frac{d_x}{\lambda} \leq \frac{1}{1 + |\sin(\vartheta_{0_{max}})|}, \tag{9.28}$$

and

$$\frac{d_y}{\lambda} \leq \frac{1}{1 + |\sin(\vartheta_{0_{max}})|}, \tag{9.29}$$

where $\vartheta_{0_{max}}$ is the maximum scan angle in which direction the beam may be pointed without the occurrence of grating lobes in visible space.

$$-90° \leq \vartheta_{0_{max}} \leq 90°. \tag{9.30}$$

For a rectangular lattice, phased array antenna, capable of steering the beam everywhere in visible space, the maximum inter-element distance - measured along the two orthogonal directions - is thus equal to half a wavelength.

$$\frac{d_x}{\lambda} \leq \frac{1}{2}, \tag{9.31}$$

and

$$\frac{d_y}{\lambda} \leq \frac{1}{2}. \tag{9.32}$$

In every cut φ, the beam broadening as discussed in the chapter on linear phased array antennas will occur - in that plane - when the beam is scanned away from broadside in that plane. Beam broadening with respect to the angle φ may - in first order - be neglected [1, 2].

REFERENCES

1. R.C. Hansen, *Phased Array Antennas*, John Wiley & Sons, New York, 1998.
2. N. Amitay, R.G. Pecina and C.P. Wu, Radiation Properties of Large Planar Arrays, *Monograph 5047*, Bell Telephone Laboratories, Inc., February 1965.

10

Special Array Antenna Configurations

Along with the linear and planar array and phased array antennas that we have discussed in detail, other array and phased array antenna configurations exist. Some of these will also be treated briefly in this chapter and will be designated *special array antenna configurations.* We use the adjective *special* since these configurations are not the most commonly encountered ones in practice. Of these special configurations we will discuss curved or *conformal* array and phased array antennas and volume array and phased array antennas that use a three-dimensional subspace for positioning the elements. Next to these non-planar array configurations we will also discuss planar array antennas and a way to create circular polarisation, using linearly polarised elements through a technique called *sequential rotation and phasing.* Finally we will discuss the technique of *reactive loading* to control the beam direction of an array antenna with only one element driven.

10.1 CONFORMAL ARRAY AND PHASED ARRAY ANTENNAS

A conformal (phased) array antenna may be defined as: 'a (phased) array antenna whose shape is dictated by the contours of a vehicle' [1]. Strictly speaking this definition implies that the linear and planar (phased) array antennas of the preceding chapters are antennas conformal to a plane. Most often though the term *conformal (phased) array antenna* is used in the sense of a one- or two-dimensionally *curved* (phased) array antenna. An example of a one-dimensionally curved (phased) array antenna is a cylindrical (phased) array antenna. Examples of a two-dimensionally curved (phased) array antenna are a spherical and a conical (phased) array antenna. Nearly all conformal array antennas share the property that the array antenna shape is not at the disposal of the antenna design engineer.

Array and Phased Array Antenna Basics Hubregt J. Visser

Conformal array antennas are required whenever an antenna must be located on a vehicle, e.g. the skin of an aircraft, missile or superstructure of a ship. Conforming the array antenna to the existing structure avoids compromising aerodynamic or stealth characteristics. This conforming to the existing structure occurs at the cost of an increased effort in designing the antenna. However, an increased field of view is often one of the benefits that comes with the employment of a conformal array antenna.

The increased design effort is caused by a number of significant differences in curved phased array antennas in comparison to planar phased array antennas [1, 2]: array factor and element pattern are no longer separable since the elements point in different directions. Not all the radiating elements contribute equally to the main beam, some radiating elements do not contribute to the main beam at all. The element patterns on a non point-symmetric surface will be different since they are embedded in different environments and may give cause to high side lobes. For a non point-symmetric surface, the polarisation vectors are in general - for an arbitrary pointing angle - not lined up, giving cause to a high cross-polarisation level.

The circular (ring) array antenna [2, 3] may be considered as one of the building blocks of a whole family of conformal array antennas. They may be stacked (each having its distinctive radius) to create a conformal array antenna on any body of revolution. In the following, a simplified analysis of a circular sector (incomplete circle) array and phased array antenna will be given, demonstrating some of the characteristics of conformal array and phased array antennas.

10.1.1 Circular Sector Array and Phased Array Antenna

To demonstrate some characteristics of conformal array and phased array antennas, we will start with considering a circular array antenna, consisting of a single row (ring) of elements, flush mounted on an electrically conducting circular cylinder, see figure 10.1.

The elements may be microstrip patch radiators or slot radiators. The element radiation pattern is considered to be cosine-like, having its maximum in the direction perpendicular to the patch or slot and thus perpendicular to the cylinder tangent. The radiators - being small with respect to the circle radius - may be thought of as being locally positioned in a plane as shown in the inset of figure 10.1. The radiation in endfire directions (and beyond) is therefore assumed to be zero in this simplified analysis.

We see in figure 10.1 that for an observation angle ϑ, half the number of elements do not radiate into that direction and that these elements are shielded by the electrically conducting cylinder. In reality, these elements can still contribute through the mechanism of mutual coupling, but in our simplified analysis we will assume mutual coupling effects to be negligible.

To further simplify our analysis, we will consider now a circular sector array, wherein the elements are positioned over one-quarter (90°) of the circumference, see figure 10.2. The far-field radiation will be evaluated over an angular range of 90° (±45° with respect to broadside, i.e. in the positive z-axis direction), where all elements still contribute to the far-field, see figure 10.1.

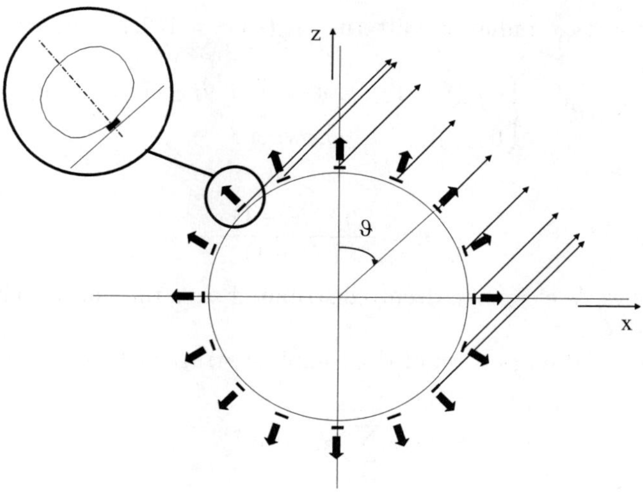

Fig. 10.1 Circular array antenna, flush mounted on an electrically conducting circular cylinder. The element radiation patterns are cosine-like.

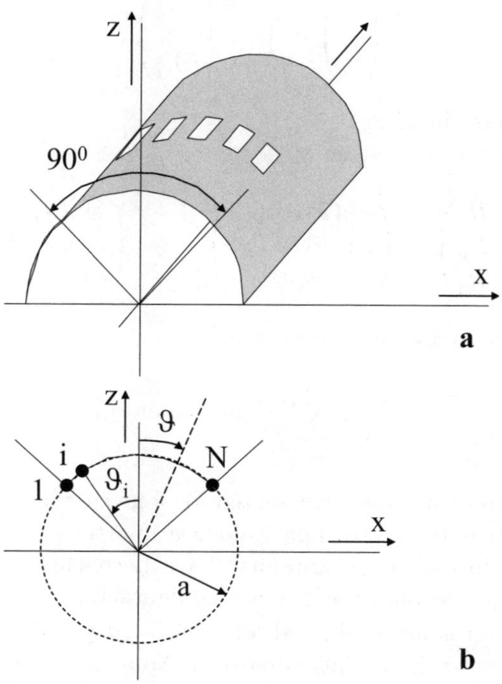

Fig. 10.2 Circular sector array antenna. a. Layout linear conformal array antenna. b. Element positions.

The element voltage radiation patterns, $S_{e_i}(\vartheta)$, $i = 1, 2, \ldots, N$, are given by

$$S_{e_i}(\vartheta) = \begin{cases} \cos(\vartheta - \vartheta_i) & \text{for } -\frac{\pi}{2} + \vartheta_i \le \vartheta \le \frac{\pi}{2} + \vartheta_i \\ 0 & \text{elsewhere} \end{cases}, \tag{10.1}$$

where

$$\vartheta_i = \frac{(2i - N - 1)\pi}{4(N-1)}, \tag{10.2}$$

assuming we have N elements, evenly distributed over the quarter circumference of the circular cylinder.

The voltage radiation pattern of the complete array, $S(\vartheta)$, is given by

$$S(\vartheta) = \sum_{i=1}^{N} S_{e_i}(\vartheta) e^{jk_0 \mathbf{r}_i \cdot \hat{\mathbf{R}}}, \tag{10.3}$$

where $k_0 = \frac{2\pi}{\lambda}$ is the free space wavenumber, $\mathbf{r}_i$ is the element position vector and $\hat{\mathbf{R}}$ is the unit vector in the direction of observation, ϑ.

The element position is given by

$$\begin{pmatrix} x_i \\ y_i \\ z_i \end{pmatrix} = \begin{pmatrix} a\sin(\vartheta_i) \\ 0 \\ a\cos(\vartheta_i) \end{pmatrix}, \tag{10.4}$$

where a is the radius of the circle.

The unit direction vector is given by

$$\begin{pmatrix} \hat{R}_x \\ \hat{R}_y \\ \hat{R}_z \end{pmatrix} = \left.\begin{pmatrix} \sin(\vartheta)\cos(\varphi) \\ \sin(\vartheta)\sin(\varphi) \\ \cos(\vartheta) \end{pmatrix}\right|_{\varphi=0} = \begin{pmatrix} \sin(\vartheta) \\ 0 \\ \cos(\vartheta) \end{pmatrix}, \tag{10.5}$$

so that we find for the array radiation pattern

$$S(\vartheta) = \sum_{i=1}^{N} S_{e_i} e^{jk_0 a \cos(\vartheta - \vartheta_i)}. \tag{10.6}$$

This radiation pattern has been calculated for four circular sector array antennas, having eight elements equally positioned along a quarter of the circumference. The radii of the four conformal array antennas have been chosen such that the array antenna lengths, projected onto the x-axis (see figure 10.2) are, respectively, $\frac{7\lambda}{4}$, $\frac{7\lambda}{2}$, 7λ and $\frac{35\lambda}{4}$. The patterns are evaluated for $-45^\circ \le \vartheta \le 45^\circ$, see figure 10.2, where all elements contribute to the radiation pattern. Note that it is not possible to apply the principle of pattern multiplication, since every element is differently oriented.

This particular choice of radii enables us to compare the radiation patterns of the curved array antennas - for the given angular range - with those of the non-curved linear array antennas - having the same number of elements and the same projected lengths - that have been shown in figures 4.4, 4.5, 4.6 and 4.7.

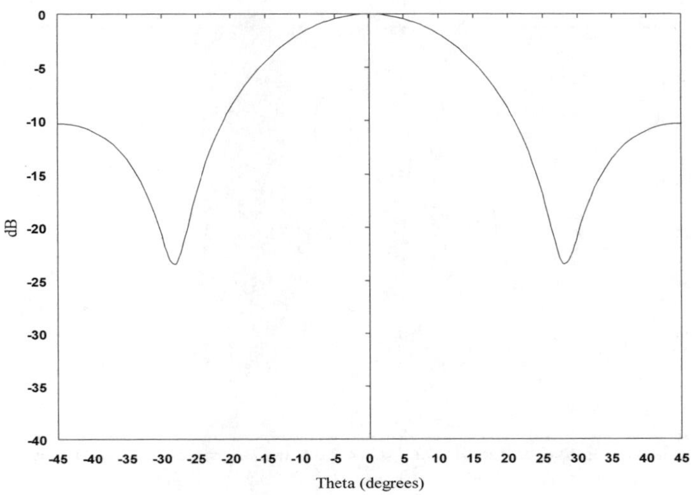

Fig. 10.3 Power radiation pattern of 8-element, 90° circular sector array antenna for $a = \frac{7\lambda}{4\sqrt{2}}$.

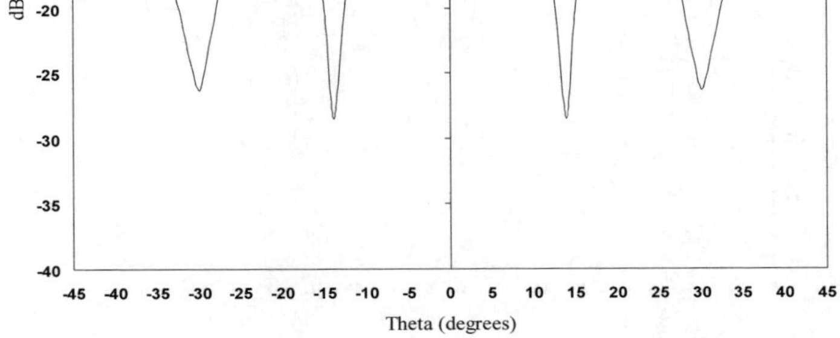

Fig. 10.4 Power radiation pattern of 8-element, 90° circular sector array antenna for $a = \frac{7\lambda}{2\sqrt{2}}$.

The conformal array antenna radiation patterns are shown in figures 10.3, 10.4, 10.5 and 10.6. The power radiation patterns are calculated as $20\log\left(\frac{|S(\vartheta)|}{8}\right)$.

We see that in comparison with the non-curved linear array antennas, the side lobe level has increased.

Next, we will scan the beam in the xz-plane. Thereto we apply a phase taper over the elements just as we did in the case of the non-conformal linear array antenna. We

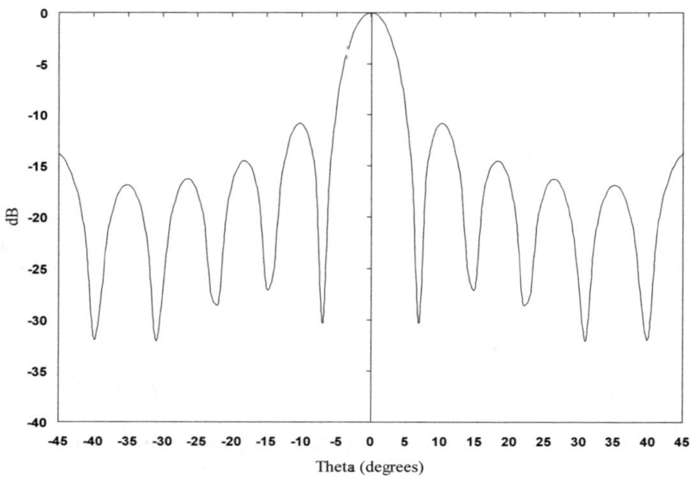

Fig. 10.5 Power radiation pattern of 8-element, 90° circular sector array antenna for $a = \frac{7\lambda}{\sqrt{2}}$.

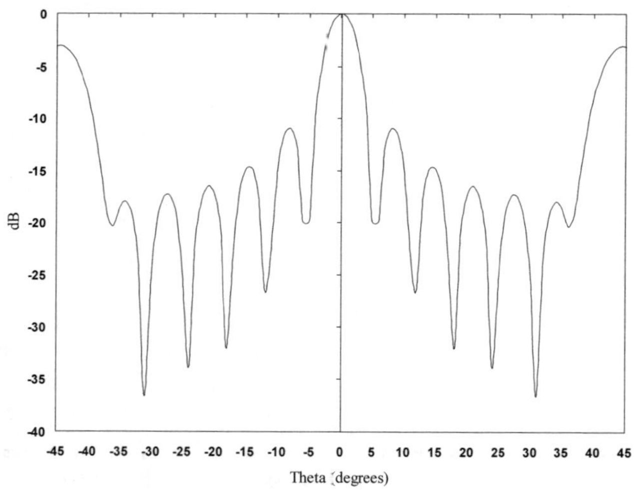

Fig. 10.6 Power radiation pattern of 8-element, 90° circular sector array antenna for $a = \frac{35\lambda}{4\sqrt{2}}$.

obtain for the conformal phased array antenna voltage radiation pattern

$$S(\vartheta) = \sum_{i=1}^{N} S_{e_i} e^{jk_0 a[\cos(\vartheta - \vartheta_i) - \cos(\vartheta_0 - \vartheta_i)]}, \tag{10.7}$$

where ϑ_0 is the desired beam-pointing direction.

For a beam scanned to the direction $\vartheta_0 = 30°$, the conformal array antenna power radiation patterns for the four different radii are shown in figures 10.7, 10.8, 10.9 and 10.10. The power radiation patterns are calculated as $20 \log\left(\frac{|S(\vartheta)|}{8}\right)$. These radiation patterns should be compared with those shown in, respectively, figures 7.2, 7.3, 7.4 and 7.5, for the scanned, non-conformal linear phased array antennas.

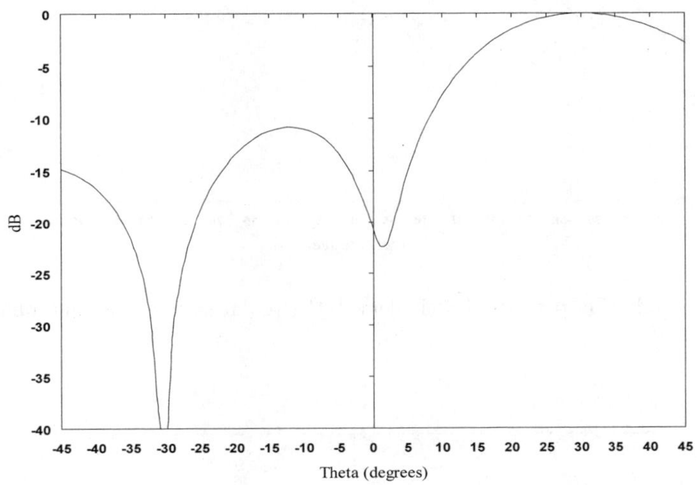

Fig. 10.7 Power radiation pattern of 8-element, 90° circular sector array antenna, scanned to $\vartheta_0 = 30°$ for $a = \frac{7\lambda}{4\sqrt{2}}$.

Again, we see an increase of the side lobe level and also a distortion of the radiation pattern.

In this and the previous example we do not see a benefit from using the conformal (phased) array antenna over a planar linear (phased) array antenna. But then we have to remember that a conformal antenna is mostly used when the form of the antenna is dictated by the carrier of the antenna. When viewed from this perspective, the conformal (phased) array antenna degradation - compared to its non-curved counterpart - seems to be acceptable in our example. When the radius of the circular cylinder is not too small, we may even approximate a circular cylindrical sector (phased) array antenna by a planar (phased) array antenna.

Situations may occur though where the use of a conformal (phased) array antenna will be beneficial over the use of a linear or planar (phased) array antenna.

It is not common to phase scan a cylindrical array antenna in the azimuth plane [2]. It is a more common practice to commutate an illuminated sector of a full ring around the array. This means that the unscanned array patterns are - through the process of switching elements on and off - rotated, see figure 10.11. Here we see a clear benefit of using a conformal array antenna. A full 360 scan is possible without mechanically rotating the antenna and with maintaining the broadside radiation pattern.

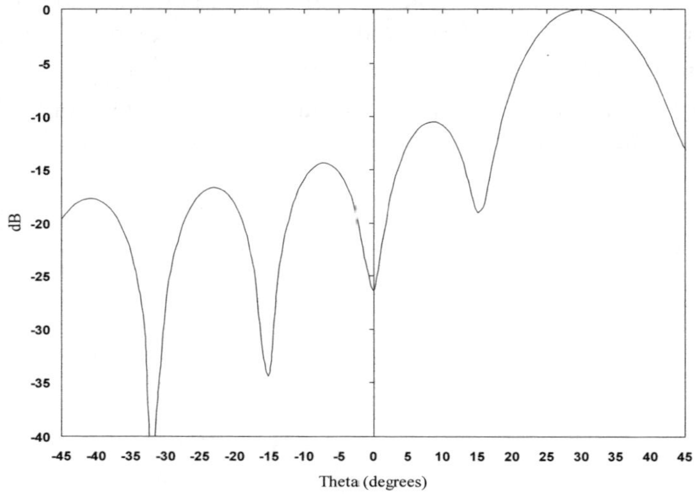

Fig. 10.8 Power radiation pattern of 8-element, 90° circular sector array antenna, scanned to $\vartheta_0 = 30°$ for $a = \frac{7\lambda}{2\sqrt{2}}$.

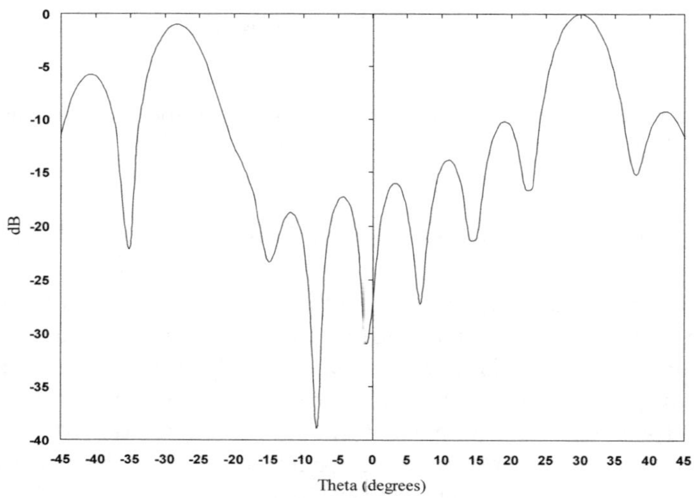

Fig. 10.9 Power radiation pattern of 8-element, 90° circular sector array antenna, scanned to $\vartheta_0 = 30°$ for $a = \frac{7\lambda}{\sqrt{2}}$.

10.2 VOLUME ARRAY AND PHASED ARRAY ANTENNAS

The concept of conformal array and phased array antennas may be extended beyond the surface to create *volume array and phased array antennas.*

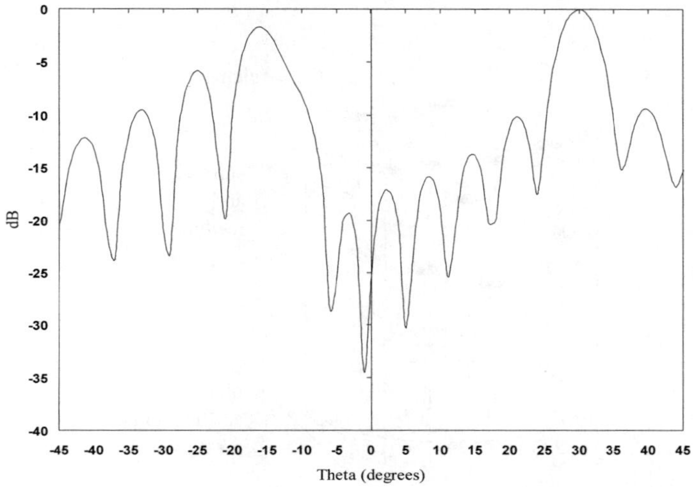

Fig. 10.10 Power radiation pattern of 8-element, 90° circular sector array antenna, scanned to $\vartheta_0 = 30°$ for $a = \frac{35\lambda}{4\sqrt{2}}$.

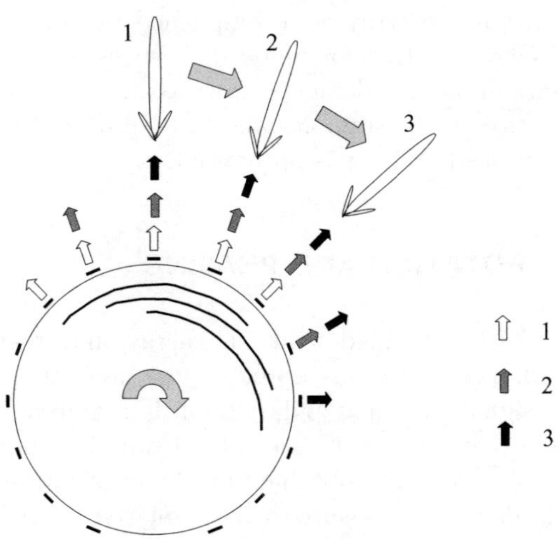

Fig. 10.11 Commutating. An illuminated region of the array is formed and switched around the array, thus providing a full 360° azimuth scan. 1. Illuminated elements to form beam 1, 2. Illuminated elements to form beam 2, 3. Illuminated elements to form beam 3.

In a volume (phased) array antenna, elements are randomly distributed within a spherical volume. The main benefit of such a (phased) array antenna is that it enables a full 360° azimuthal coverage where - as opposed to the situation for the cylindrical array antenna - all radiating elements are participating simultaneously. Furthermore, the projected aperture size will be equal in all scan directions [4].

The schematic of a volume array is shown in figure 10.12.

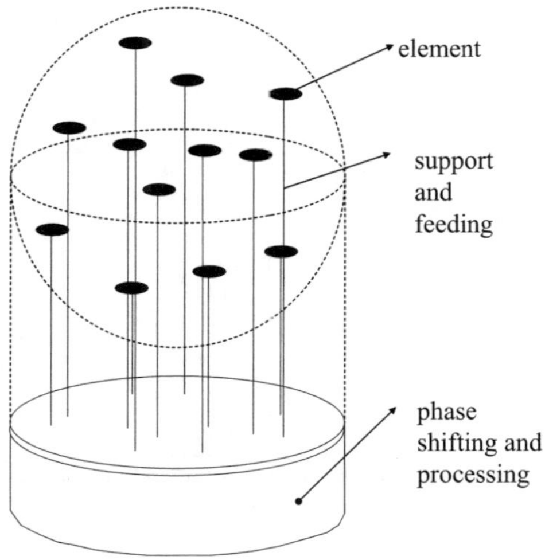

Fig. 10.12 Schematic of a volume array.

Beam scanning in a volume array is accomplished by connecting a phase shifter to every element and compensating for phase differences between the elements for a desired scan direction. The array elements are chosen such that the polarisation is orthogonal to the direction of the supports and feeding lines. In this way interference from these supports and feeding lines is minimised.

10.3 SEQUENTIAL ROTATION AND PHASING

So far, we have implicitly assumed that our array and phased array antennas were linearly polarised. For numerous applications, however, e.g. the reception of satellite broadcasting signals, circular polarisation is required. Although single-feed circularly polarised elements do exist, like - for example - perturbated microstrip patch antennas, the best results are obtained by the application of two separate and spatially orthogonal feeds that are excited with a relative phase difference of 90°.

The implementation of dual-feed elements in an array or phased array antenna will put high constraints on the feeding network. Housing the required 'real estate' can become a serious problem. The high constraints may be relaxed a bit by maintaining the spatially orthogonal, 90° out of phase feeds, but using separate elements instead of having both feeds at one element. In this way we may use linearly polarised elements, in a *sequentially rotated and phased* fashion to create circular polarisation [6, 7].

To explain the principle of sequential rotation and phasing, let's start with two linearly polarised elements, placed apart a distance d on the x-axis of a Cartesian coordinate system, see figure 10.13.

The polarisations are indicated in the figure by arrows. For a broad range of angles ϑ relative to broadsight in the yz-plane, the polarisation is circular, see figure 10.13.

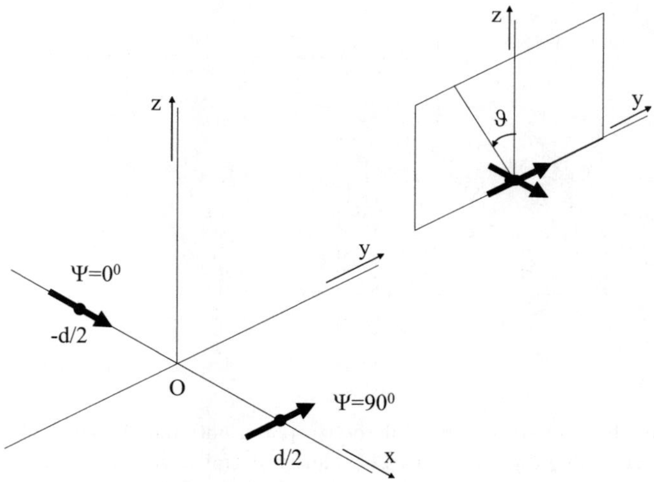

Fig. 10.13 Phased array antenna consisting of two linearly polarised elements, phased for a 90° phase difference. In the yz-plane a circularly polarised element is created by virtue of perpendicular projection.

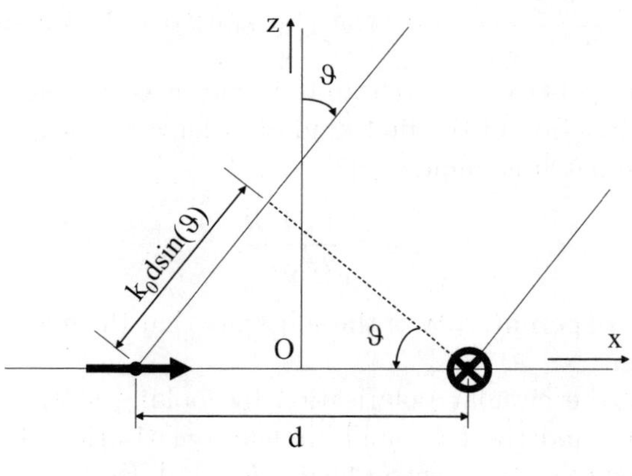

Fig. 10.14 Additional spatial phase delay in the xz-plane of a phased array antenna consisting of two linearly polarised elements, phased for a 90° phase difference as shown in figure 10.13.

In the xz-plane, however, the phase difference starts to deviate from 90° for angles ϑ increasing from broadside. This deviation is due to the additional spatial phase delay, $\Delta\psi = k_0 d \sin(\vartheta)$, as shown in figure 10.14 [7].

To demonstrate the effect of sequential rotation and phasing, we will from here on work with linearly polarised, circular microstrip patch antenna elements, see figure 10.15.

We choose for this element, since the electric far-field components, $\mathbf{E}_\vartheta = E_\vartheta \mathbf{u}_\vartheta$ and $\mathbf{E}_\varphi = E_\varphi \mathbf{u}_\varphi$ are very easily described, as is shown by the equations for the dominant

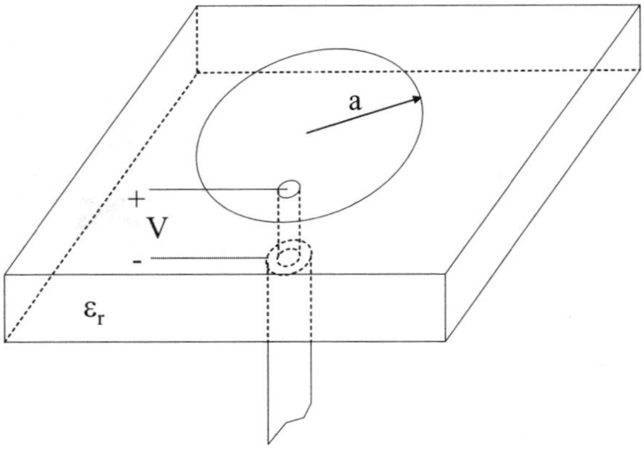

Fig. 10.15 Coaxially excited, circular microstrip patch antenna element. The patch of radius a is placed on a grounded dielectric sheet of relative permittivity ε_r.

TM_{11}-mode [5]

$$E_\varphi = j\frac{Vak_0}{2}\frac{e^{-jk_0 r}}{r}\cos(\vartheta)\sin(\varphi)\left[J_2\left(k_0 a\sin(\vartheta)\right) + J_0\left(k_0 a\sin(\vartheta)\right)\right], \tag{10.8}$$

$$E_\vartheta = j\frac{Vak_0}{2}\frac{e^{-jk_0 r}}{r}\cos(\varphi)\left[J_2\left(k_0 a\sin(\vartheta)\right) - J_0\left(k_0 a\sin(\vartheta)\right)\right], \tag{10.9}$$

where V is the voltage between patch and ground, a is the radius of the patch and $J_n(x)$ is the Bessel function of the first kind of order n and argument x.

The radius of the patch is found as, [5]

$$a = \frac{1.841\lambda}{2\pi\sqrt{\varepsilon_r}}, \tag{10.10}$$

where ε_r is the relative permittivity of the substrate that the microstrip patch radiator is photo etched from.

We will evaluate the circular polarisation by looking at the right hand circular polarisation (RHCP) and the left hand circular polarisation (LHCP). The RHCP- and LHCP-components of the electric field - E_R and E_L respectively - follow from the ϑ and φ components by

$$E_R = \frac{1}{\sqrt{2}}\left(E_\vartheta + jE_\varphi\right), \tag{10.11}$$

and

$$E_L = \frac{1}{\sqrt{2}}\left(E_\vartheta - jE_\varphi\right). \tag{10.12}$$

The dominant part of both polarisations is called the *co-polarisation*, the other component is called the *cross-polarisation*. The purity of the polarisation is best expressed by means of the *axial ratio* (AR)

$$AR = \left|\frac{|E_L| + |E_R|}{|E_L| - |E_R|}\right|. \tag{10.13}$$

For a pure LHCP or pure RHCP wave, the axial ratio is equal to one or, on a logarithmic scale, 0dB.

For the two-element array of figure 10.13, taking circular microstrip patch radiators for the elements and choosing the element distance $d = 0.8\lambda_0$, where λ_0 is the free space wavelength, the principal plane radiation patterns are calculated and plotted, together with the axial ratio as a function of angle ϑ in figure 10.16. The frequency of operation is f=1.55GHz and the relative permittivity of the substrate is $\varepsilon_r = 2.33$. These values are used to determine the radius of the circular patch element.

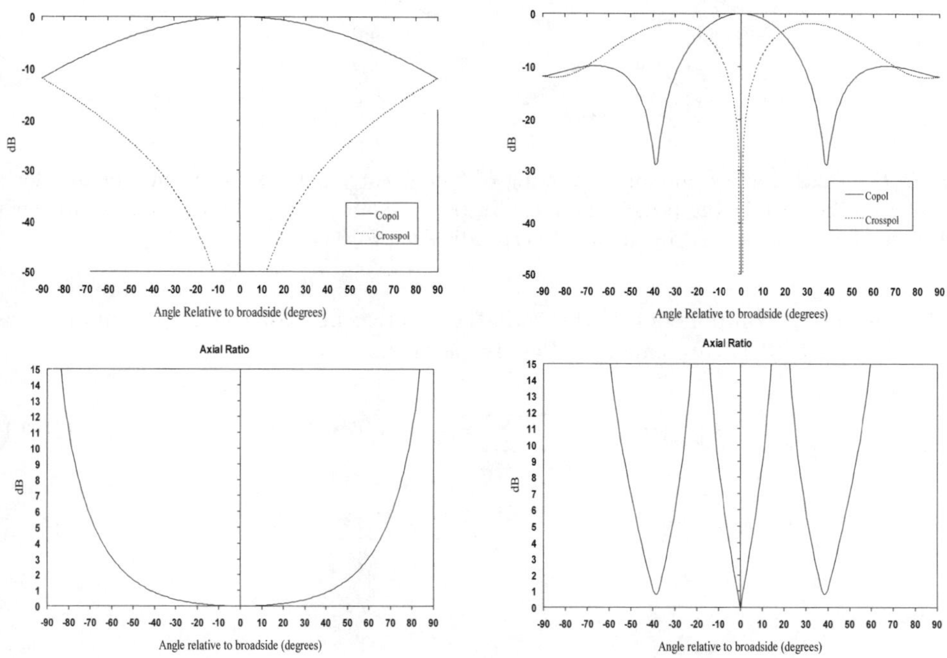

a **b**

Fig. 10.16 Co- and cross-polarised power radiation patterns and axial ratio in the principal planes for a two-element, sequentially rotated and phased, array antenna of circular, linearly polarised microstrip patch elements, displaced $0.8\lambda_0$ (see figure 10.13). a. Plane $\varphi = 90°$ (yz-plane). b. Plane $\varphi = 0°$ (xz-plane).

The degradation in circular polarisation off broadside in the plane containing the two radiating elements (xz-plane) can be eliminated by expanding the linear array into a planar array as shown in figure 10.17 [6, 7].

Now, in both principal planes we have - by virtue of orthogonal projection - created a circularly polarised element.

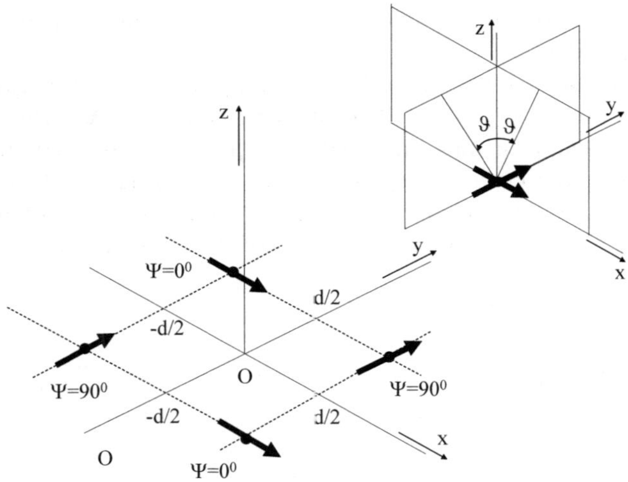

Fig. 10.17 Phased array antenna consisting of four linearly polarised elements, phased for a 90° phase difference in the principal planes. In the xz-plane and in the yz-plane a circularly polarised element is created by virtue of perpendicular projection.

The ϑ and φ components of the radiated electric field of the 2 × 2 subarray as shown in figure 10.17, are calculated as, respectively

$$E_{\vartheta_{sub}}(\vartheta,\varphi) = \sum_{p=0}^{1}\sum_{q=0}^{1} E_{\vartheta_{pq}} e^{j\frac{2\pi}{\lambda}(pd_x u + qd_y v)}, \tag{10.14}$$

$$E_{\varphi_{sub}}(\vartheta,\varphi) = \sum_{p=0}^{1}\sum_{q=0}^{1} E_{\varphi_{pq}} e^{j\frac{2\pi}{\lambda}(pd_x u + qd_y v)}, \tag{10.15}$$

where

$$u = \sin(\vartheta)\cos(\varphi), \tag{10.16}$$

$$v = \sin(\vartheta)\sin(\varphi), \tag{10.17}$$

and

$$E_{\vartheta_{00}} = E_{\vartheta_{11}} = -\cos(\varphi)F_1(\vartheta), \tag{10.18}$$

$$E_{\varphi_{00}} = E_{\varphi_{11}} = \cos(\vartheta)\sin(\varphi)F_2(\vartheta), \tag{10.19}$$

$$E_{\vartheta_{01}} = E_{\vartheta_{10}} = -j\sin(\varphi)F_1(\vartheta), \tag{10.20}$$

$$E_{\varphi_{01}} = E_{\varphi_{10}} = -j\cos(\vartheta)\cos(\varphi)F_2(\vartheta). \tag{10.21}$$

Herein,

$$F_1(\vartheta) = J_0(k_0 a\sin(\vartheta)) - J_2(k_0 a\sin(\vartheta)), \tag{10.22}$$

and

$$F_2(\vartheta) = J_0(k_0 a\sin(\vartheta)) + J_2(k_0 a\sin(\vartheta)). \tag{10.23}$$

Furthermore, we have allowed for different spacings in the x- and y-directions and we have placed the origin of the coordinate system on one of the subarray elements, see figure 10.18.

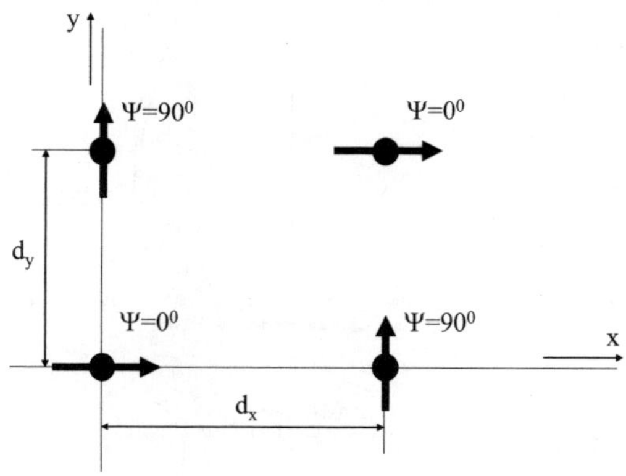

Fig. 10.18 2×2 sequentially rotated subarray in coordinate system.

The quality of the circular polarisation is now worst in the diagonal plane, as shown in figure 10.19.

When we compare the axial ratio in the diagonal plane of this 2×2 subarray with the axial ratio in the plane containing the elements of the 2×1 array (figure 10.16b), we see a slight improvement of the axial ratio for angles ϑ off broadsight.

This improvement may be explained by the smaller element distance in the projected linear array, see figure 10.20.

Larger arrays may be constructed using the 2×2 sequentially rotated and phased subarray as a basic building block, see figure 10.21.

For the array antenna as shown in figure 10.21, the ϑ and φ components of the radiated electric field are calculated as

$$E_\vartheta(\vartheta, \varphi) = E_{\vartheta_{sub}}(\vartheta, \varphi) \sum_{m=0}^{\frac{M-2}{2}} \sum_{n=0}^{\frac{N-2}{2}} e^{j\frac{4\pi}{\lambda}(md_x u + nd_y v)}, \tag{10.24}$$

$$E_\varphi(\vartheta, \varphi) = E_{\varphi_{sub}}(\vartheta, \varphi) \sum_{m=0}^{\frac{M-2}{2}} \sum_{n=0}^{\frac{N-2}{2}} e^{j\frac{4\pi}{\lambda}(md_x u + nd_y v)}, \tag{10.25}$$

where M is the number of elements in the x-direction and N is the number of elements in the y-direction.

For a square array consisting of 2×2, sequentially rotated and phased 2×2 subarrays, with inter-element distances of $0.8\lambda_0$ and inter-subarray distances of $1.6\lambda_0$, the radiation patterns and axial ratio in the diagonal plane are shown in figure 10.22.

We see a smaller beam due to the fact that the array antenna has become larger. The cross-polarisation as a function of the angle relative to broadside has not changed, as can be seen from the axial ratio as a function of the angle relative to broadside.

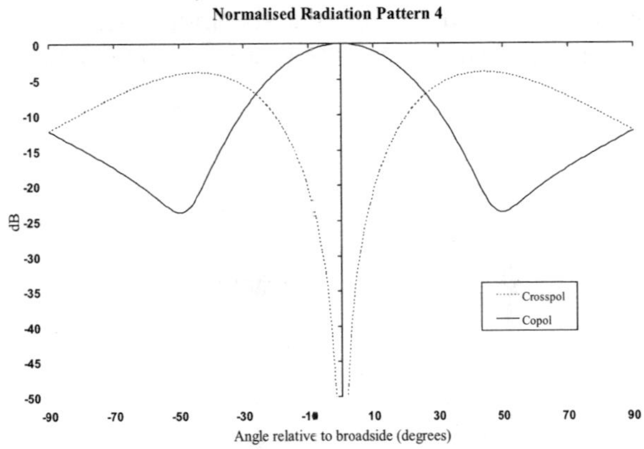

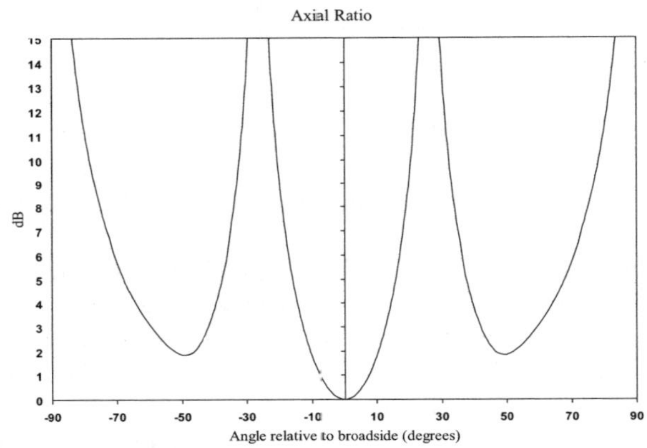

Fig. 10.19 Co- and cross-polarised power radiation patterns and axial ratio in the diagonal plane for a four-element, sequentially rotated and phased, array antenna consisting of circular, linearly polarised microstrip patch elements, displaced $0.8\lambda_0$ (see figure 10.17).

Since the main lobe has become smaller, the need for a low axial ratio over a *broad* angular range no longer exists.

If, however, we do need to reduce the axial ratio outside the principal planes, we may accomplish this by reducing the element spacing as shown in figure 10.23 which shows a polar plot of the axial ratio of a square array antenna for different element spacings. The array antenna consists of sequentially rotated and phased, circular, linearly polarised microstrip patch elements, resonant at 1.55GHz and evaluated at an angle 20° off broadside.

The figure clearly shows that the polarisation is purely circular in both principal planes and worst in or near the diagonal planes. Upon decreasing the element distances, the axial ratio outside the principal planes improves.

Normally, placing array antenna elements closer together will increase the mutual coupling effects. Since the elements are linearly polarised and orthogonally oriented

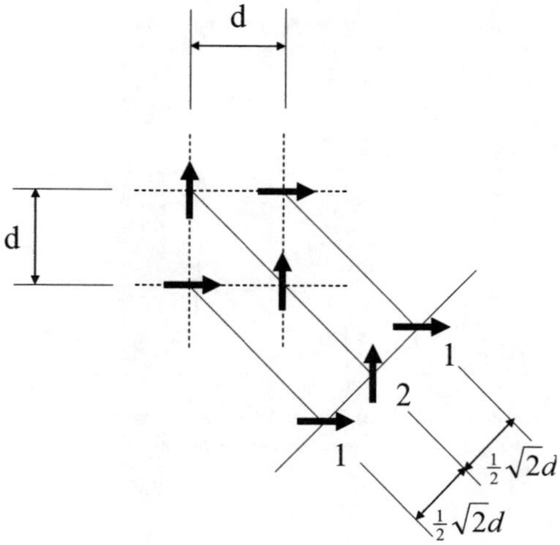

Fig. 10.20 Projected linear array with amplitude weighting in the diagonal plane of a four-element, sequentially rotated, array antenna of linearly polarised elements.

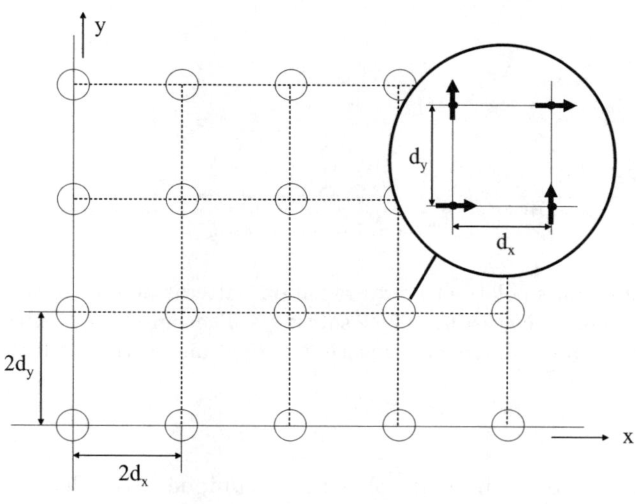

Fig. 10.21 Array antenna consisting of 2 × 2, sequentially rotated and phased subarrays.

towards each other, mutual coupling is already low. However, when using microstrip patch antenna elements and employing them in a *complete* sequential rotation fashion instead of the until now used *limited* sequential rotation fashion, the mutual coupling effects due to higher order modes will be suppressed.

In the *complete* sequential rotation fashion [6, 7], the elements and phasings of one subarray will undergo sequential rotations and phasings equal to 360° divided by the number of elements, see figure 10.24.

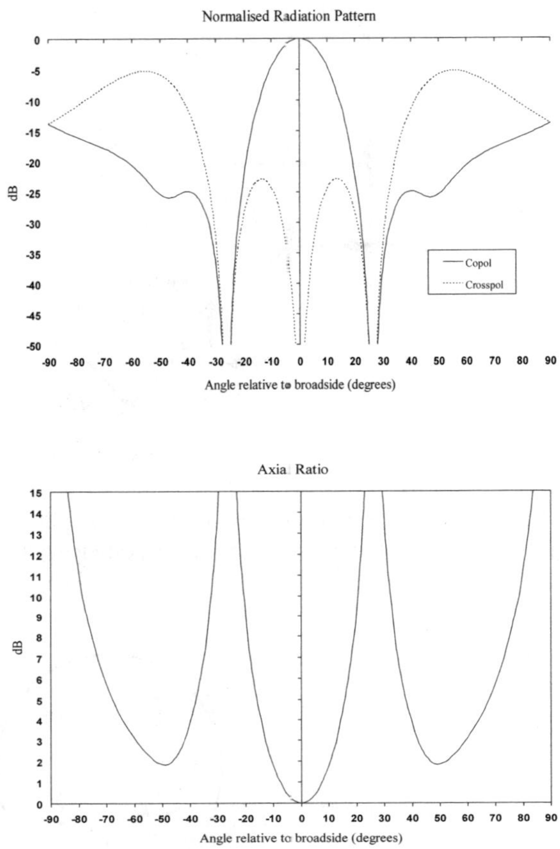

Fig. 10.22 Co- and cross-polarised power radiation patterns and axial ratio in the diagonal plane for an array antenna consisting of 2×2 subarrays. The four-element subarrays consist each of 2×2, sequentially rotated, circular, linearly polarised microstrip patch elements, displaced $0.8\lambda_0$.

The sequential rotation and phasing technique also leads to bandwidth improvement [7], due to mismatch cancellation in the subarray feeding network [8]. Of all the possible subarray configurations, the three-element subarray (as shown in figure 10.24a) will lead to the largest frequency bandwidth [8].

10.4 REACTIVE LOADING

We have already mentioned that the feeding network of a (phased) array antenna - due to its size and complexity - will contribute considerably to the costs of the complete (phased) array antenna. In the chapter on endfire array antennas we have already seen an alternative feeding structure that is simpler and thus less expensive

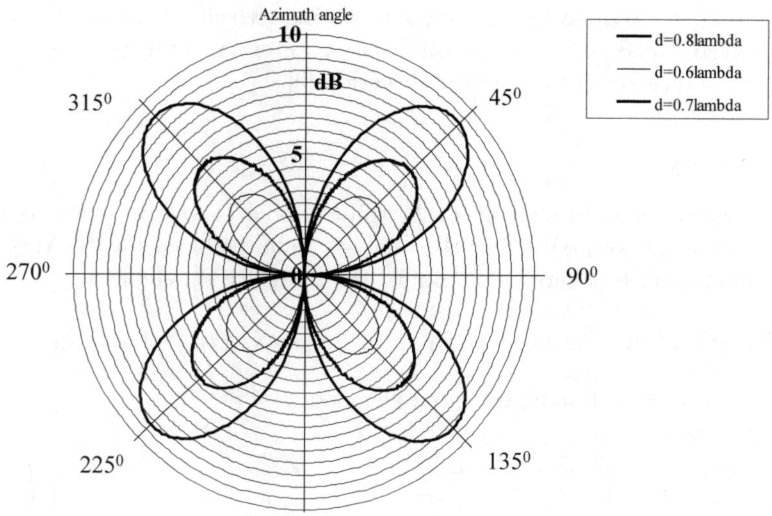

Fig. 10.23 Axial ratio as function of azimuth angle for elevation angle $\vartheta = 20°$ for a square array of sequentially rotated and phased elements for different element spacings.

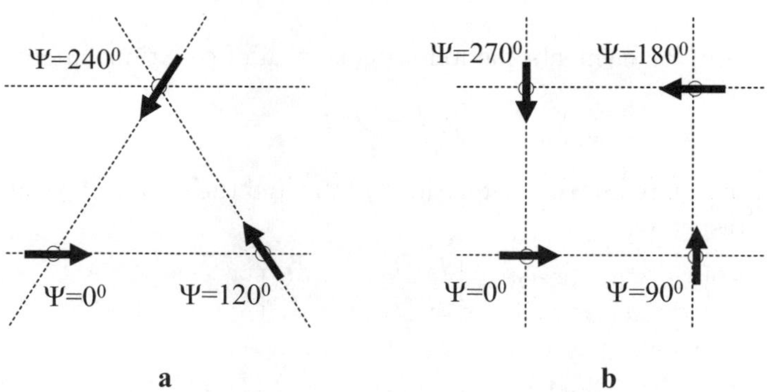

Fig. 10.24 Examples of complete sequential rotation and phasing. a. Triangular lattice subarray. b. Rectangular lattice subarray.

than the series or corporate feeding structure. In a Yagi–Uda array antenna, only one element is driven, all other elements act as parasitics. The driver in a Yagi–Uda array antenna is a (half-wave) dipole radiator, the parasitic elements are short-circuited dipole radiators.

The idea of having one (or only a few) array antenna element(s) driven and the other ones acting as parasitic elements may be taken one step further by not necessarily having the parasitic elements short circuited. Instead, the parasitic elements may be reactively (capacitively or inductively) loaded. By choosing the correct reactive loading of the parasitic array antenna elements, the array antenna beam may be directed into a desired direction [9].

10.4.1 Theory

For an array antenna consisting of K thin[1] dipole elements, where only the first element is driven,[2] we may follow the analysis we introduced for the Yagi–Uda array antenna in chapter 6. There we found for the first, driven, element

$$V_1 = Z_{11}I_1 + Z_{12}I_2 + \cdots + Z_{1i}I_i + \cdots + Z_{1,K-1}I_{K-1} + Z_{1K}I_K, \tag{10.26}$$

and for the next $K-1$ short circuited elements

$$\begin{pmatrix} -Z_{21} \\ -Z_{31} \\ \vdots \\ -Z_{i1} \\ \vdots \\ -Z_{K-1,1} \\ -Z_{K1} \end{pmatrix} = \begin{pmatrix} Z_{22} & Z_{23} & \dots & Z_{2i} & \dots & Z_{2K} \\ Z_{32} & Z_{33} & \dots & Z_{3i} & \dots & Z_{3K} \\ \vdots & \vdots & \vdots & \vdots & \vdots & \vdots \\ Z_{i2} & Z_{i3} & \dots & Z_{ii} & \dots & Z_{iK} \\ \vdots & \vdots & \vdots & \vdots & \vdots & \vdots \\ Z_{K-1,2} & Z_{K-1,3} & \dots & Z_{K-1,i} & \dots & Z_{K-1,K} \\ Z_{K2} & Z_{K3} & \dots & Z_{Ki} & \dots & Z_{KK} \end{pmatrix} \begin{pmatrix} \frac{I_2}{I_1} \\ \frac{I_3}{I_1} \\ \vdots \\ \frac{I_i}{I_1} \\ \vdots \\ \frac{I_{K-1}}{I_1} \\ \frac{I_K}{I_1} \end{pmatrix}. \tag{10.27}$$

For the non-driven elements the following system of equations apply

$$[V] = [Z][I], \tag{10.28}$$

where $[Z]$ is the square matrix in equation (10.27) and the $[I]$ and $[V]$ column matrices are given by, respectively

$$[I] = \left[\frac{I_2}{I_1}, \frac{I_3}{I_1}, \dots, \frac{I_K}{I_1}\right]^T, \tag{10.29}$$

and

$$[V] = [-Z_{21}, -Z_{31}, \dots, -Z_{K1}]^T. \tag{10.30}$$

Since the non-driven elements will now be reactively loaded, we have to add a series impedance to every element. The system of equations therefore changes into

$$[V] = ([Z] + [Z_L])\,[I], \tag{10.31}$$

[1] For thin dipoles we may - in a first-order approximation - assume the currents on the dipoles to be sinusoidal.

[2] We have arranged the numbering of the elements such that element no. 1 is the driven element. This does *not* have consequences for the position of this element in the array antenna.

where $[Z_L]$ is a diagonal matrix containing the element load impedances jX_k, for $k = 2, 3, \ldots, K$

$$[Z_L] = \begin{pmatrix} jX_2 & 0 & 0 & \ldots & 0 \\ 0 & jX_3 & 0 & \ldots & 0 \\ 0 & 0 & jX_4 & \ldots & 0 \\ \vdots & \vdots & \vdots & \ldots & \vdots \\ 0 & 0 & 0 & \ldots & jX_K \end{pmatrix}. \tag{10.32}$$

Once the correct impedances jX_k are known, the currents on the non-driven dipole elements, relative to the current on the driven element (I_1), may be found from

$$[I] = ([Z] + [Z_L])^{-1} [V]. \tag{10.33}$$

10.4.2 Circular Dipole Array Antenna

As an example of the preceding, we will analyse a seven-element circular array antenna of one driven and six reactively loaded dipole elements as discussed in [9].

The configuration is shown in figure 10.25.

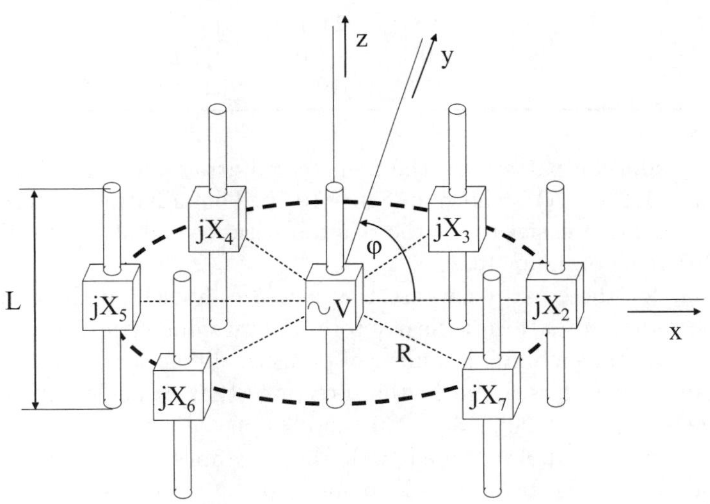

Fig. 10.25 Seven-element circular array antenna of one driven and six reactively loaded dipole elements. The driven element is positioned in the centre of a circle. The six reactively loaded elements are equally displaced over a circle with radius R. All element lengths are equal to $L = \frac{\lambda}{2}$.

The element positions (x_k, y_k), for $k = 1, 2, \ldots, K$ ($K = 7$) are given by

$$(x_k, y_k) = \begin{cases} (0, 0) & k = 1 \\ \left(R\cos\left[\dfrac{(k-2)2\pi}{K-1}\right], R\sin\left[\dfrac{(k-2)2\pi}{K-1}\right] \right) & k > 1 \end{cases} \tag{10.34}$$

The inter-element distances d_{kl}, for $k, l = 1, 2, \ldots, K$ ($K = 7$) are

$$d_{kl} = \begin{cases} 0 & k = l \\ R & k = 1 \vee l = 1 \\ \sqrt{(x_k - x_l)^2 + (y_k - y_l)^2} & k \neq 1 \wedge l \neq 1 \wedge k \neq l \end{cases} \tag{10.35}$$

By optimisation [9], the reactive loads are found for steering the beam of a seven-elements circular array, see figure 10.25, into the directions $\varphi = 0°$, $\varphi = 10°$, $\varphi = 20°$ and $\varphi = 30°$. We will use these results of Harrington [9] that are stated in table 10.1 for a radius and element spacings $R = d_{23} = d_{34} = d_{45} = d_{56} = d_{67} = \frac{\lambda}{4}$.

Table 10.1 Reactive loads (Ω), [9], for beam-steering a 7-elements, circular, reactively loaded dipole array antenna.

element number	$\varphi = 0°$	$\varphi = 10°$	$\varphi = 20°$	$\varphi = 30°$
2	–j63.6	–j64.3	–j64.7	–j57.4
3	–j94.6	–j85.2	–j71.0	–j57.4
4	j9.0	j34.8	180.0	–j398.1
5	j10.8	j11.4	j9.7	j1.7
6	j9.0	–j1.0	–j4.1	j1.7
7	–j94.6	–j105.6	–j.134.1	–j401.0

The power radiation pattern for the non-steered array antenna (all reactive loads $X_k = 0$, for $k = 1, 2, \ldots, 7$) is shown in figure 10.26. The radiation pattern is shown in a polar plot, making it easier to see the relation between radiation and the geometry of the circular array of figure 10.25.

We see that for the situation where the parasitic dipoles on the circle are short circuited, the radiation pattern is nearly rotationally symmetric. The small ripple is caused by the fact that a limited number of parasitic dipoles are placed on the circle. The maximum radiation is found in the directions that coincide with the parasitic dipole positions (0°, 60°, 120°, 180°, 240° and 300°).

With the parasitic dipoles loaded with the reactances as stated in the second column of table 10.1, the beam should point into the direction $\varphi = 0°$. The power radiation pattern is shown in figure 10.27.

The radiation pattern reveals that the beam is indeed steered to $\varphi = 0°$.

For the reactances as stated in the third to fifth column of table 10.1, the power radiation patterns are shown in respectively figures 10.28, 10.29 and 10.30.

The radiated maximum for the beam that should be steered to $\vartheta = 10°$ is found at $\vartheta = 7°$, see figure 10.28. For the beam that should be steered to $\vartheta = 20°$, the maximum is found at $\vartheta = 17°$, see figure 10.29 and for the beam that should be steered to $\vartheta = 30°$, the maximum is found at $\vartheta = 21°$, see figure 10.30. Since the beams are rather broad and the levels in the desired beam-pointing directions are not that much different from the maximum levels, we may conclude that the concept of reactive loading works well.

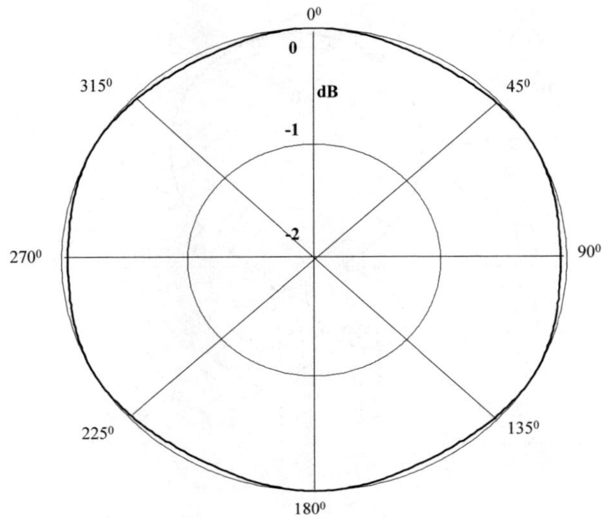

Fig. 10.26 Power radiation pattern of a seven-element circular array antenna of one driven and six reactively loaded dipole elements. All reactive loads are equal to zero. All dipole lengths are equal to half a wavelength, the circle radius and inter-element distances are equal to a quarter of a wavelength.

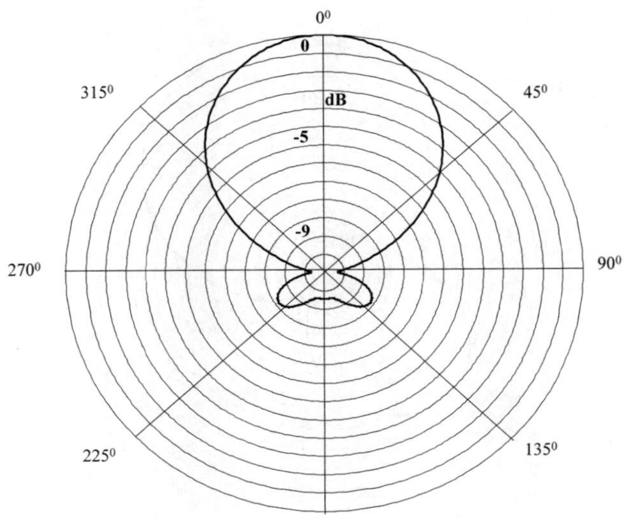

Fig. 10.27 Power radiation pattern of a seven-element circular array antenna of one driven and six reactively loaded dipole elements. The reactive loads are as stated in table 10.1 for a beam steered to $\varphi = 0°$. All dipole lengths are equal to half a wavelength, the circle radius and inter-element distances are equal to a quarter of a wavelength.

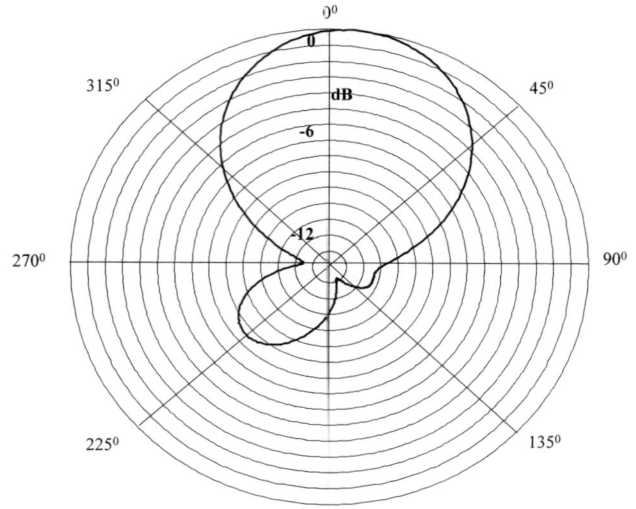

Fig. 10.28 Power radiation pattern of a seven-element circular array antenna of one driven and six reactively loaded dipole elements. The reactive loads are as stated in table 10.1 for a beam steered to $\varphi = 10°$. All dipole lengths are equal to half a wavelength, the circle radius and inter-element distances are equal to a quarter of a wavelength.

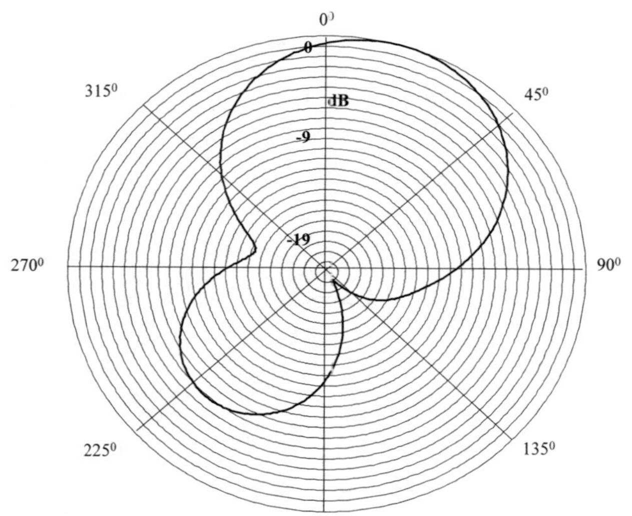

Fig. 10.29 Power radiation pattern of a seven-element circular array antenna of one driven and six reactively loaded dipole elements. The reactive loads are as stated in table 10.1 for a beam steered to $\varphi = 20°$. All dipole lengths are equal to half a wavelength, the circle radius and inter-element distances are equal to a quarter of a wavelength.

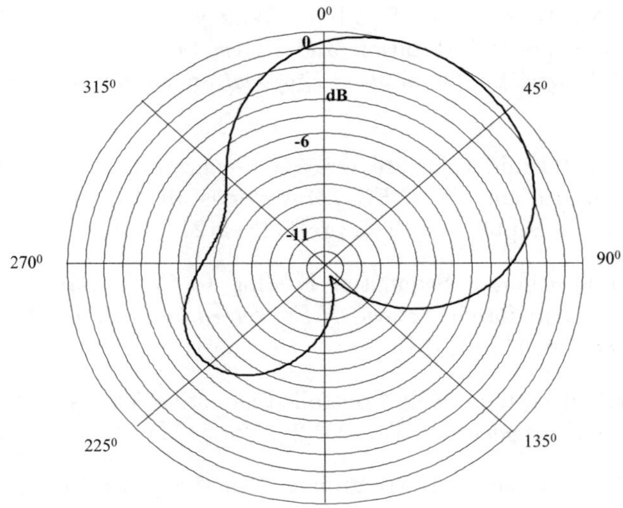

Fig. 10.30 Power radiation pattern of a seven-element circular array antenna of one driven and six reactively loaded dipole elements. The reactive loads are as stated in table 10.1 for a beam steered to $\varphi = 30^\circ$. All dipole lengths are equal to half a wavelength, the circle radius and inter-element distances are equal to a quarter of a wavelength.

We have seen earlier that capacitive or inductive loads may be realised by lengths of short-circuited transmission line. So by (electronically) switching pieces of short-circuited transmission line, the reactive loading may be changed and thus different beam-pointing directions may be selected.

The reactive loading may be used in conjunction with array antenna elements other than dipoles. The technique has been used with (dielectric filled) waveguide radiators [10, 11] and microstrip patch radiators [12, 13].

REFERENCES

1. Wolfgang H. Kummer, Preface of the Special Issue on Conformal Arrays, *IEEE Transactions on Antennas and Propagation*, Vol. AP-22, No. 1, January 1974, pp. 1–2.

2. Robert J. Mailloux, *Phased Array Antenna Handbook*, Artech House, Boston, 1994.

3. R.C. Hansen, *Phased Array Antennas*, John Wiley & Sons, New York, 1998.

4. H. Wilden and J. Ender, The Crow's Nest Antenna - Experimental Results, *Proceedings IEEE International Radar Conference*, 1990, pp. 280–285.

5. I.J. Bahl and P. Bhartia, *Microstrip Antennas*, Artech House, Boston, 1980.

6. T. Teshirogi, M. Tanaka and W. Chujo, Wideband Circularly Polarised Array Antenna with Sequential Rotation and Phase Shift of Elements, *Proceedings International Symposium Antennas & Propagation*, Japan, August 1985, pp. 117–120.

7. John Huang, A Technique for an Array to Generate Circular Polarization with Linearly Polarized Elements, *IEEE Transactions on Antennas and Propagation*, Vol. AP-34, No. 9, September 1986, pp. 1113–1124.

8. P.S. Hall, Application of Sequential Feeding to Wide Bandwidth, Circularly Polarised Microstrip Patch Arrays, *IEE Proceedings, Pt. H.*, Vol. 136, No. 11, November 1988, pp. 1510–1518.

9. Roger F. Harrington, Reactively Controlled Directive Arrays, *IEEE Transactions on Antennas and Propagation*, Vol. AP-26, No. 3, May 1978, pp. 390–395.

10. John Luzwick and Roger F. Harrington, A Reactively Loaded Aperture Antenna Array, *IEEE Transactions on Antennas an Propagation*, Vol. AP-26, No. 4, July 1978, pp. 543–547.

11. Fritz Arndt, Karl-Heinz Wolff, Lür Brünjes, Reiner Heyen, Frank Siefken-Herrlich, Wolfgang Bothmer and Ernst Forgber, Generalized Moment Method Analysis of Planar Reactively Loaded Rectangular Waveguide Arrays, *IEEE Transactions on Antennas and Propagation*, Vol. 37, No. 3, March 1989, pp. 329–338.

12. Robert J. Dinger, Reactively Steered Adaptive Array Using Microstrip Patch Elements at 4 GHz, *IEEE Transactions on Antennas and Propagation*, Vol. AP-32, No. 8, August 1984, pp. 848–856.

13. John Huang, Microstrip Yagi Array Antenna for Mobile Satellite Vehicle Application, *IEEE Transactions on Antennas and Propagation*, Vol. 39, No. 7, July 1991, pp. 1024–1030.

11

Array and Phased Array Antenna Measurement

The basic parameters that describe an antenna are the input impedance and the radiation pattern. This applies equally well to a single radiator as to an array or phased array antenna. Measurement of the input impedance is - with modern measurement equipment - relatively straightforward. Measurement of the radiation pattern requires some additional effort. For large array and phased array antennas, input impedance measurement and radiation pattern measurement would require the construction of one or more costly feeding networks. Besides, the size of the antenna would put extreme demands on the antenna radiation pattern measurement site. These problems may be circumvented by measuring the mutual coupling between radiators in an array *without a feeding network*. Mathematical manipulation of this measurement data will then reveal the information necessary for evaluating the array or phased array antenna.

11.1 INPUT IMPEDANCE, SELF-COUPLING AND MUTUAL COUPLING

The input impedance of an antenna can be easily obtained by measuring the reflection or self-coupling of the antenna, using a *vector network analyser* (VNA). A VNA is a two-port microwave receiver that is capable of determining the complex (amplitude and phase) ratios of reflected and incident voltage wave amplitudes. We remember that a two-port network as shown in figure 11.1 may be described by scattering parameters according to

$$b_1 = S_{11}a_1 + S_{12}a_2, \tag{11.1}$$

$$b_2 = S_{21}a_1 + S_{22}a_2, \tag{11.2}$$

Array and Phased Array Antenna Basics Hubregt J. Visser

where b_1 is the complex voltage wave amplitude of the reflected wave at port 1, b_2 is the complex voltage wave amplitude of the reflected wave at port 2, a_1 is the complex voltage wave amplitude of the incident wave at port 1 and a_2 is the complex voltage wave amplitude of the incident wave at port 2.

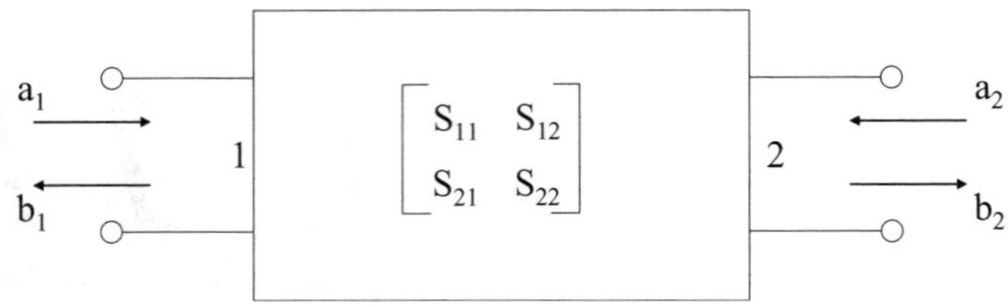

Fig. 11.1 Two-port microwave network and incident and reflected complex voltage wave amplitudes.

The basic VNA arrangement for the measurement of S_{11} and S_{21} of a device under test (DUT) is shown in figure 11.2 [1].

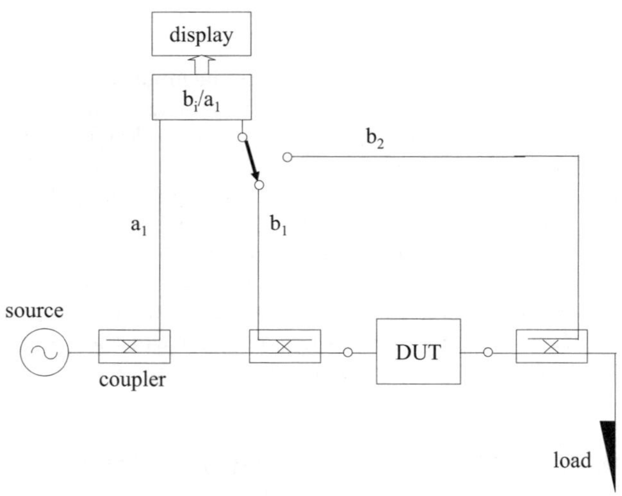

Fig. 11.2 Basic VNA arrangement for the measurement of S_{11} and S_{21}.

In this arrangement. $a_2 = 0$.

If the DUT is reversely connected into the network of figure 11.2, the remaining scattering coefficients, S_{21} and S_{22}, can be measured. The same can be accomplished by keeping the DUT in its position and use switches to reconnect source and load ($a_1 = 0$) and at the same time redefine the complex ratio.

To determine the input impedance of an antenna, we only need to determine the reflection coefficient, ρ, at the input. We may disregard the second port of the VNA and perform a one-port measurement to obtain $S_{11} = \rho$.

From transmission line theory we have already established a relation between reflection coefficient and input impedance, Z_{in}

$$S_{11} = \frac{Z_{in} - Z_0}{Z_{in} + Z_0}, \tag{11.3}$$

where Z_0 is the characteristic impedance of the VNA measurement system (usually 50Ω). From the above equation, we find for the input impedance

$$Z_{in} = Z_0 \frac{1 + S_{11}}{1 - S_{11}}, \tag{11.4}$$

where Z_0 is known and S_{11} is obtained from measurement.

Now that we are discussing two ports, we may as well take a preliminary sidestep towards mutual coupling that will be discussed in detail later on in this chapter. We will take a look at a two-antenna system and the signal flows, see figure 11.3a.

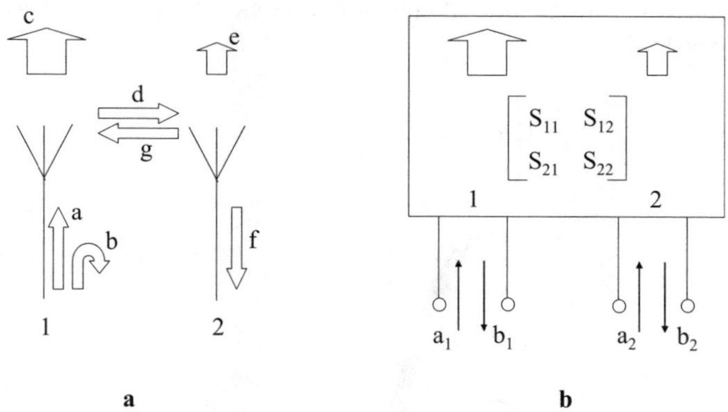

Fig. 11.3 Mutual coupling in an array antenna consisting of two radiators. a. Coupling mechanism. b. Scattering matrix representation.

Let's assume that antenna 1 is energised or excited (a). Due to a mismatch, a small part of the incident wave will be reflected (b) and most of the incident energy will be radiated (c, d). A part of the radiated energy will arrive at the aperture of antenna 2 (d). By the mechanisms of reflection and diffraction, a part of this energy incident upon antenna 2 is reradiated (e, g). A part of the energy incident upon antenna 2 will be received and transported into the feeding network (f). Due to internal reflections in the feeding network, a part of this energy (f) will be reradiated by both antenna 1 and 2. This is not shown in the figure. Part of the reradiated energy by antenna 2 will arrive at antenna 1 (g), where a similar process as described for antenna 2 will be taking place. The signal levels, however, will by now have decreased considerably.

We see that mutual coupling will affect both the radiator input impedance and the radiation pattern of the radiator. The radiation pattern of an element in an array environment will - in general - differ from the one in an isolated (no other elements present) situation. Furthermore, mutual coupling will also affect the polarisation characteristics (direction of the radiated electric field vector) of an element. We will not discuss this effect though.

The mutual coupling process for these two radiators may also be described by means of the 2×2 scattering matrix, see figure 11.3b. The two-element array antenna is considered as a microwave two-port network. The mutual coupling in an N-element linear or planar or curved array antenna may be described by the $N \times N$ scattering matrix of a microwave N-port. One of the advantages of such a description is that the elements of the scattering matrix can be obtained directly by measurement, using a VNA. How the obtained scattering matrix helps in obtaining the array radiation pattern will be explained after we have discussed the measurement of antenna radiation patterns in general.

11.2 RADIATION PATTERN MEASUREMENT

The radiation power or field pattern of an antenna is a graphical representation of the radiated power or field amplitude of that antenna as a function of direction angles ϑ and φ, see figure 11.4.

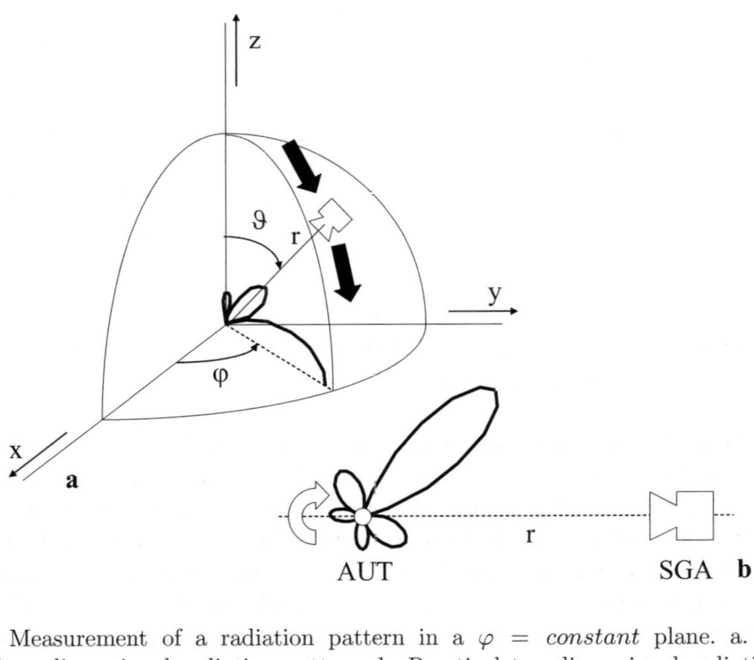

Fig. 11.4 Measurement of a radiation pattern in a φ = *constant* plane. a. Probing a cut in a three-dimensional radiation pattern. b. Practical two-dimensional radiation pattern measurement.

Normally, we are interested in two-dimensional cuts taken from the three-dimensional pattern and most often these cuts take the form of the radiated power (field amplitude) as a function of ϑ for a constant angle φ as shown in figure 11.4a for one-eight of a sphere of radius r. Here we assume that the antenna to be evaluated, i.e. the antenna under test (AUT) is placed in the origin of the coordinate system, that this antenna is acting as a transmitting antenna and that on a sphere with radius r, a probe is moved that receives the signal transmitted by the AUT. This received signal is plotted as a function of direction and thus gives the radiation pattern.[1] [2]

A practical implementation of measuring two-dimensional cuts of the three-dimensional radiation pattern is shown in figure 11.4b. The AUT is placed on a pedestal and rotated around its axis, while a standard gain antenna (SGA) is placed on a distance r from the AUT and receives the signal transmitted by the AUT while being kept in a fixed position. An SGA is an antenna with a known gain as a function of direction and frequency. Provided that the distance r satisfies the far-field condition that we stated in chapter 3,

$$r \geq \frac{2D^2}{\lambda}, \tag{11.5}$$

where D is the largest dimension of AUT and SGA and λ is the used wavelength, the radio equation (chapter 3) gives us the possibility to calculate the gain function from the received power as a function of direction.

When the far-field condition is met exactly, the spherical wavefront (transmitted by the SGA, assuming that the SGA is smaller than the AUT) deviates from a planar wavefront (over the aperture of the AUT) maximally 22.5° [2]. For most measurements, this deviation from a plane wave is acceptable.

In the following, we will very briefly outline the most common ways in which antenna radiation patterns are measured nowadays.

11.2.1 Far-Field Antenna Measurement Range

The obvious way to perform an antenna radiation pattern measurement is to build a set-up as shown in figure 11.4b and make sure that this set-up is constructed in such a way that possible sources of error are reduced to an acceptable level. Since equation (11.5) tells us that for large antennas (large in terms of wavelengths) the far field distance may become considerable, an outdoor antenna range seems to be a good solution.

The most important error source is formed then by reflections from ground and surrounding objects. The influence of these error sources can be reduced by elevating the antennas above the ground and possibly above the reflecting surrounding at ground level. An *elevated far-field antenna range* is shown in figure 11.5.

In fact, the outdoor far-field antenna range as shown in the figure is a combination of a true elevated antenna range (AUT and SGA on the same horizontal level) and a so-called *slant range* [2, 3]. A slant range is a far-field antenna range where either the AUT or the known antenna is elevated.

[1] When the antenna only, i.e. without active transmitting or receiving circuitry is measured, the antenna is reciprocal, meaning that transmit and receive patterns are identical. Therefore the AUT may also be used in receive mode, while the probe is being used in transmit mode.

[2] In fact, any antenna is reciprocal. However, when a transmitter and/or a receiver are/is directly connected to the antenna and the antenna clamps are no longer physically accessible, the whole system of antenna and active circuitry is no longer reciprocal. Although - in the author's opinion - the whole system should not be called an antenna, it has become common to do so and these 'antennas' are nowadays known as *active antennas*.

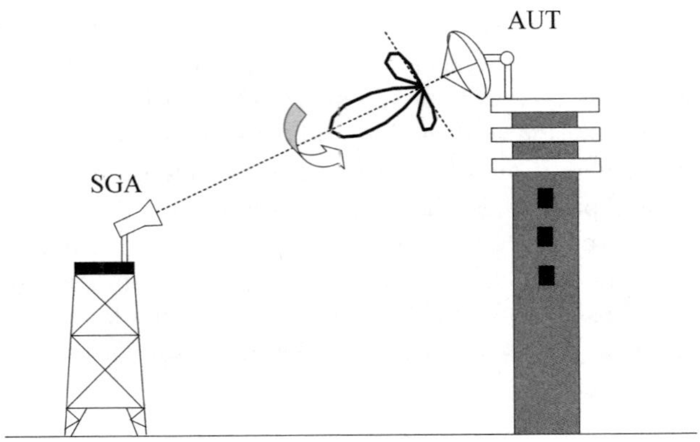

Fig. 11.5 Elevated far-field antenna range.

The disadvantages of any outdoor range are - besides the possible disturbance caused by scattering from objects in the surrounding - disturbances due to electromagnetic interference caused by other services using the same part of the electromagnetic spectrum as being used in the measurements, the inability to perform measurements during rain and a lack of security.

For antennas that are small with respect to wavelength, these problems can be overcome by moving the antenna measurement site indoors and taking special measures to reduce the reflections from walls, ceiling and floor.

11.2.2 Anechoic Chamber

Indoor reflections can be reduced by lining walls, ceiling and floor with radar absorbing material (RAM). Thus a so-called *anechoic chamber* is created. RAM basically consists of a carbon-impregnated foam that is produced in tiles. These tiles are often shaped as arrays of pyramids. These shapes ensure a minimisation of front face reflection and work well up to angles of approximately 50° from normal incidence [4].

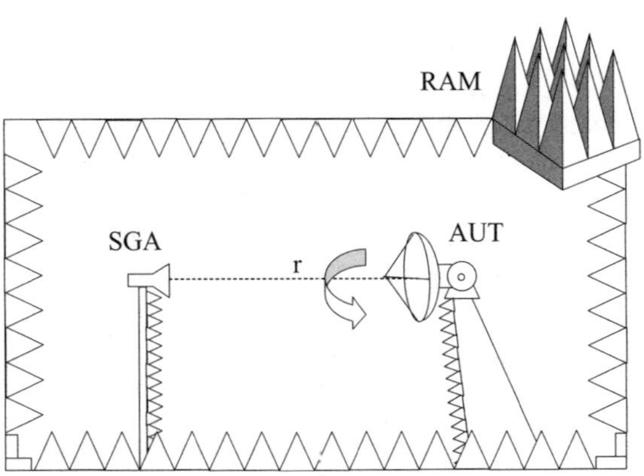

Fig. 11.6 Anechoic chamber and typical RAM tile, used to line walls, ceiling and floor.

Figure 11.6 shows an anechoic chamber and a typical RAM tile.

Since the radiation pattern measurement performed inside an anechoic room is a far-field measurement, the distance between the AUT and SGA must satisfy the far-field condition as stated in equation (11.5). This will put a limit on the maximum allowable size of the antennas to be tested, since indoor space will - in general - be limited.

If an antenna is too large for the far-field condition to be met indoors, the wavefront at the AUT will deviate from a planar wavefront too much. This statement in itself contains the solution for this problem: if we want to measure the radiation pattern of an antenna that has dimensions such that the far-field distance is beyond the maximum allowed antenna separation in an anechoic chamber, we need to *synthesise* a planar wavefront over the AUT.

11.2.3 Compact Antenna Test Range

In a *compact antenna test range* (CATR), planar wavefronts are generated in a very short range, hence its name. The planar wavefronts are generated by virtue of one or two curved metal reflectors. In figure 11.7, a so-called *dual parabolic-cylinder CATR* is shown.

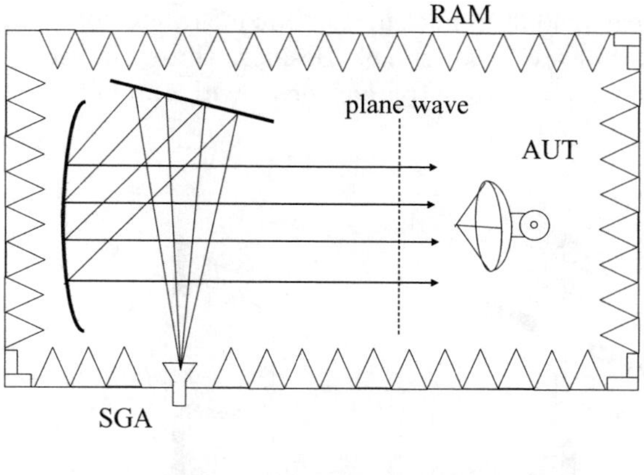

Fig. 11.7 Compact antenna test range using two parabolic-cylinder reflectors.

In this set-up, two parabolic-cylinder reflectors are used, having their one-dimensional curvature positioned perpendicular to one another. A spherical wavefront, created by the SGA, hits the first reflector whose curvature is - in figure 11.7 - directed inside the paper. The reflected wavefront, that is now cylindrical, then hits the second reflector with curvature as shown in the figure. The doubly reflected wavefront at the position of the AUT has become planar. Of course the wavefront is not perfectly planar everywhere upon leaving the second reflector. The area where the wavefront is planar 'enough' for antenna measurement purposes is called the *quiet zone*. This quiet zone is about 50%–60% of the dimensions of the reflector. To reduce the disturbance in the quiet zone caused by diffracted fields from the reflector edges, these edges are in practice either serrated or rolled.

Another compact type of antenna radiation pattern measurement site is formed by the family of near-field ranges.

11.2.4 Near-Field Antenna Measurement Range

A type of antenna radiation pattern measurement range that is gaining in popularity is the so-called *near-field antenna measurement range*. In a near-field range the AUT is kept in a fixed position and the radiated field is probed in a plane, on a cylinder or on a sphere close to and in front of or surrounding the AUT. Next, the sampled near-field data is Fourier-transformed to obtain the (three-dimensional) radiation pattern [5]. The far-field thus obtained may be inverse-Fourier-transformed backwards beyond the sampling plane onto the antenna aperture to obtain the aperture distribution for diagnostic purposes. If the AUT is an array antenna for example, this backward transformation may be used to find defective array antenna elements. With the exception of the spherical near-field antenna measurement range, near-field antenna ranges only sample a part of the near-field surrounding the antenna. Therefore, a planar near-field antenna measurement range is best suited for highly directive antennas where most of the radiated energy will be directed into the forward direction. The spherical near-field antenna measurement range is best suited for near-omnidirectional radiators. The cylindrical near-field antenna measurement range forms a good alternative for measuring low-directivity antennas. Besides, the cylindrical near-field antenna range is easily combined with a planar near-field antenna range as is shown in figure 11.8.

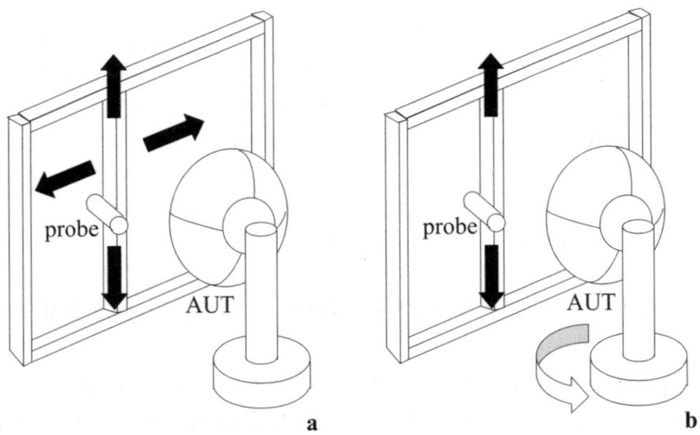

Fig. 11.8 Near-field antenna radiation pattern measurement range. a. Planar near-field range. b. Cylindrical near-field range.

Figure 11.8a shows a planar near-field antenna measurement range. The scanner moves the probe around to take samples of the near-field in amplitude and phase over a regular grid. This scanner is shown here without its lining of RF absorbers (RAM). The probe is a small antenna (often taking the form of an open-ended waveguide), that is kept as small as possible to minimise reflections between the AUT and probe. As figure 11.8b shows, the planar scanner when operated as a vertical linear scanner, combined with the AUT being rotated on its pedestal, allow - in a relatively easy and cost-effective way - the operation of the planar near-field range as a cylindrical near-field range.

Although the (planar) near-field antenna measurement range offers the opportunity to obtain the radiation pattern of a large phased array antenna in a limited-size indoor environment, its value within the design path of a phased array antenna lies mostly in the final, verification stage of the complete antenna with feeding network and phase shifters.

Since this feeding network with phase shifters will contribute substantially to the costs of a phased array antenna, it is not cost-effective at all to construct different complete phased array antennas in the design process. To minimise costs, it is preferred to separate array antenna and feeding network and optimise the array antenna first before adding the feeding network in the final stage. So what is needed is an effective way to obtain radiation patterns of the array antenna without a feeding network. The concept of the *scan element pattern* [6], also known as the *active element pattern*, offers a way to obtain this desired array antenna pattern in a cost-effective way. Waveguide simulator measurements also offer a cost-effective means of obtaining scanned array antenna characteristics without the need of constructing a feeding network.

11.3 SCAN ELEMENT PATTERN

The characterisation of a large array antenna may be accomplished on the basis of the assumption that in a large array nearly all elements encounter a similar array environment and therefore also encounter identical mutual coupling effects from the surrounding [6–8]. Next to the mutual coupling, also the concept of the *scan element pattern*, i.e. the radiation pattern of a singly excited element in its array environment - where all other elements are terminated into matched loads - will prove to be useful.

The *scan element pattern* is also known as the *active element pattern*, but we choose the first term, following [6], in order to avoid confusion. The scan element pattern will provide the phased array antenna gain at the position of the scanned beam as a function of scan angle. For a large phased array antenna, all scan element patterns will be nearly identical and the phased array antenna performance may be approximated by applying pattern multiplication. In this pattern multiplication, the common scan element pattern is multiplied with the array factor. All coupling effects are accounted for then in the scan element pattern [6, 8].

We will discuss the mutual coupling and scan element pattern, using a K-element linear (phased) array antenna as an example in order to keep the discussion as clear as possible. We have seen that the translation of the analysis results for a linear to a planar (phased) array antennas is straightforward. Therefore - after the linear array antenna analysis - we will simply state the results for a planar array antenna.

11.3.1 Mutual Coupling

We assume the linear array antenna to consist of K identical elements, displaced a distance d with respect to one another and being matched to a voltage source as shown in figure 11.9 [8].

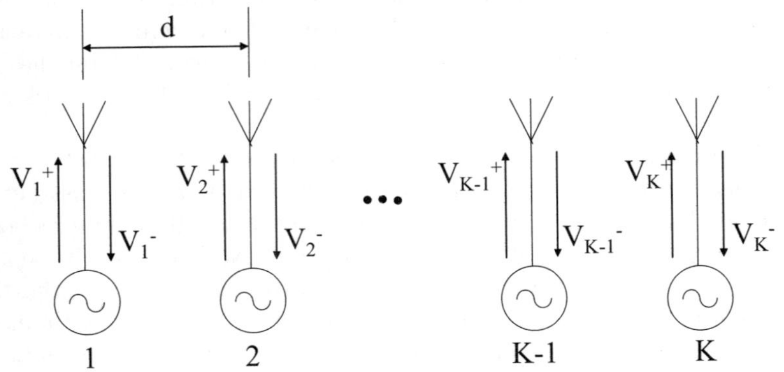

Fig. 11.9 K-element linear array antenna. Every element is match-connected to a voltage source.

Since the radiators are match-connected to their sources, the voltage waves travelling in negative directions are due to mutual coupling from surrounding elements.

To obtain the scan element pattern, only one of the K elements will be excited, the other elements will be disconnected from their sources and will be terminated in matched (reflectionless) loads. This situation is shown for element k, where $k = 1, 2, \ldots, K$, in figure 11.10.

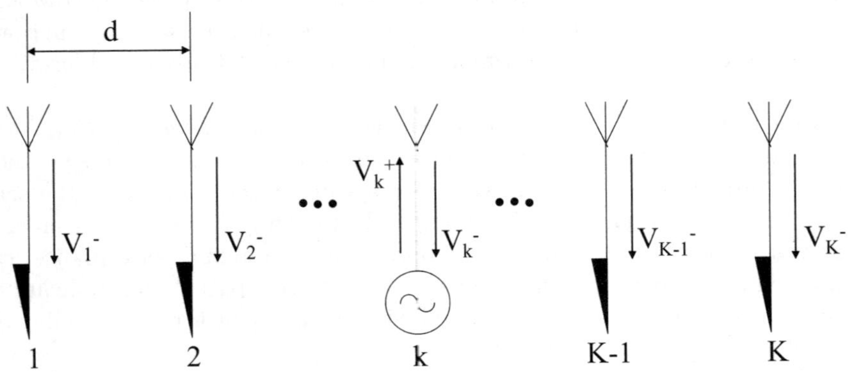

Fig. 11.10 K-element linear array antenna. Feeding for obtaining the scan element pattern for element k.

The effect of the mutual coupling is shown in figure 11.10 by the presence of waves travelling from the not-excited elements into the loads.

We relate the voltage wave amplitude to the radiated electric and magnetic fields, E_e and H_e, of an isolated element through

$$E_e(\vartheta) = V_e S_e(\vartheta)\frac{e^{-jk_0 r}}{r}, \tag{11.6}$$

$$H_e(\vartheta) = \frac{V_e}{\eta} S_e(\vartheta)\frac{e^{-jk_0 r}}{r}, \tag{11.7}$$

where V_e is the voltage wave amplitude of the excitation, $k_0 = \frac{2\pi}{\lambda}$ and $S_e(\vartheta)$ is the normalised isolated element radiation pattern. η is the free space characteristic impedance and is given by

$$\eta = \sqrt{\frac{\mu_0}{\varepsilon_0}}, \tag{11.8}$$

where μ_0 is the free space permeability and ε_0 is the free space permittivity.

Implicitly we assume here that the element radiates a single mode and that a single mode propagates through the feed network, an assumption that has given us fairly good results in the preceding chapters.

The K-element linear array may be seen as a K-port, characterised by a $K \times K$ scattering matrix, where the scattering coefficients relate the element voltage wave amplitudes according to

$$S_{kl} = \left.\frac{V_k^-}{V_l^+}\right|_{V_m^+=0;m\neq l} \quad \text{for } k, l, m = 1, 2, \ldots, K. \tag{11.9}$$

This equation tells us that the scattering coefficient S_{kl} is obtained by comparing the voltage wave amplitude of the wave entering element k from free space with the voltage wave amplitude of the wave exciting element l. For this comparison, none of the elements, with the exception of element l is being excited. All these elements are terminated in matched loads. This process is visualised in figure 11.11.

With V_k^+ and V_k^- the voltage wave amplitude of, respectively, the incident and reflected wave at element k and with use of equation (11.9), we find for the voltage wave amplitude of the reflected wave at element k

$$V_k^- = \sum_{l=1}^{K} S_{kl} V_l^+. \tag{11.10}$$

The total voltage over the antenna clamps of element l, V_l, is the sum of incident and reflected voltage wave amplitudes

$$V_l = V_l^+ + V_l^-. \tag{11.11}$$

The total current is given by

$$I_l = I_l^+ + I_l^- = \frac{1}{Z_0}\left(V_l^+ - V_l^-\right), \tag{11.12}$$

where Z_0 is the characteristic impedance of the transmission line.

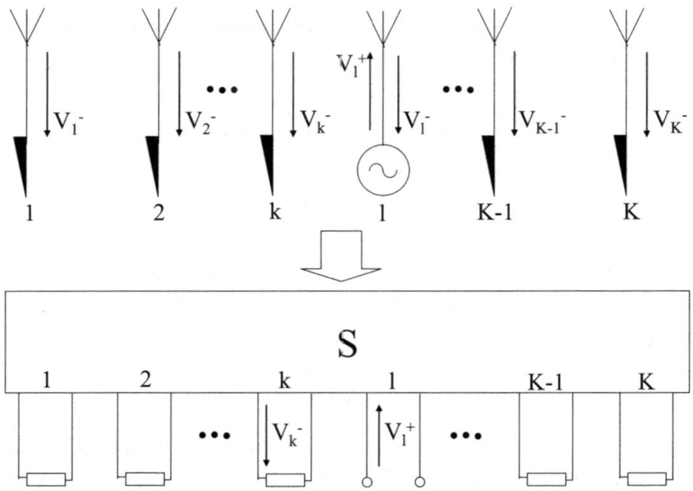

Fig. 11.11 Definition of scattering coefficient S_{kl}.

For reasons of keeping the discussion simple, we will assume that the characteristic impedance of the transmission line connected to the antenna element, Z_0, is equal to one, i.e. we will work with a normalised characteristic impedance. Thus

$$I_l = V_l^- - V_l^-. \tag{11.13}$$

If we want the antenna beam of the K-element linear array to be directed to $\vartheta = \vartheta_0$, the previous chapters have told us to apply a phase taper that satisfies

$$V_l^+ = e^{-jk_0(l-1)d\sin(\vartheta_0)}, \tag{11.14}$$

where we have put the phase reference on element 1.

The *scan reflection coefficient* of element k, $\Gamma_k(\vartheta_0)$, i.e. the reflection coefficient of element k in its (scanned) array environment with all elements excited and properly phased, is found with the aid of equations (11.10) and (11.14)

$$\Gamma_k(\vartheta_0) = \frac{V_k^-}{V_k^+} = e^{jk_0(k-1)d\sin(\vartheta_0)} \sum_{l=1}^{K} S_{kl} e^{-jk_0(l-1)d\sin(\vartheta_0)}. \tag{11.15}$$

The radiated electric field of the complete K-element linear array antenna, with all elements excited, $E_a(\vartheta)$, is found to be, applying pattern multiplication

$$E_a(\vartheta) = E_e(\vartheta) \sum_{k=1}^{K} V_k e^{jk_0(k-1)d\sin(\vartheta)}. \tag{11.16}$$

Note that we have simply multiplied the isolated element pattern with the array factor, so mutual coupling effects are not taken into account.

To bring in the mutual coupling effects, we first relate the voltage wave amplitude, V_k, to the scan reflection coefficient, $\Gamma(\vartheta)$, by using equations (11.11) and (11.15)

$$V_k = V_k^+ + V_k^- = [1 + \Gamma_k(\vartheta_0)]\, V_k^+ . \tag{11.17}$$

Substituting this result, together with equation (11.14) into equation (11.16) gives for the radiated electric field of the completely excited array antenna

$$E_a(\vartheta) = E_e(\vartheta) \sum_{k=1}^{K} [1 + \Gamma_k(\vartheta_0)]\, e^{jk_0(k-1)d[\sin(\vartheta)-\sin(\vartheta_0)]}, \tag{11.18}$$

and, phased for $\vartheta = \vartheta_0$

$$E_a(\vartheta_0) = E_e(\vartheta_0) \sum_{k=1}^{K} [1 + \Gamma_k(\vartheta_0)] . \tag{11.19}$$

Substitution of equation (11.6) into equation (11.19) gives

$$E_a(\vartheta_0) = V_e S_e(\vartheta_0) \sum_{k=1}^{K} [1 + \Gamma_k(\vartheta_0)] \frac{e^{-jk_0 r}}{r} . \tag{11.20}$$

In a similar way we find for the radiated magnetic field of the fully excited array

$$H_a(\vartheta_0) = \frac{V_e}{\eta} S_e(\vartheta_0) \sum_{k=1}^{K} [1 - \Gamma_k(\vartheta_0)] \frac{e^{-jk_0 r}}{r} . \tag{11.21}$$

In the chapter on antenna parameters we have seen that the gain function $G(\vartheta)$ is given by

$$G(\vartheta) = \frac{P(\vartheta)}{P_{in}/4\pi}, \tag{11.22}$$

where $P(\vartheta)$ is the radiated power in the direction ϑ and P_{in} is the totally incident available power.

The time-average power density $\mathbf{S}$ (watts per square metre) of the radiated fields can be calculated as [2]

$$\mathbf{S}(\mathbf{r}) = \frac{1}{2} \Re \{\mathbf{E}_a(\mathbf{r}) \times \mathbf{H}_a^*(\mathbf{r})\} = \frac{1}{2} \Re \{E_a H_a^*\}\, \mathbf{u}_r . \tag{11.23}$$

The radiated power, $P(\vartheta)$, per solid angle unit $d\Omega = \sin(\vartheta) d\vartheta d\varphi$, see figure 11.12, is then given by

$$P(\mathbf{r}) = P(\vartheta) = \left| r^2 \mathbf{S}(\mathbf{r}) \right| . \tag{11.24}$$

The radiated power into the direction ϑ_0 thus becomes

$$P(\vartheta_0) = V_e^2 S_e^2(\vartheta_0) \frac{1}{2\eta} \left\{ \sum_{k=1}^{K} [1 + \Gamma_k(\vartheta_0)] \right\} \left\{ \sum_{k=1}^{K} [1 - \Gamma_k(\vartheta_0)] \right\} . \tag{11.25}$$

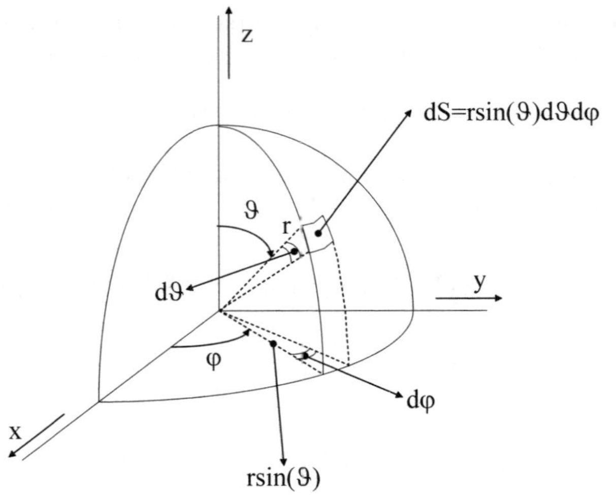

Fig. 11.12 Solid angle $d\Omega = \sin(\vartheta)d\vartheta d\varphi$.

The incident available power is, since we assumed a normalised characteristic impedance,

$$P_{in} = K\frac{V_e^2}{2}, \tag{11.26}$$

so that we find for the gain of the fully excited array in the direction ϑ_0

$$G_a(\vartheta_0) = \frac{4\pi S_e^2(\vartheta_0)}{K\eta}\left(K^2 - \left|\sum_{k=1}^{K}\Gamma_k(\vartheta_0)\right|^2\right). \tag{11.27}$$

So, based on the scan reflection coefficient, Γ_k, that can be obtained from pair-wise mutual coupling measurements between the array elements, and the isolated element pattern - that in general is relatively easy to obtain - we may find the gain of the complete array antenna in the direction of the scanned beam.

Recapitulating: The *isolated element pattern* is the radiation pattern of a single element, measured in an empty environment, i.e. *not* in its array environment. The *scan element pattern* is the radiation pattern of a singly excited element in its array environment, where all other array antenna elements are terminated into matched loads. The *scan reflection coefficient* is the reflection coefficient of a single element in its array environment, where all elements are excited and properly phased.

The scan element pattern is difficult, although not impossible to measure directly. The scan reflection coefficient is difficult to measure directly. Mutual coupling coefficients are relatively easy to measure. These coupling coefficients may be used to determine the scan reflection coefficient. As we will see in the next paragraph, the scan element pattern may be obtained from the scan reflection coefficient.

We return now to the expression for the gain of a fully excited array, evaluated at the direction of the scanned beam.

We may simplify this equation for large array antennas.

In a large one- or two-dimensional array antenna, nearly all elements experience an identical array environment. Only the elements at and near the edges of the array experience an environment that differs substantially from those in the centre. When the array antenna is sufficiently large,[3] the number of centre elements outweighs the number of edge elements and approximately all elements experience the same mutual coupling effects. Therefore, in this situation, all element scan reflection coefficients are approximately equal, $\Gamma_k \approx \Gamma$, so that

$$G_a(\vartheta_0) = KG_e(\vartheta_0)\left(1 - |\Gamma(\vartheta_0)|^2\right), \tag{11.28}$$

where G_e is the gain of an isolated element, that is given by

$$G_e(\vartheta) = \frac{4\pi S_e^2(\vartheta)}{\eta}. \tag{11.29}$$

So, if - for a large array - we measure all scattering coefficients relative to a centre element,[4] of which we know the isolated gain pattern, we may calculate the scan reflection coefficient, Γ, using equation (11.15) and calculate the gain of the fully excited array in the direction ϑ_0, using equation (11.28).

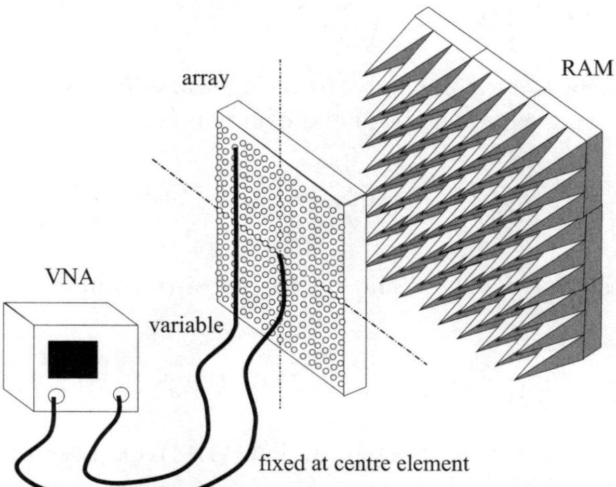

Fig. 11.13 Measuring coupling coefficients relative to the centre element in one-half or one-quarter of the array. All elements not involved in the current coupling measurements are terminated into matched loads. The array antenna is facing an anechoic environment.

[3]How large 'sufficiently large' is, is determined by the desired accuracy in calculating the antenna parameters using the large array antenna approximations.
[4]This means that we keep one port of a two-port VNA connected to the centre element and connect the other port to every other element, one after the other, while keeping all other elements terminated into matched loads. If the array antenna is symmetric, measurements may be restricted to one-half or one-quarter of the array. While measuring the pair-wise coupling coefficients, the array antenna aperture should be directed into an RF absorbing surrounding like an anechoic room, see figure 11.13.

11.3.2 Scan Element Pattern

To obtain the scan element pattern, i.e. the pattern of one excited element in its array environment where all other elements are terminated into matched loads, we take a closer look at our K-element (phased) array antenna. We assume that element k is excited. To calculate the radiated electric far-field, $E_a^s(\vartheta)$, we start with equation (11.16)

$$E_a^s(\vartheta) = E_e(\vartheta) \sum_{l=1}^{K} V_l e^{jk_0(l-k)d\sin(\vartheta)}, \tag{11.30}$$

where we have moved the phase reference to element k.

Since only element k is excited, equation (11.11) reveals that $V_k = V_k^+ + V_k^-$ and $V_l = V_l^-$ for $l \neq k$. This leads to

$$E_a^s(\vartheta) = E_e(\vartheta) \left[V_k^+ + \sum_{l=1}^{K} V_l^- e^{jk_0(l-k)d\sin(\vartheta)} \right], \tag{11.31}$$

and since $V_l^- = \sum_{m=1}^{K} S_{lm} V_m^+$, (equation (11.10))

$$E_a^s(\vartheta) = E_e(\vartheta) \left[V_k^+ + \sum_{l=1}^{K} \left(\sum_{m=1}^{K} S_{lm} V_m^+ \right) e^{jk_0(l-k)d\sin(\vartheta)} \right]. \tag{11.32}$$

Since only element k is excited, only the term for $m = k$ remains in the summation over m. Further, using equation (11.6) and equation (11.14) ($V_k^+ = 1$), leads to

$$E_a^s(\vartheta) = V_e S_e(\vartheta) \left[1 + \sum_{l=1}^{K} S_{lk} e^{jk_0(l-k)d\sin(\vartheta)} \right] \frac{e^{-jk_0 r}}{r}. \tag{11.33}$$

From equations (11.10), (11.14) and (11.15), substituted in the above equation, we finally obtain

$$E_a^s(\vartheta) = V_e S_e(\vartheta) \left[1 + e^{-jk_0(k-1)d\sin(\vartheta)} \Gamma_k(-\vartheta) \right] \frac{e^{-jk_0 r}}{r}, \tag{11.34}$$

where the phase reference, by virtue of equation (11.14), has been put back to element 1 and where use has been made of $S_{kl} = S_{lk}$.

For the radiated magnetic field, H_a^s, we find

$$H_a^s(\vartheta) = \frac{V_e S_e(\vartheta)}{\eta} \left[1 - e^{-jk_0(k-1)d\sin(\vartheta)} \Gamma_k(-\vartheta) \right] \frac{e^{-jk_0 r}}{r}. \tag{11.35}$$

The gain of the singly excited element then follows from

$$G_k^s(\vartheta) = \frac{P(\vartheta)}{P_{in}/4\pi} = \frac{4\pi r^2}{2P_{in}} \Re\left\{ E_a^s H_a^{s^*} \right\} = G_e(\vartheta) \left[1 - |\Gamma_k(-\vartheta)|^2 \right]. \tag{11.36}$$

For a very large array, $\Gamma_k(-\vartheta) = \Gamma_k(\vartheta) = \Gamma(\vartheta)$ and the gain function reduces to

$$G_k^s(\vartheta) = G_e(\vartheta) \left[1 - |\Gamma(\vartheta)|^2 \right]. \tag{11.37}$$

So the scan element pattern (of a singly excited element) is simply related to the scan reflection coefficient of the completely excited array antenna. This scan reflection coefficient can be calculated from the mutual coupling coefficients between the array elements, that can be relatively easy obtained by measurement, see figure 11.13.

11.3.3 Blind Scan Angles

Upon comparing equation (11.37), the gain function of a singly excited element, with equation (11.28), the gain of a fully exited array in the direction $\vartheta = \vartheta_0$, we see that

$$G_a(\vartheta_0) = KG_k^s(\vartheta_0) . \tag{11.38}$$

The gain of a large, K-element, fully excited phased array antenna, in the direction $\vartheta = \vartheta_0$, i.e. in the direction of the scanned beam, is equal to K times the scanned element pattern into that direction.

This means that if we observe anomalies in the scanned element pattern - like 'dips' or 'nulls' - these anomalies will also result in the fully excited phased array antenna pattern. A 'null' in the scanned element pattern means that for the phased array antenna scanned into the direction of this 'null' the radiation pattern will also show a 'null', i.e. energy will not be radiated into that direction. The specific angle ϑ where such a phenomenon occurs is called a *blind scan angle.*

A blind scan angle may occur in any array that has a periodic structure and in which materials are used in such a way that a leaky wave may propagate over the array face [6]. The leaky wave emits radiation that in certain directions (the blind scan angles) cancels the array antenna radiation. The effect is in general more serious for large array antennas.

A blind scan angle occurs before the appearance of a grating lobe. Therefore, by changing the inter-element distances in an array, blind scan angles may be removed from the scan area of interest. For printed antennas, blind scan angle positions may be calculated, using the approximate equations of [9]. An elaborate discussion and literature survey on scan blindness and measures to avoid blindness - that is beyond the scope of this book - may be found in [6].

The measurement of mutual coupling between array antenna elements has made it possible to predict the scan behaviour of a phased array antenna without having to construct one or more costly feeding networks. Although this has resulted in a substantial reduction of development costs, still one or more array antennas need to be realised. A means to further reduce development costs and bypass the need for full-scale array antenna realisation is offered by the *waveguide simulator.*

11.4 WAVEGUIDE SIMULATOR

Let's assume that we have an infinite, regularly spaced, planar array antenna of identical elements. This means that in the two transverse directions there are no edge elements, only centre elements exist. The infinite array assumption allows us to assume all elements to be identical with respect to mutual coupling effects.

Assume now that the array antenna is phased in such a way that a beam is directed towards a certain direction ϑ_0, different from broadside and that the polarisation is such that the electric field is directed perpendicular to the plane of scan, see figure 11.14a.

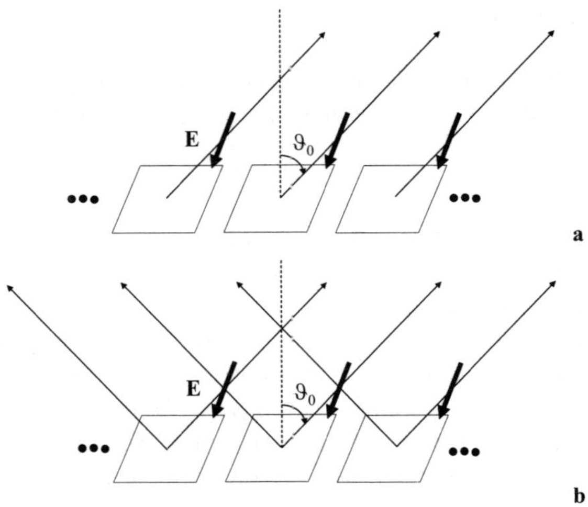

Fig. 11.14 Scanned beams for an infinite array antenna. a. Plane wavefront direction for a beam scanned to $\vartheta = \vartheta_0$. b. Plane wavefront direction for a beam scanned to $\vartheta = -\vartheta_0$. The polarisation is perpendicular to the plane of scan.

By measuring the reflection coefficient of an arbitrary element for the beam scanned to $\vartheta = \vartheta_0$, we can calculate the phased array antenna characteristics as explained in the previous section.

Next we assume that a phase taper is applied for scanning the beam into the direction $\vartheta = -\vartheta_0$ and we assume that both phase tapers exist simultaneously. Then we get the situation as shown in figure 11.14b.

By superposition, the reflection coefficient of an arbitrary element in the array, emitting both plane wavefronts is identical to that of an element in an array emitting a single beam [10].

In the planes where the two planar wavefronts cross, the net electric field is zero due to the opposite phasing of the waves. Since on a perfect electric conductor the tangential electric field is zero, we may place metallic walls in these planes without this wall placing affecting the array antenna characteristics. If we choose the plane wave directions such that the positions of these metallic walls coincide with planes of symmetry of the array antenna (as is the case in figure 11.14), we may analyse a single waveguide containing a repeatable cell of the array antenna, see figure 11.15.

The walls of the waveguide simulator act as mirrors and by reflection of the array cell and its images an infinite array antenna is created. The waveguide simulator needs to be sufficiently large and needs to be terminated into a waveguide load to prevent waves to be incident upon the array cell.

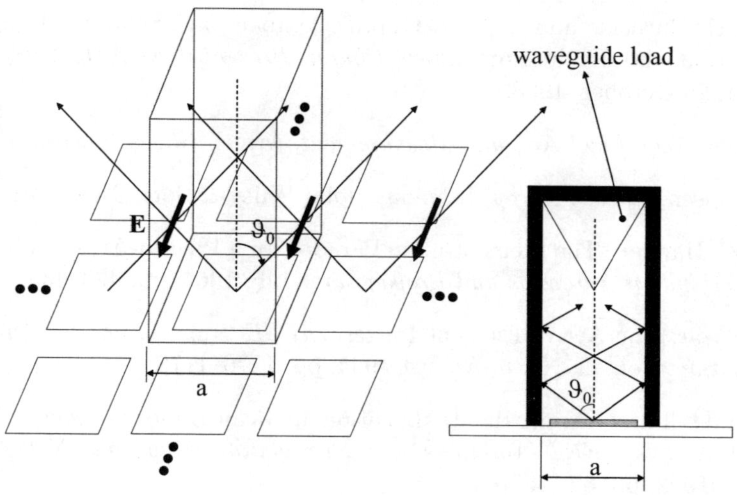

Fig. 11.15 Simulation of an infinite array antenna, phased for scanning to $\vartheta = \vartheta_0$, using a waveguide simulator.

The array cell as shown in figure 11.15 contains a single element which is the smallest cell that repeats itself to form the infinite array. Cells containing more elements, even parts of elements are also allowed [10, 11], as long as by reflection in the waveguide walls the original infinite array antenna is created. The choice of the waveguide dimensions then depends on the particular scan angle ϑ_0 that needs to be analysed. This angle depends on frequency (wavelength) and waveguide simulator dimensions, see figure 11.15

$$\sin(\vartheta_0) = \frac{\lambda}{2a}. \tag{11.39}$$

This equation and the fact that a repeatable cell needs to be created means that waveguide simulators can be employed only for a discrete set of scan angles.

REFERENCES

1. Roger D. Pollard, Microwave Network Characterisation, in *Microwave Subsystem Design*, Short course organised by the Microwave Solid-State Group Dept. of Electrical & Electronic Engineering, University of Leeds, UK, 1989.

2. Constantine E. Balanis, *Antenna Theory, Analysis and Design, second edition*, John Wiley & Sons, New York, 1997.

3. Warren L. Stutzman and Gary A. Thiele, *Antenna Theory and Design, second edition*, John Wiley & Sons, New York, 1998.

4. Walter D. Burnside and Inder J. Gupta, Compact and Spherical Range Design, Application and Evaluation, *Short Course Presented at AMTA'98*, Montreal, Canada, 25 October, 1998.

5. D. Slater, *Near-Field Antenna Measurement*, Artech House, Boston, MA, 1993.

6. R.C. Hansen, *Phased Array Antennas*. John Wiley & Sons, New York, 1998.

7. Peter W. Hannan, The Element-Gain Paradox for a Phased-Array Antenna, *IEEE Transactions on Antennas and Propagation*, July 1964, pp. 423–433.

8. D.M. Pozar, The Active Element Pattern, *IEEE Transactions on Antennas and Propagation*, Vol. 42, No. 8, August 1994, pp. 1176–1178.

9. Parbhu D. Patel, Approximate Location of Scan-Blindness Angle in Printed Phased Arrays, *IEEE Antennas and Propagation Magazine*, Vol. 34, No. 5, October 1992, pp. 53–54.

10. P.W. Hannan and M.A. Balfour, Simulation of a Phased-Array Antenna in Waveguide, *IEEE Transactions on Antennas and Propagation*, May 1965, pp. 342–353.

11. David M. Pozar, Analysis of an Infinite Phased Array of Aperture Coupled Microstrip Patches, *IEEE Transactions on Antennas and Propagation*, Vol. 37, No. 4, April 1989, pp. 418–425.

Appendix A *Complex Analysis*

The use of complex variables in solving problems in the applied sciences, like in electrical engineering, appears to be a very valuable tool. Especially when dealing with sinusoidal excitations, the introduction of complex variables will simplify the solution process. Before this simplification is fully appreciated though, one has to deal first with the somewhat awkward concept of complex numbers.

A.1 COMPLEX NUMBERS

We are all familiar with the real numbers and the permitted and non-permitted operations on real numbers. So, it is for example permitted to calculate the square root of the number 3.79 ($\sqrt{3.79} = 1.95$), but the square root of -4 does not exist.

The *complex numbers* allow for the latter square root to exist, through the introduction of so-called *imaginary numbers* next to the real numbers. Any complex number consists of a real part and an imaginary part and is generally denoted as

$$c = a + jb, \tag{A.1}$$

where c is a complex number, a is the real part of the complex number

$$a = \Re(c), \tag{A.2}$$

Array and Phased Array Antenna Basics Hubregt J. Visser

and b is the imaginary part of the complex number

$$b = \Im(c). \tag{A.3}$$

j is the imaginary unit[1] that exhibits the special characteristic

$$j^2 = -1. \tag{A.4}$$

Before we move on to the arithmetic concerning complex numbers, we first define the *complex conjugate* of a complex number.

The complex conjugate of a complex number c, denoted c^*, is defined by replacing j by $-j$ everywhere in the complex number. So, if

$$c = a + jb, \tag{A.5}$$

then

$$c^* = a - jb, \tag{A.6}$$

The addition and subtraction of complex numbers is straightforward. If $c = a + jb$ and $d = e + jf$, then

$$c + d = a + jb + e + jf = (a + e) + j(b + f), \tag{A.7}$$
$$c - d = a + jb - (e + jf) = (a - e) + j(b - f). \tag{A.8}$$

Multiplication makes use of the special characteristic of the imaginary unit

$$e \cdot (a + jb) = ea + jeb, \tag{A.9}$$

$$\begin{aligned} (e + jf) \cdot (a + jb) &= \\ ea + jeb + jaf + j^2 fb &= \\ (ea - fb) + j(eb + af). \end{aligned} \tag{A.10}$$

Multiplication of a complex number with its complex conjugate results in

$$\begin{aligned} cc^* = (a + jb)(a - jb) &= \\ a^2 + jab - jab - j^2b^2 &= \\ a^2 + b^2 &= \\ &|c|^2. \end{aligned} \tag{A.11}$$

For division, use is made of the complex conjugate. If $c = a + jb$ and $d = e + jf$, then

$$\begin{aligned} \frac{c}{d} = \frac{cd^*}{dd^*} &= \\ \frac{(a + jb)(e - jf)}{e^2 + f^2} &= \\ \frac{ae + bf}{e^2 + f^2} + j\frac{be - af}{e^2 + f^2} \end{aligned} \tag{A.12}$$

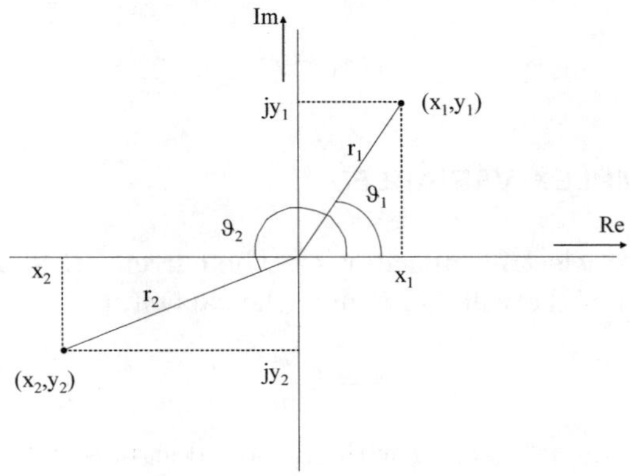

Fig. A.1 Complex numbers (x_1, y_1) and (x_2, y_2) in the complex plane.

Complex numbers may be graphically represented in the *complex plane*, see figure A.1.

The real part of a complex number is plotted along the horizontal axis, the imaginary part of the complex number, multiplied by j, is plotted along the vertical axis.

Another way of representing complex numbers instead of using Cartesian coordinates is using polar coordinates. For $z_1 = x_1 + jy_1$, see figure A.1, the polar form is

$$z_1 = r_1 \left[\cos(\vartheta_1) + j \sin(\vartheta_1)\right], \tag{A.13}$$

where

$$r_1 = |z_1|, \tag{A.14}$$

and

$$\vartheta_1 = \arctan\left(\frac{\Im(z_1)}{\Re(z_1)}\right). \tag{A.15}$$

If we differentiate the expression of z_1 to ϑ_1, we get

$$\begin{aligned}\frac{dz_1}{d\vartheta_1} = r_1 \left[-\sin(\vartheta_1) + j\cos(\vartheta_1)\right] = \\ jr_1 \left[\cos(\vartheta_1) + j\sin(\vartheta_1)\right] = \\ jz_1,\end{aligned} \tag{A.16}$$

and therefore

$$\begin{aligned}z_1 = r_1 \left[\cos(\vartheta_1) + j\sin(\vartheta_1)\right] = \\ r_1 e^{j\vartheta_1}.\end{aligned} \tag{A.17}$$

As a last check, we verify that

$$z_1^* = r_1 e^{-j\vartheta_1}, \tag{A.18}$$

[1]Mathematicians and physicists use the symbol i for the imaginary unit, but since in electrical engineering this symbol is already reserved for current, electrical engineers use the symbol j instead.

and therefore

$$z_1^* z_1 = r_1^2. \tag{A.19}$$

A.2 USE OF COMPLEX VARIABLES

Let us first look at an electric capacitor, C. The current 'through' a capacitor, i, is described as function of the voltage, v, over the capacitor

$$i = C\frac{dv}{dt}. \tag{A.20}$$

Now, assume that we are dealing with sinusoidal signals (currents and voltages)

$$v = V_1 \cos(\omega t), \tag{A.21}$$

where $\omega = 2\pi f$ is the angular frequency.

The current 'through' the capacitor is now given by

$$\begin{aligned} i = C\frac{d}{dt}\left[V_1 \cos(\omega t)\right] &= \\ -V_1 C\omega \sin(\omega t). \end{aligned} \tag{A.22}$$

Next, we look at an inductor, L. The voltage over the inductor, v, is described as a function of the current, i, through the inductor

$$v = L\frac{di}{dt}. \tag{A.23}$$

For sinusoidal signals, $i = I_1 \cos(\omega t)$,

$$\begin{aligned} v = L\frac{d}{dt}\left[I_1 \cos(\omega t)\right] &= \\ -I_1 L\omega \sin(\omega t). \end{aligned} \tag{A.24}$$

If we work with complex signals, we get for the capacitor

$$v = \Re\left(V_1 e^{j\omega t}\right), \tag{A.25}$$

$$\begin{aligned} i = C\Re\left(\frac{d}{dt}V_1 e^{j\omega t}\right) &= \\ C\Re\left(j\omega V_1 e^{j\omega t}\right) &= \\ -V_1 C\omega \sin(\omega t), \end{aligned} \tag{A.26}$$

and for the inductor

$$i = \Re\left(I_1 e^{j\omega t}\right), \tag{A.27}$$

$$\begin{aligned} v = L\Re\left(\frac{d}{dt}I_1 e^{j\omega t}\right) &= \\ L\Re\left(j\omega I_1 e^{j\omega t}\right) &= \\ -I_1 L\omega \sin(\omega t), \end{aligned} \tag{A.28}$$

We see that for the complex signals the operator $\frac{d}{dt}$ is replaced by the multiplication with $j\omega$.

Therefore, we can define a complex impedance for the capacitor as

$$Z_C = \frac{V_1 e^{j\omega t}}{jC\omega V_1 e^{j\omega t}} = \frac{1}{j\omega C}. \tag{A.29}$$

Similarly, we may define a complex impedance for the inductor as

$$Z_L = \frac{jL\omega I_1 e^{j\omega t}}{I_1 e^{j\omega t}} = j\omega L. \tag{A.30}$$

To show the ease of working with complex signals, we will calculate the input impedance of a parallel circuit consisting of a resistor, R, an inductor, L, and a capacitor, C, see figure A.2.

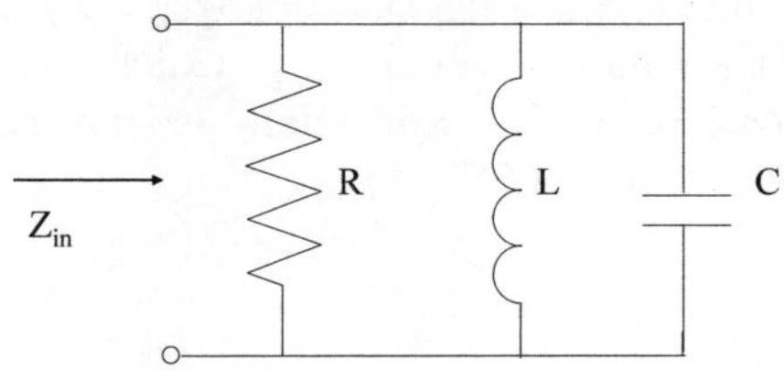

Fig. A.2 Parallel electric circuit consisting of resistor R, inductor L and capacitor C.

The complex input admittance Y_{in} is

$$Y_{in} = \frac{1}{R} + \frac{1}{j\omega L} + j\omega C = \frac{1}{R} - j\frac{1}{\omega L}\left(1 - \omega^2 LC\right), \tag{A.31}$$

and the complex input impedance is therefore found to be

$$Z_{in} = \frac{1}{Y_{in}} = \frac{1}{\frac{1}{R} - j\frac{1}{\omega L}\left(1 - \omega^2 LC\right)} = \frac{R}{1 - j\frac{R}{\omega L}\left(1 - \omega^2 LC\right)} = \frac{R\left[1 + j\frac{R}{\omega L}\left(1 - \omega^2 LC\right)\right]}{1 + \frac{R^2}{\omega^2 L^2}\left(1 - \omega^2 LC\right)^2}. \tag{A.32}$$

Finally, the 'real-world' input resistance is found as the real part of the complex input impedance

$$R_{in} = \Re\{Z_{in}\} = \frac{R}{1 + \frac{R^2}{\omega^2 L^2}\left(1 - \omega^2 LC\right)^2}. \tag{A.33}$$

The phase is found to be, see also figure A.2

$$\Psi_{R_{in}} = \arctan\left(\frac{\Im\{Z_{in}\}}{\Re\{Z_{in}\}}\right) = \frac{1}{\omega L}\left(1 - \omega^2 LC\right). \tag{A.34}$$

We see that the calculation of the input resistance of the parallel circuit, using complex numbers, is really straightforward, while a direct solution based on time-derivatives of real sinusoidal currents and voltages would have been far more complicated.

So, the general idea in dealing with a sinusoidal excitation, S_{in} (voltage or current) is to create a complex number, S_{in}^c, such that $\Re\{S_{in}^c\} = S_{in}$ (or $\Im\{S_{in}^c\} = S_{in}$), then calculate the response of the complex signal, S_{out}^c - which is easy since time-derivatives are replaced by multiplications by $j\omega$ - and finally extract the real response, S_{out} by taking $S_{out} = \Re\{S_{out}^c\}$ (or $S_{out} = \Im\{S_{out}^c\}$).

Appendix B
Vector Analysis

Scalars are defined by an amplitude only (e.g. charge or temperature). *Vectors* are not only defined by their amplitude, but also by their *direction*. This means that operations familiar for scalars, like addition and subtraction, multiplication and division and taking the derivative become more complicated for vectors.

B.1 NOTATION

A *scalar* in printed text is represented by a normal, though often cursive, letter (e.g. charge q). A vector in printed text is often represented by a bold face letter (e.g. electric field $\mathbf{E}$). In writing (like on a blackboard), scalars pose no problem, but vectors do if the above stated convention is followed. Therefore, vectors are also written as letters with an arrow on top of them. This arrow may be deformed into a half arrow or even a bar and this bar may also be found underneath the letter. The vector $\mathbf{a}$ may therefore be represented as

$$\mathbf{a}, \vec{a}, \bar{a}, \underline{a}. \tag{B.1}$$

The amplitude of this vector $\mathbf{a}$, represented by $|\mathbf{a}|$, is a scalar (a). For a vector in two-dimensional space, the relation between amplitude and direction is given by, see figure B.1,

$$\mathbf{a} = a\cos(\varphi)\hat{\mathbf{u}}_x + a\sin(\varphi)\hat{\mathbf{u}}_y, \tag{B.2}$$

where $\hat{\mathbf{u}}_x$ and $\hat{\mathbf{u}}_y$ are *unit vectors* in respectively the x- and y-direction. Unit vectors have a length *1* and are mutually perpendicular. In three dimensions, a vector may be decomposed into Cartesian coordinates as shown in figure B.2.

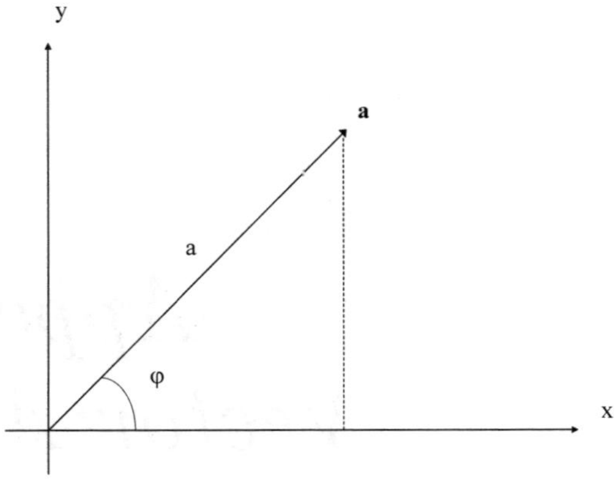

Fig. B.1 Two-dimensional vector **a** having amplitude a and direction φ.

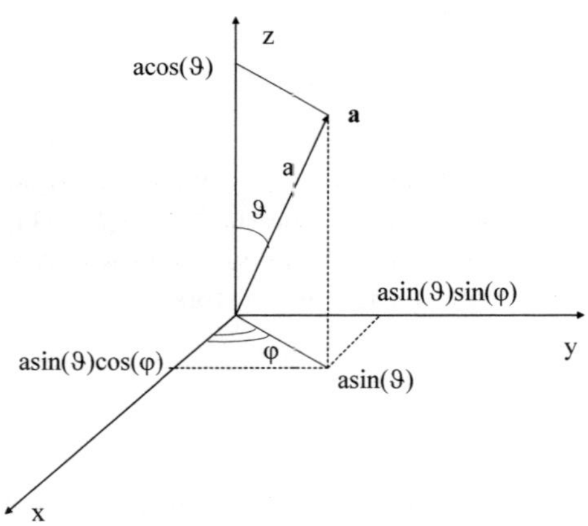

Fig. B.2 Three-dimensional vector **a** having amplitude a and direction ϑ, φ.

The vector **a**, decomposed into unit vectors, is given by

$$\mathbf{a} = a\sin(\vartheta)\cos(\varphi)\hat{\mathbf{u}}_x + a\sin(\vartheta)\sin(\varphi)\hat{\mathbf{u}}_y + a\cos(\vartheta)\hat{\mathbf{u}}_z, \tag{B.3}$$

were, again, use is made of $a = |\mathbf{a}|$.

The terms $\sin(\vartheta)\cos(\varphi)$, $\sin(\vartheta)\sin(\varphi)$ and $\cos(\vartheta)$ are often referred to as *direction cosines*. This terminology is not quite correct and finds its roots into a decomposition of the vector into unit vectors where other than the usual spherical coordinate angles are being used, see figure B.3.

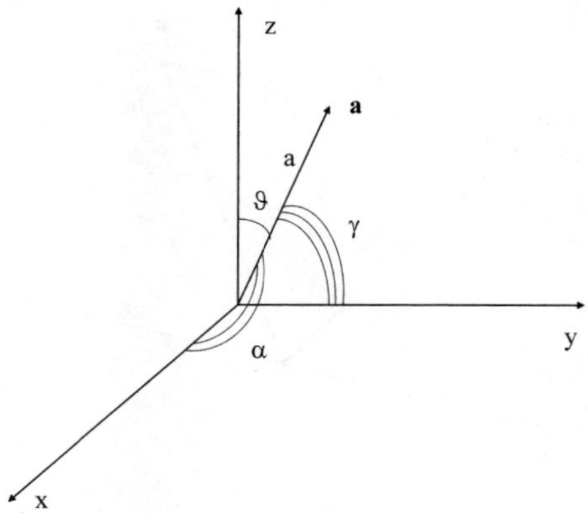

Fig. B.3 Three-dimensional vector **a** having amplitude a and direction ϑ, α, γ.

For the angles shown in figure B.3, we find

$$\mathbf{a} = a\cos(\alpha)\hat{\mathbf{u}}_x + a\cos(\gamma)\hat{\mathbf{u}}_y + a\cos(\vartheta)\hat{\mathbf{u}}_z, \tag{B.4}$$

explaining the name *direction cosines.*

B.2 ADDITION AND SUBTRACTION

The addition of vectors may be graphically represented using the 'parallelogram rule' as shown in figure B.4 for the addition of the vectors **a** and **b**.

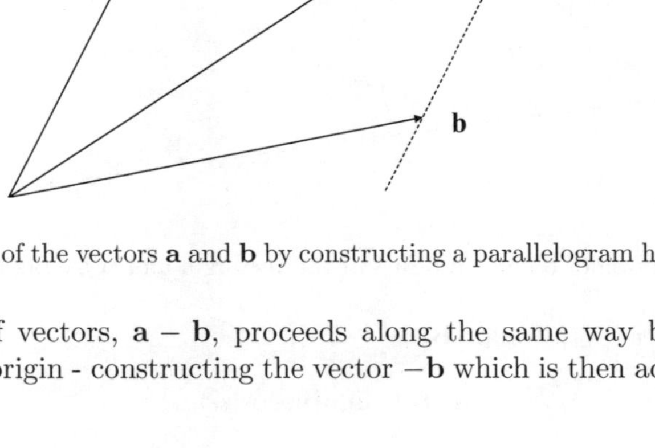

Fig. B.4 Addition of the vectors **a** and **b** by constructing a parallelogram having sides **a** and **b**.

Subtraction of vectors, $\mathbf{a} - \mathbf{b}$, proceeds along the same way by first - through reflection in the origin - constructing the vector $-\mathbf{b}$ which is then added to the vector **a**, see figure B.5.

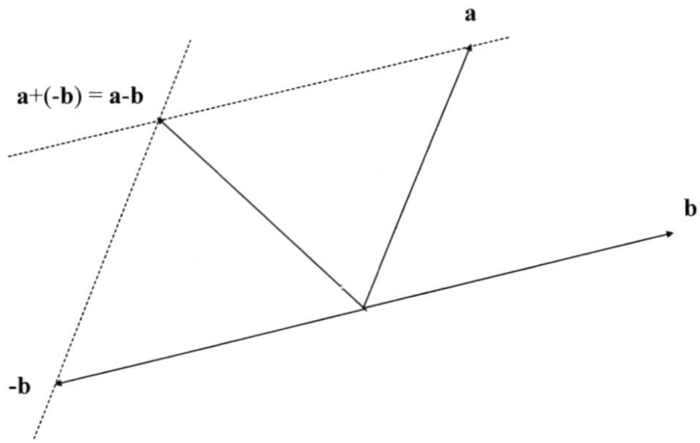

Fig. B.5 Subtraction of the vectors **a** and **b** by first constructing -**b**, followed by the addition of **a** and -**b**.

B.3 PRODUCTS

For the scalar quantities a and b only one product exists: $a \cdot b$. With vectors we have two possibilities: The so-called *scalar product* or *dot product* and the *vector product* or *cross product.*

B.3.1 Scalar Product or Dot Product

The dot product of two vectors results in a scalar (thus having only an amplitude and not a direction). The dot product of two vectors is obtained by projecting one of the vectors onto the other one, see figure B.6.

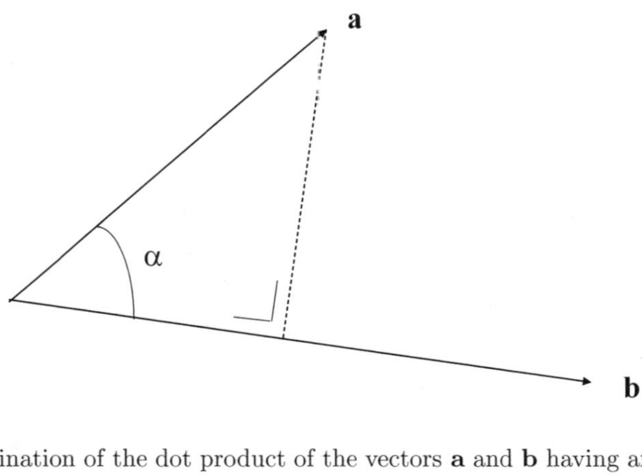

Fig. B.6 Determination of the dot product of the vectors **a** and **b** having an internal angle α.

The dot product (notation $\mathbf{a} \cdot \mathbf{b}$) is

$$\mathbf{a} \cdot \mathbf{b} = |\mathbf{a}|.|\mathbf{b}|. \cos(\alpha). \tag{B.5}$$

Note that for $\alpha = \pi/2$, the dot product becomes zero. It does not matter which one of the two vectors is projected onto the other one, or

$$\mathbf{a} \cdot \mathbf{b} = \mathbf{b} \cdot \mathbf{a}. \tag{B.6}$$

B.3.2 Vector Product or Cross Product

The cross product of two vectors, **a** and **b**, results in another vector (thus having amplitude *and* direction). The direction of the cross product follows that of a right-handed screw, see figure B.7. The amplitude of the cross product is equal to the area of the parallelogram defined by both vectors, **a** and **b**.

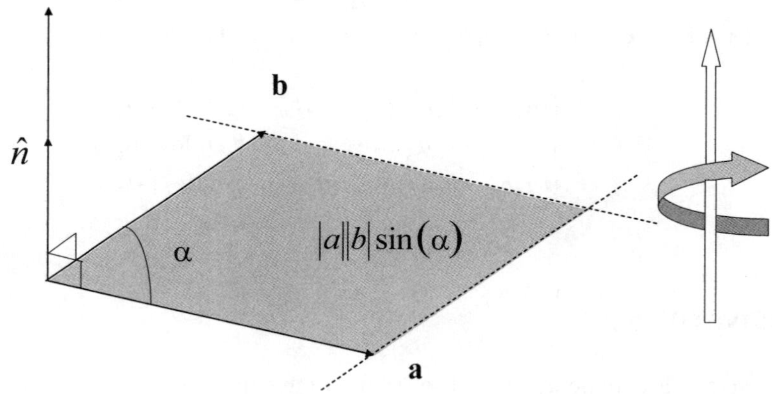

Fig. B.7 Determination of the cross product of the vectors **a** and **b**, having an internal angle α.

The cross product of the vectors **a** and **b** (notation $\mathbf{a} \times \mathbf{b}$) is

$$\mathbf{a} \times \mathbf{b} = |\mathbf{a}|.|\mathbf{b}|.\sin(\alpha)\hat{\mathbf{n}}, \tag{B.7}$$

where $\hat{\mathbf{n}}$ is the so-called *normal* (length *1*) on the plane defined by the vectors **a** and **b**. Unlike the situation in the dot product, in the cross product the order of the vectors does matter,

$$\mathbf{a} \times \mathbf{b} = -\mathbf{b} \times \mathbf{a}. \tag{B.8}$$

B.3.3 Threefold Product

With the previously defined dot product and cross product, two possibilities arise for the threefold vector product of the three vectors **a**, **b** and **c**, $(\mathbf{a} \times \mathbf{b}) \cdot \mathbf{c}$ and $(\mathbf{a} \times \mathbf{b}) \times \mathbf{c}$.

First, we determine the cross product of the vectors **a** and **b**. Thereto we decompose the vectors into Cartesian components

$$\begin{aligned} \mathbf{a} &= a_x \hat{\mathbf{u}}_x + a_y \hat{\mathbf{u}}_y + a_z \hat{\mathbf{u}}_z, \\ \mathbf{b} &= b_x \hat{\mathbf{u}}_x + b_y \hat{\mathbf{u}}_y + b_z \hat{\mathbf{u}}_z, \end{aligned} \tag{B.9}$$

and apply the 'determinant rule' for determining the cross product

$$(\mathbf{a}\times\mathbf{b}) = \begin{vmatrix} \hat{\mathbf{u}}_x & \hat{\mathbf{u}}_y & \hat{\mathbf{u}}_z \\ a_x & a_y & a_z \\ b_x & b_y & b_z \end{vmatrix} = (a_y b_z - b_y a_z)\hat{\mathbf{u}}_x - (a_x b_z - b_x a_z)\hat{\mathbf{u}}_y + (a_x b_y - b_x a_y)\hat{\mathbf{u}}_z. \quad \text{(B.10)}$$

Next, we take the dot product with $\mathbf{c} = c_x\hat{e}_x + c_y\hat{e}_y + c_z\hat{e}_z$, which results in

$$(\mathbf{a}\times\mathbf{b})\cdot\mathbf{c} = (a_y b_z - b_y a_z)c_x - (a_x b_z - b_x a_z)c_y + (a_x b_y - b_x a_y)c_z. \quad \text{(B.11)}$$

With equation (B.10), the 'determinant rule' and some calculation and regrouping of terms, we find

$$\begin{aligned}(\mathbf{a}\times\mathbf{b})\times\mathbf{c} = &\begin{vmatrix} \hat{\mathbf{u}}_x & \hat{\mathbf{u}}_y & \hat{\mathbf{u}}_z \\ a_y b_z - b_y a_z & a_z b_x - b_z a_x & a_x b_y - b_x a_y \\ c_x & c_y & c_z \end{vmatrix} = \\ &\{(a_z b_x - b_z a_x)c_z - (a_x b_y - b_x a_y)c_y\}\,\hat{\mathbf{u}}_x - \\ &\{(a_y b_z - b_y a_z)c_z - (a_x b_y - b_x a_y)c_x\}\,\hat{\mathbf{u}}_y + \\ &\{(a_y b_z - b_y a_z)c_y - (a_z b_x - b_z a_x)c_x\}\,\hat{\mathbf{u}}_z = \\ &(\mathbf{a}\cdot\mathbf{c})\mathbf{b} - (\mathbf{b}\cdot\mathbf{c})\mathbf{a}. \end{aligned} \quad \text{(B.12)}$$

B.4 DERIVATIVES

Suppose a vector $\mathbf{a}$ is dependent on the (scalar) parameter t

$$\mathbf{a} = \mathbf{a}(t). \quad \text{(B.13)}$$

For a small change δt the vector transforms into, see figure B.8:

$$\mathbf{a}(t+\delta t) = \mathbf{a} + \delta\mathbf{a}. \quad \text{(B.14)}$$

The derivative of $\mathbf{a}$ to t we define, in the usual way, as

$$\frac{d\mathbf{a}}{dt} = \lim_{\delta t\to 0}\frac{\delta\mathbf{a}}{\delta t}. \quad \text{(B.15)}$$

The derivative to t of the product of two scalars $a(t)$ and $b(t)$ is found by applying the 'chain rule'

$$\frac{d}{dt}(ab) = \frac{da}{dt}b + a\frac{db}{dt}. \quad \text{(B.16)}$$

The dot product of two vectors $\mathbf{a}(t)$ and $\mathbf{b}(t)$ may be written, using equation (B.5), as $ab\cos(\alpha)$, where α is the internal angle. Applying the 'chain rule' results in

$$\frac{d}{dt}(\mathbf{a}\cdot\mathbf{b}) = \frac{da}{dt}b\cos(\alpha) + a\frac{db}{dt}\cos(\alpha). \quad \text{(B.17)}$$

When δt in the limit approaches zero, also $\delta\mathbf{a}$ and $\delta\mathbf{b}$ will approach zero and therefore $\frac{d\mathbf{a}}{dt}$ and $\frac{d\mathbf{b}}{dt}$ will have the same direction as, respectively, $\mathbf{a}$ and $\mathbf{b}$.

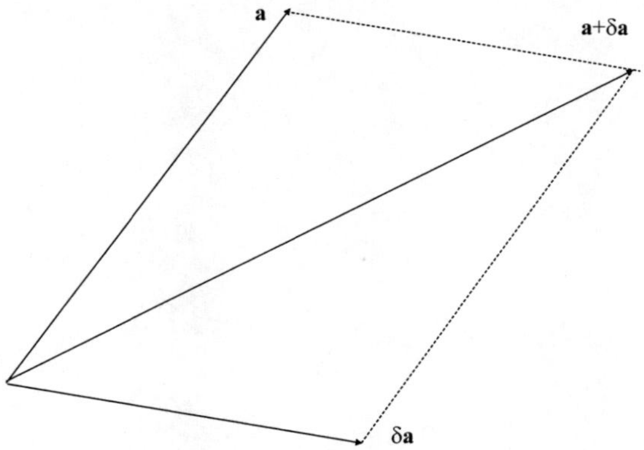

Fig. B.8 Translation of the vector $\mathbf{a}$(t) into the vector $\mathbf{a}(t + \delta t) = \mathbf{a} + \delta\mathbf{a}$.

Equation (B.17) then transforms into

$$\frac{d}{dt}(\mathbf{a} \cdot \mathbf{b}) = \frac{d\mathbf{a}}{dt} \cdot \mathbf{b} + \mathbf{a} \cdot \frac{d\mathbf{b}}{dt}. \tag{B.18}$$

In a similar way we find

$$\frac{d}{dt}(\mathbf{a} \times \mathbf{b}) = \frac{d\mathbf{a}}{dt} \times \mathbf{b} + \mathbf{a} \times \frac{d\mathbf{b}}{dt}. \tag{B.19}$$

Appendix C
Effective Aperture and Directivity

The effective aperture of an antenna is uniquely related to its directivity. By using the directivity and effective aperture of a short dipole, which are relatively easy calculated, a general interrelation between effective aperture and directivity, valid for any antenna may be derived.

Consider the two-antenna communication system of figure C.1. System *1* may be the transmitter, while system *2* is the receiver, or the other way around. The antennas are displaced a distance R and are assumed to be lined up with respect to polarisation and directivity. The directivity of antenna *1* is D_T, its effective aperture is A_{eT}. Directivity and effective area of antenna *2* are, respectively D_R and A_{eR}. We start by considering the first option; antenna *1* is transmitting and antenna *2* is receiving.

The totally radiated power by antenna *1* is P_T. If antenna *1* were an isotropic radiator, the power density, S_0, at distance R from antenna *1* would be

$$S_0 = \frac{P_T}{4\pi R^2}. \tag{C.1}$$

Due to the directive properties of antenna *1*, the power density, S_T at distance R is

$$S_T = S_0 D_T = \frac{P_T D_T}{4\pi R^2}. \tag{C.2}$$

Array and Phased Array Antenna Basics Hubregt J. Visser

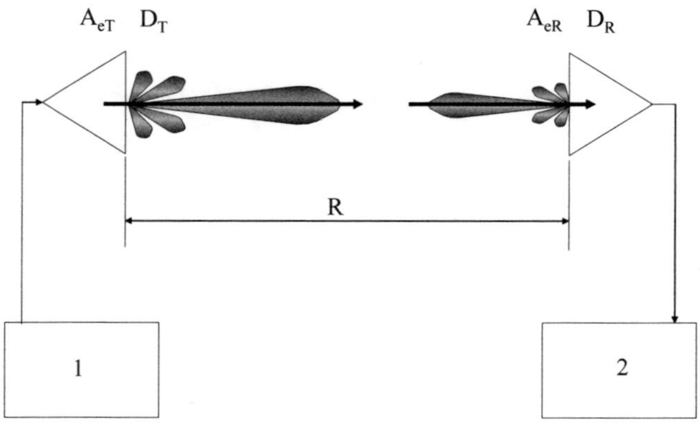

Fig. C.1 Two-antenna communication system. The antennas are displaced a distance R and are assumed to be lined up with respect to polarisation and directivity. The directivity of antenna *1* is D_T, its effective aperture is A_{eT}. Directivity and effective area of antenna *2* are, respectively D_R and A_{eR}.

The power received by antenna *2*, P_R is then

$$P_R = S_T A_{eR} = \frac{P_T D_T A_{eR}}{4\pi R^2}, \tag{C.3}$$

where A_{eR} is the effective aperture of antenna *2*. Rearranging this equation gives

$$D_T A_{eR} = \frac{P_R}{P_T}\left(4\pi R^2\right). \tag{C.4}$$

If we now let antenna *2* transmit P_T and we look at the received power at antenna *1*, which - by virtue of reciprocity - is equal to P_R, we find

$$D_R A_{eT} = \frac{P_R}{P_T}\left(4\pi R^2\right), \tag{C.5}$$

so

$$\frac{D_T}{A_{eT}} = \frac{D_R}{A_{eR}}. \tag{C.6}$$

If we now assume that in the two-antenna system, the transmitting antenna is an isotropic radiator, then $D_T = 1$ and the above equation transforms into

$$A_{eT_{ISO}} = \frac{A_{eR}}{D_R}, \tag{C.7}$$

which means that

the effective aperture of an isotropic radiator is equal to the ratio of effective aperture and directivity of any *antenna.*

If we take, for example, a short dipole, we may relatively easy calculate the effective area and directivity as

$$A_e = \frac{3}{8\pi}\lambda^2, \tag{C.8}$$

$$D = \frac{3}{2}, \tag{C.9}$$

where λ is the used wavelength. Therefore the effective area of an isotropic radiator is

$$A_{e_{ISO}} = \frac{\lambda^2}{4\pi}, \tag{C.10}$$

and thus for any antenna

$$\frac{\lambda^2}{4\pi} = \frac{A_e}{D}. \tag{C.11}$$

This gives us the sought after relation between effective aperture, A_e, and directivity, D, of an antenna

$$D = \frac{4\pi A_e}{\lambda^2}. \tag{C.12}$$

Appendix D
Transmission Line Theory

At microwave frequencies, the wavelengths have become so small that the physical dimensions of transmission lines and even those of lumped elements, like resistors, capacitors and inductors, are in the order of these wavelengths. This means that at these frequencies we have to consider effects of waves, like standing waves and reflections. Depending on the type of transmission line under consideration, these effects may be best characterised employing a field description or employing a circuit description. In this appendix we will limit ourselves to a circuit description of transmission lines.
The microwave frequency range is somewhat arbitrary, but in practice, frequencies between 300MHz and 30GHz may be considered as being in the microwave spectrum.

D.1 DISTRIBUTED PARAMETERS

A general long - i.e. *long* with respect to wavelength - two-wire transmission line may be characterised by distributed transmission line parameters, see figure D.1.

Herein R is the sum of resistances in both conductors per unit of length, G is the conductivity per unit of length, L is the self-inductance per unit of length and C is the capacity per unit of length.

Array and Phased Array Antenna Basics Hubregt J. Visser

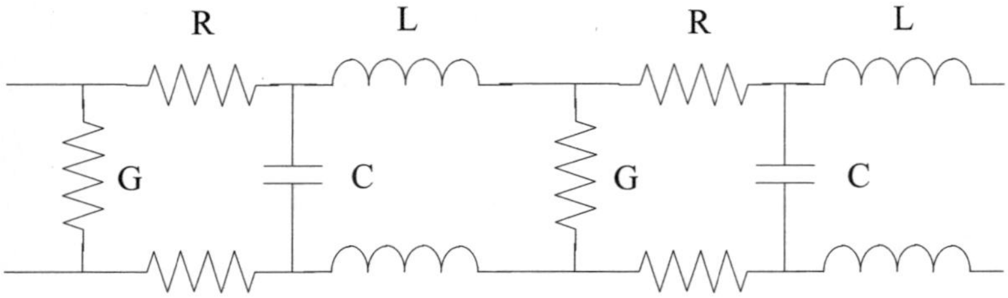

Fig. D.1 Distributed transmission line parameters in a general long two-wire transmission line.

When the distributed transmission line parameters are known, the characteristic impedance, Z_0, and propagation constant, γ_0, may be calculated as

$$Z_0 = \sqrt{\frac{R + j\omega L}{G + j\omega C}}, \tag{D.1}$$

$$\gamma_0 = \sqrt{(R + j\omega L)(G + j\omega C)}. \tag{D.2}$$

We will demonstrate this, as well as state the definitions for characteristic impedance and propagation constant, using an infinitesimal length, Δz, of transmission line as shown in figure D.2.

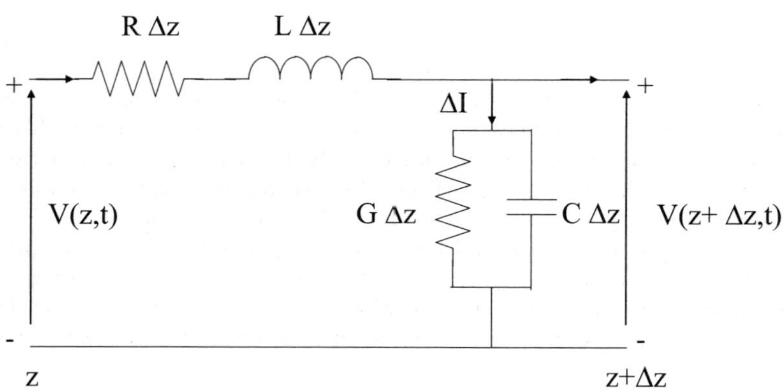

Fig. D.2 Equivalent network for an infinitesimal length of transmission line.

Applying the Kirchhoff *voltage* law to the circuit of figure D.2 gives

$$V(z,t) = R\Delta z I(z,t) + L\Delta z \frac{\partial I(z,t)}{\partial t} + V(z + \Delta z, t), \tag{D.3}$$

or, after rearranging terms,

$$-\frac{V(z + \Delta z, t) - V(z,t)}{\Delta z} = RI(z,t) + L\frac{\partial I(z,t)}{\partial t}. \tag{D.4}$$

In the limit $\Delta z \to 0$, this equation reduces to

$$-\frac{\partial V(z,t)}{\partial z} = RI(z,t) + L\frac{\partial I(z,t)}{\partial t}. \tag{D.5}$$

Next, applying the Kirchhoff *current* law to the circuit of figure D.2 gives

$$I(z,t) = I(z+\Delta z,t) + \Delta I = I(z+\Delta z,t) + G\Delta z V(z+\Delta z,t) + C\Delta z\frac{\partial V(z+\Delta z,t)}{\partial t}, \tag{D.6}$$

which may be written as

$$-\frac{I(z+\Delta z,t) - I(z,t)}{\Delta z} = GV(z+\Delta z,t) + C\frac{\partial V(z+\Delta z,t)}{\partial t}. \tag{D.7}$$

In the limit $\Delta z \to 0$, this equation reduces to

$$-\frac{\partial I(z,t)}{\partial z} = GV(z,t) + C\frac{\partial V(z,t)}{\partial t}. \tag{D.8}$$

If we now suppose time harmonic signals, i.e. signals having a (co)sinusoidal time-dependency, we may describe the voltages and currents using complex quantities:

$$V(z,t) = \Re\left\{V_s(z)e^{j\omega t}\right\}, \tag{D.9}$$
$$I(z,t) = \Re\left\{I_s(z)e^{j\omega t}\right\}, \tag{D.10}$$

where $\Re\{x\}$ means the real part of complex argument x. The parameter $\omega = 2\pi f$ is the angular frequency, where f is the frequency.

Substitution of equations (D.9) and (D.10) into equations (D.5) and (D.8) gives

$$-\frac{dV_s}{dz} = (R + j\omega L)I_s, \tag{D.11}$$
$$-\frac{dI_s}{dz} = (G + j\omega C)V_s. \tag{D.12}$$

Taking the derivative to z of equation (D.11) and substituting equation (D.12) into that equation yields

$$\frac{d^2V_s}{dz^2} = (R + j\omega L)(G + j\omega C)V_s. \tag{D.13}$$

This equation may be written as

$$\frac{d^2V_s}{dz^2} - \gamma^2 V_s = 0, \tag{D.14}$$

where

$$\gamma = \alpha + j\beta = \sqrt{(R + j\omega L)(G + j\omega C)}. \tag{D.15}$$

Equation (D.14) is known as the *wave equation* or *Helmholtz equation*, γ is known as the *propagation constant.* The propagation constant consists of an *attenuation constant*, α, and a *phase constant*, β, where $\beta = \frac{2\pi}{\lambda}$, λ being the wavelength.

For the current, a wave equation of identical form may be derived

$$\frac{d^2 I_s}{dz^2} - \gamma^2 I_s = 0. \tag{D.16}$$

Solutions of the Helmholtz equations for voltage and current are

$$V_s(z) = V_0^+ e^{-\gamma z} + V_0^- e^{+\gamma z}, \tag{D.17}$$

$$I_s(z) = I_0^+ e^{-\gamma z} + I_0^- e^{+\gamma z}, \tag{D.18}$$

where V_0^+ and I_0^+ are the amplitudes of, respectively, voltage and current waves travelling in positive z-direction and V_0^- and I_0^- are the amplitudes of, respectively, voltage and current waves travelling in negative z-direction.

The characteristic impedance, Z_0, of the transmission line is defined as the ratio of voltage and current amplitude of in positive direction travelling voltage and current waves:

$$Z_0 = \frac{V_0^+}{I_0^+}. \tag{D.19}$$

With use of equations (D.11), (D.12), (D.17) and (D.18), we find for the characteristic impedance

$$Z_0 = \frac{V_0^+}{I_0^+} = \frac{R + j\omega L}{\gamma} = \frac{\gamma}{G + j\omega C}, \tag{D.20}$$

and upon substitution of equation (D.15)

$$Z_0 = \sqrt{\frac{R + j\omega L}{G + j\omega C}}. \tag{D.21}$$

D.2 GUIDED WAVES

We have seen in the previous section that by introducing complex quantities, we may describe voltage and current at any place on a transmission line as a superposition of a (voltage or current) wave travelling in positive direction and one travelling in negative direction. We used this concept, in combination with that of distributed transmission line parameters, to derive expressions for the propagation constant and characteristic impedance of a transmission line.

In this section we will further explore this guided wave[1] property of transmission lines to derive practical parameters like *voltage standing wave ratio* (VSWR), *reflection factor*, *impedance* and *input impedance*. These parameters prove to be very useful in designing microwave networks, as they allow us to design subsystems and predict the behaviour of the interconnected subsystems on basis of the values of these parameters.

In order to analyse transmission lines, we will take the general two-wire transmission line of figure D.3 and look in detail at the voltage between and the current through the two wires.

[1] *Guided* waves as opposed to *unguided* waves as we encounter in a radio link.

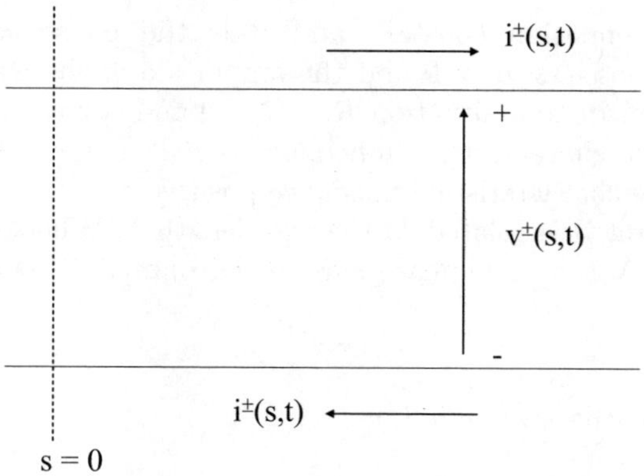

Fig. D.3 Voltage and current of a two-wire transmission line.

Voltage and current in the long transmission line can propagate as a wave going in positive s-direction and as a wave going in negative s-direction. For the wave propagating in positive s-direction

$$v^+(s,t) = \Re\left\{V^+(s)e^{j\omega t}\right\}, \tag{D.22}$$

where

$$V^+(s) = Ae^{-\gamma s}, \tag{D.23}$$

and

$$i^+(s,t) = \Re\left\{I^+(s)e^{j\omega t}\right\}, \tag{D.24}$$

where

$$I^+(s) = \frac{A}{Z_0}e^{-\gamma s}, \tag{D.25}$$

Z_0 being the characteristic impedance of the transmission line. A is for the moment an unknown complex amplitude coefficient.

For the wave propagating in negative s-direction

$$v^-(s,t) = \Re\left\{V^-(s)e^{j\omega t}\right\}, \tag{D.26}$$

where

$$V^-(s) = Be^{+\gamma s}, \tag{D.27}$$

and

$$i^-(s,t) = \Re\left\{I^-(s)e^{j\omega t}\right\}, \tag{D.28}$$

where

$$I^-(s) = -\frac{B}{Z_0}e^{+\gamma s}. \tag{D.29}$$

B is for the moment an unknown complex amplitude coefficient.

The propagation constant is also in general complex,

$$\gamma = \alpha + j\beta, \tag{D.30}$$

where α is the attenuation constant and β is the phase constant. When the transmission line is lossless, $\alpha = 0$ and the amplitude of the wave is constant over the transmission line. In that situation $R = G = 0$ and equation (D.21) reveals that for that situation the characteristic impedance is real. Using the sign convention of figure D.3 makes the characteristic impedance positive.

The phase constant β is related to the wavelength λ. Whenever s increases with an amount equal to λ, the same phase must be encountered: $\beta\lambda = 2\pi$ and thus

$$\beta = \frac{2\pi}{\lambda}. \tag{D.31}$$

β is also known as the *wave number*.

D.2.1 VSWR and Reflection Factor

When a travelling wave at some point is totally or partially reflected, a standing wave is created. The ratio of the absolute values of the complex voltage wave amplitude at maximum and at minimum is known as the *voltage standing wave ratio* (VSWR), S.

$$S = \frac{|V|_{max}}{|V|_{min}} = \frac{|V^+| + |V^-|}{|V^+| - |V^-|}. \tag{D.32}$$

The reflection factor, ρ, is defined as

$$\rho = \frac{V^-}{V^+} = \frac{B}{A} e^{2\gamma s}. \tag{D.33}$$

From equations (D.32) and (D.33) follows

$$S = \frac{1 + |\rho|}{1 - |\rho|}, \tag{D.34}$$

and

$$|\rho| = \frac{S - 1}{S + 1}. \tag{D.35}$$

D.2.2 Impedance and Relative Impedance

The impedance, Z - which is a function of the position s along the transmission line, just like ρ - is defined as

$$Z = \frac{V}{I} = \frac{V^+ + V^-}{I^+ + I^-}. \tag{D.36}$$

The relative impedance, z, is the impedance Z normalised to the characteristic impedance of the transmission line

$$z = \frac{Z}{Z_0} = \frac{1 + \frac{B}{A} e^{2\gamma s}}{1 - \frac{B}{A} e^{2\gamma s}}. \tag{D.37}$$

In determining the impedance we suppose the transmission line being cut at position s. The impedance is related to the part of the transmission line to the right of the cut. An excitation voltage $V(s)$ then results in a current $I(s)$.

Substitution of equation (D.33) into equation (D.37) leads to

$$z = \frac{1+\rho}{1-\rho}, \tag{D.38}$$

and

$$\rho = \frac{z-1}{z+1}. \tag{D.39}$$

D.3 INPUT IMPEDANCE OF A TRANSMISSION LINE

Assume a transmission line of length l, meaning $0 \leq s \leq l$. We will add a subscript $_i$ to the parameters Z, z, S and ρ when they refer to the input of the transmission line ($s = 0$). When they refer to the output of the transmission line ($s = l$), we will use a subscript $_u$. For an arbitrary position s ($0 < s < l$) we will use no subscript.

We now want to express the input impedance, Z_i, as a function of the load impedance, Z_u, at the end of the transmission line.

With use of equation (D.33) we find

$$\rho_i = \rho_u e^{-2\gamma l}. \tag{D.40}$$

Upon substitution of this result into equation (D.38) and using equation (D.39), we find for the normalised input impedance

$$z_i = \frac{1+\rho_u e^{-2\gamma l}}{1-\rho_u e^{-2\gamma l}} = \frac{1+\frac{z_u-1}{z_u+1}e^{-2\gamma l}}{1-\frac{z_u-1}{z_u+1}e^{-2\gamma l}}. \tag{D.41}$$

After some straightforward, though lengthy calculations,[2] this equation may be rewritten into

$$z_i = \frac{z_u \cosh(\gamma l) + \sinh(\gamma l)}{z_u \sinh(\gamma l) + \cosh(\gamma l)}. \tag{D.42}$$

For a lossless transmission line, $\gamma = j\beta = j\frac{2\pi}{\lambda}$, so that

$$z_i = \frac{z_u + j\tan\left(2\pi\frac{l}{\lambda}\right)}{1 + jz_u \tan\left(2\pi\frac{l}{\lambda}\right)}. \tag{D.43}$$

D.4 TERMINATED LOSSLESS TRANSMISSION LINE

Using equation (D.43), that relates the normalised input impedance to the normalised load impedance of a lossless transmission line of length l, we will now look into some special situations.

[2]Start with multiplying the numerator and denominator of the equation with $\frac{1}{2}(z_u + 1)e^{\gamma l}$.

D.4.1 Matched Load

A transmission line is terminated into a *matched load*, if the load impedance is the complex conjugate of the characteristic impedance of the transmission line. Since - by virtue of equation (D.43) - we assume our transmission line to be lossless, we have seen that the characteristic impedance is real. Therefore, the line is terminated into a matched load if the load impedance is equal to the characteristic impedance of the transmission line.

So, we have the situation that $Z_u = Z_0$ and therefore $z_u = 1$. This makes that everywhere on the transmission line $Z = Z_0$ and $\rho = 0$ (no reflection). Also, $S = 1$.

D.4.2 Short Circuit

When we terminate the transmission line into a short circuit, $Z_u = 0$ and therefore also $z_u = 0$. Substitution of $z_u = 0$ into equation (D.43) gives

$$z_i = j\tan\left(\frac{2\pi l}{\lambda}\right). \qquad \text{(D.44)}$$

Upon a closer inspection of this equation we see that - going from the short circuit over the transmission line to the input - we alternately, at intervals of a quarter wavelength, encounter an impedance that is either purely inductive or purely capacitive, see also figure D.4.

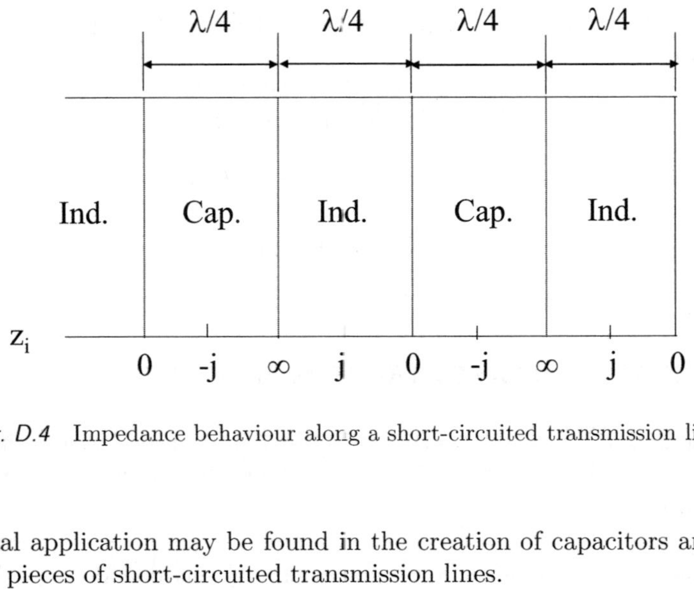

Fig. D.4 Impedance behaviour along a short-circuited transmission line.

A practical application may be found in the creation of capacitors and inductors by means of pieces of short-circuited transmission lines.

D.4.3 Open Circuit

If we leave the transmission line open at the end, $Z_u = \infty$ and thus $z_u = \infty$. Substitution of $z_u = \infty$ into equation (D.43) gives

$$z_i = -j\frac{1}{\tan\left(\frac{2\pi l}{\lambda}\right)}. \tag{D.45}$$

We find an input impedance behaviour over the transmission line analogous to the situation depicted in figure D.4, but shifted over a quarter of a wavelength.

D.4.4 Imaginary Unit Termination

When $Z_u = jZ_0$, remembering that Z_0 is real, $z_u = j$. Substitution of $z_u = j$ into equation (D.43) gives

$$z_i = j\tan\left(2\pi\frac{l}{\lambda} + \frac{\pi}{4}\right). \tag{D.46}$$

Again, we find an input impedance behaviour similar to the situation as depicted in figure D.4, but now shifted over one-eighth of a wavelength.

D.4.5 Real Termination

If Z_u is real, $z_u = r_u$. We may distinguish two situations:

D.4.5.1 $\boldsymbol{r_u < 1}$ For this situation

$$|\rho| = \frac{1 - r_u}{1 + r_u}, \tag{D.47}$$

and

$$S = \frac{1}{r_u} > 1. \tag{D.48}$$

D.4.5.2 $\boldsymbol{r_u > 1}$ For this situation

$$|\rho| = \frac{r_u - 1}{r_u + 1}, \tag{D.49}$$

and

$$S = r_u > 1. \tag{D.50}$$

D.5 QUARTER WAVELENGTH IMPEDANCE TRANSFORMER

We have seen that if the termination of a lossless transmission line, i.e. the impedance connected at the end of the transmission line, is not identical to the characteristic impedance of that transmission line, reflections will occur.

Not only are these reflections unwanted due to the fact that they prohibit a complete signal transfer, they are also unwanted since they distort the quality of the signal transferred. When a *mismatch* exists not only at the end of the transmission line but also at the beginning, the generator side, part of the signal power will reach the load (at the end of the line) after a certain time delay.

To overcome the negative effects of a mismatch, an impedance transformer may be placed in between transmission line or microwave circuit and load to make a reflection free transition between transmission line or microwave circuit and load.

Such an impedance transformer may be realised very easily, employing a piece of transmission line with the right characteristic impedance.

Starting with equation (D.43), we take a piece of transmission line of characteristic impedance Z_c and length equal to a quarter of a wavelength at the frequency of operation. The load impedance is Z_L. The unnormalised input impedance is then found to be

$$Z_{in} = Z_c z_{in} = Z_c \left[\frac{Z_L + jZ_c \tan\left(\frac{\pi}{2}\right)}{Z_c + jZ_L \tan\left(\frac{\pi}{2}\right)} \right] = \frac{Z_c^2}{Z_L}. \tag{D.51}$$

So, if we take a piece of transmission line, a quarter of a wavelength long (at the operating frequency) and dimension this transmission line such that its characteristic impedance is equal to

$$Z_c = \sqrt{Z_{in} Z_L}, \tag{D.52}$$

where Z_{in} is the required input impedance (usually the impedance level of the circuit connected to the load), and place this piece of transmission line between circuit output and load, we have created a reflectionless transition from circuit to load, see figure D.5.

The equivalent of the impedance transformer and load impedance Z_L is a new load impedance equal to Z_{in}.

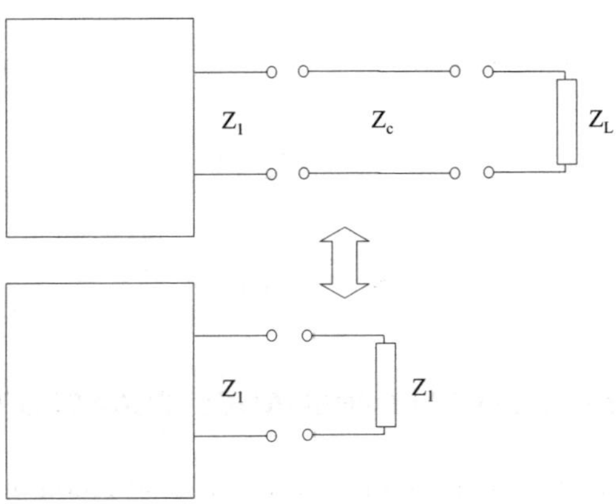

Fig. D.5 Impedance matching using a quarter wavelength impedance transformator.

Appendix E
Scattering Matrix

At microwave frequencies it is easier to work with waves than with voltages and currents. A microwave network therefore is most easily described in terms of waves. The scattering matrix relates amplitudes and phases of incoming to outgoing waves for an n-port network. The incoming and outgoing waves are suitably normalised for a network experiencing identical impedance levels at all ports. For networks experiencing different impedance levels at the ports, it is convenient to use the *unnormalised* scattering matrix. The unnormalised scattering matrix is, however, uniquely related to the normalised scattering matrix.

E.1 NORMALISED SCATTERING MATRIX

In an n-port network (see figure E.1), a wave that is incident at one port may be distributed to one or more other ports. Due to internal reflections, power from the input will be distributed not only to the intentioned ports but - to various degrees - to all other ports.

The complex amplitudes of the ingoing waves are denoted a_i, those of the outgoing waves b_i, where $i = 1, 2, \ldots, n$.

To ease the discussion of the scattering, we will look now at a two-port network as shown in figure E.2.

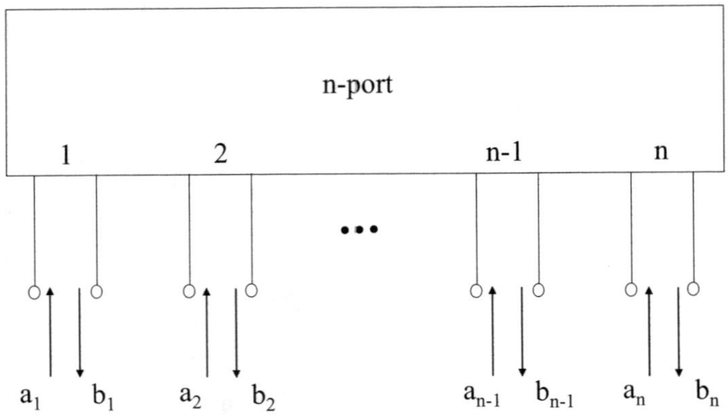

Fig. E.1 In- and outgoing waves in an n-port microwave network.

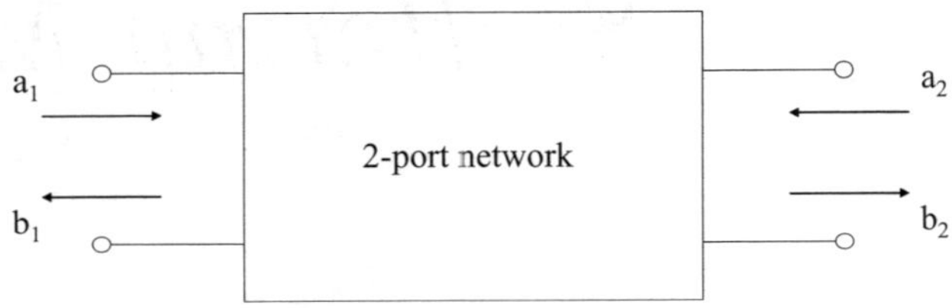

Fig. E.2 In- and outgoing waves in a two-port microwave network.

The complex amplitudes of the outgoing waves are related to those of the ingoing waves through

$$\begin{pmatrix} b_1 \\ b_2 \end{pmatrix} = \begin{pmatrix} S_{11} & S_{12} \\ S_{21} & S_{22} \end{pmatrix} \begin{pmatrix} a_1 \\ a_2 \end{pmatrix}, \tag{E.1}$$

or

$$\begin{aligned} b_1 &= S_{11}a_1 + S_{12}a_2, \\ b_2 &= S_{21}a_1 + S_{22}a_2. \end{aligned} \tag{E.2}$$

The scattering coefficients, i.e. the elements of the scattering matrix are defined as

$$S_{11} = \left. \frac{b_1}{a_1} \right|_{a_2=0}, \tag{E.3}$$

$$S_{12} = \left. \frac{b_1}{a_2} \right|_{a_1=0}, \tag{E.4}$$

$$S_{21} = \left. \frac{b_2}{a_1} \right|_{a_2=0}, \tag{E.5}$$

$$S_{22} = \left. \frac{b_2}{a_2} \right|_{a_1=0}. \tag{E.6}$$

S_{11} therefore is the reflection coefficient at port 1 when port 2 is not excited and S_{22} is the reflection coefficient at port 2 when port 1 is not excited. S_{12} is the transfer from port 2 to port 1 when port 1 is not excited and S_{21} is the transfer from port 1 to port 2 when port 2 is not excited.

The normalisation of the complex wave amplitudes is such that $\frac{1}{2}a_i a_i^*$ is the available power at port i and $\frac{1}{2}b_i b_i^*$ is the power emergent from port i, where $i = 1, 2$. This means that the wave amplitudes must satisfy

$$a_i = \frac{v_i^+}{\sqrt{Z_0}}, \tag{E.7}$$

$$b_i = \frac{v_i^-}{\sqrt{Z_0}}, \tag{E.8}$$

where $i = 1, 2$ and v_i^+ and v_i^- are, respectively, the voltage corresponding to the wave going into port i and the voltage corresponding to the wave leaving port i. The impedances looking into either port 1 or port 2 are assumed to be equal to Z_0.

This can be understood by evaluating the average power, P_i, flowing into port i. Thereto we look at the total voltage, v_i, and total current, i_i, at port i, using transmission line theory (see appendix D):

$$v_i = v_i^+ + v_i^- = \sqrt{Z_0}\,(a_i + b_i)\,, \tag{E.9}$$

$$i_i = \frac{1}{Z_0}\left(v_i^+ - v_i^-\right) = \frac{1}{\sqrt{Z_0}}\,(a_i - b_i)\,. \tag{E.10}$$

The average power flowing into port i is

$$P_i = \frac{1}{2}\Re\left\{v_i i_i^*\right\} = \frac{1}{2}\left(a_i a_i^* - b_i b_i^*\right), \tag{E.11}$$

which is the power of the incident wave at port i minus the power in the reflected wave at that same port.

Although we have demonstrated the normalisation using a two-port, the theory is general and applies to any n-port. Since we are especially interested in three ports (split-T power dividers), we will demonstrate the unnormalised scattering matrix properties for a three-port, keeping in mind that the concept is more general.

E.2 UNNORMALISED SCATTERING MATRIX

In general, the relationship between incoming and outgoing voltage waves of a three-port is given by

$$\begin{pmatrix} b_1 \\ b_2 \\ b_3 \end{pmatrix} = \begin{pmatrix} S_{11} & S_{12} & S_{13} \\ S_{21} & S_{22} & S_{23} \\ S_{31} & S_{32} & S_{33} \end{pmatrix} \begin{pmatrix} a_1 \\ a_2 \\ a_3 \end{pmatrix}, \tag{E.12}$$

where a_i and b_i, $i = 1, 2, 3$, are the (complex) amplitudes of incoming and outgoing voltage waves, that are normalised such that $\frac{1}{2}a_i a_i^*$ is the average incoming power at port i and $\frac{1}{2}b_i b_i^*$ is the average outgoing power at port i.

This *normalised* scattering matrix is used whenever the three-port exhibits equal impedance levels at all ports. However, when the three-port has unequal impedance levels at the ports - as is the case for a split-T power divider - the *unnormalised* voltage scattering matrix T is used. T relates the amplitudes of (unnormalised) incoming and outgoing voltage waves through

$$\begin{pmatrix} d_1 \\ d_2 \\ d_3 \end{pmatrix} = \begin{pmatrix} T_{11} & T_{12} & T_{13} \\ T_{21} & T_{22} & T_{23} \\ T_{31} & T_{32} & T_{33} \end{pmatrix} \begin{pmatrix} c_1 \\ c_2 \\ c_3 \end{pmatrix}. \tag{E.13}$$

The unnormalised voltage wave amplitudes are related to the normalised ones through [1]

$$d_i = g_i b_i, \tag{E.14}$$

$$c_i = g_i a_i, \tag{E.15}$$

for $i = 1, 2, 3$, where

$$g_i = \sqrt{Z_0^i}. \tag{E.16}$$

Z_0^i is the impedance level at port i.

With use of equations (E.12) to (E.16) we find

$$\begin{aligned} d_1 = g_1 b_1 = g_1 S_{11} a_1 + g_1 S_{12} a_2 + g_1 S_{13} a_3 = \\ T_{11} c_1 + T_{12} c_2 + T_{13} c_3 = \\ T_{11} g_1 a_1 + T_{12} g_2 a_2 + T_{13} g_3 a_3, \end{aligned} \tag{E.17}$$

and thus

$$S_{11} = T_{11}, \tag{E.18}$$

$$S_{12} = \frac{g_2}{g_1} T_{12}, \tag{E.19}$$

$$S_{13} = \frac{g_3}{g_1} T_{13}. \tag{E.20}$$

In the same way we find

$$S_{21} = \frac{g_1}{g_2} T_{21}, \tag{E.21}$$

$$S_{22} = T_{22}, \tag{E.22}$$

$$S_{23} = \frac{g_3}{g_2} T_{23}, \tag{E.23}$$

and

$$S_{31} = \frac{g_1}{g_3} T_{31}, \tag{E.24}$$

$$S_{32} = \frac{g_2}{g_3} T_{32}, \tag{E.25}$$

$$S_{33} = T_{33}. \tag{E.26}$$

Using the reciprocity of the S-matrix [1], we find

$$T_{12} = \left(\frac{g_1}{g_2}\right)^2 T_{21}, \tag{E.27}$$

$$T_{13} = \left(\frac{g_1}{g_3}\right)^2 T_{31}, \tag{E.28}$$

and

$$T_{23} = \left(\frac{g_2}{g_3}\right)^2 T_{32}. \tag{E.29}$$

REFERENCES

1. C.G. Montgomery, R.H. Dicke and E.M. Purcell, *Principles of Microwave Circuits*, McGraw-Hill, New York, 1948.

Appendix F *Voltage Incident at a Transmission Line*

For a transmission line connected at one side to a generator with internal impedance equal to the characteristic impedance of the transmission line and connected at the other side to a load also equal to the characteristic impedance, the incident voltage on the transmission line is equal to half the generator voltage.

In the even and odd mode analysis of a split-T power divider, we stated that the incident voltage at a transmission line is half the generator voltage. In this appendix we will show that this is true.

Therefore we start with the circuit as shown in figure F.1.

The voltage at the beginning ($s = 0$) and end of the line ($s = l$) are found by inspection from figure F.1

$$V(0) = E - Z_G I(0), \tag{F.1}$$

$$V(l) = Z_L I(l). \tag{F.2}$$

The voltage and current along the transmission line may be written as, see also appendix D

$$V(s) = Ae^{-\gamma s} + Be^{+\gamma s}, \tag{F.3}$$

$$I(s) = \frac{A}{Z_c}e^{-\gamma s} - \frac{B}{Z_c}e^{+\gamma s}, \tag{F.4}$$

Array and Phased Array Antenna Basics Hubregt J. Visser

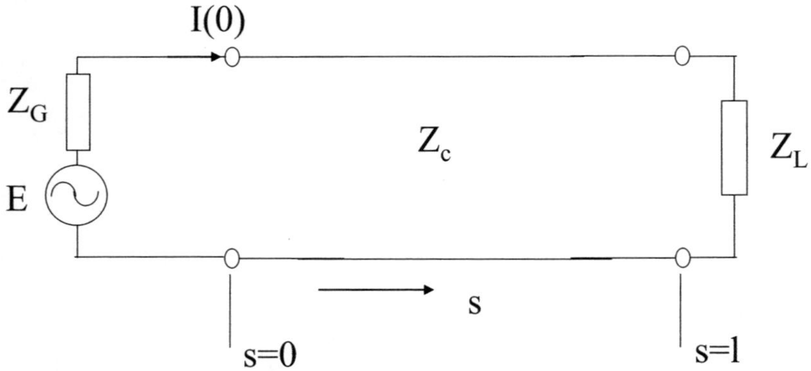

Fig. F.1 Transmission line of length l and characteristic impedance Z_c connected at one side to a generator of voltage E and internal impedance Z_G and at the other side to a load impedance Z_L.

where γ is the propagation constant, A is the complex voltage wave amplitude of the voltage wave going to the right and B is the complex voltage wave amplitude of the voltage wave going to the left.

Substitution of equations (F.3) and (F.4) for $s = 0$ into equation (F.1) gives

$$\left(1 + \frac{Z_G}{Z_c}\right) A = E - \left(1 - \frac{Z_G}{Z_c}\right) B. \tag{F.5}$$

Substitution of equations (F.3) and (F.4) for $s = l$ into equation (F.2) gives

$$\left(1 - \frac{Z_L}{Z_c}\right) Ae^{-\gamma l} = -\left(1 + \frac{Z_L}{Z_c}\right) Be^{+\gamma l}. \tag{F.6}$$

From equations (F.5) and (F.6) we obtain:

$$A = -\frac{Z_c\,(Z_c + Z_L)\,e^{+\gamma l} E}{(Z_c - Z_G)\,(Z_c - Z_L)\,e^{-\gamma l} - (Z_c + Z_G)\,(Z_c + Z_L)\,e^{+\gamma l}}, \tag{F.7}$$

$$B = \frac{Z_c\,(Z_c - Z_L)\,e^{-\gamma l} E}{(Z_c - Z_G)\,(Z_c - Z_L)\,e^{-\gamma l} - (Z_c + Z_G)\,(Z_c + Z_L)\,e^{+\gamma l}}. \tag{F.8}$$

If $Z_c = Z_L$, then

$$A = \frac{Z_c E}{Z_c + Z_G}, \tag{F.9}$$

and

$$B = 0. \tag{F.10}$$

If also $Z_c = Z_G$, we finally find that

$$A = \frac{1}{2} E. \tag{F.11}$$

Appendix G *Cascaded Scattering Matrices*

A split-T Power divider in microstrip or any other transmission line technology may be analysed as consisting of a basic power divider and two impedance transformers. The basic power divider is a transmission line of desired characteristic impedance that splits into two transmission lines of zero length with characteristic impedances determined by the desired power ratio of the two outputs. The impedance transformers are transmission lines of quarter lambda length at the centre frequency that transform these impedance levels to the desired characteristic impedance. The basic power divider may be described by a 3×3 scattering matrix, each impedance transformer may be described by a 2×2 scattering matrix. The overall scattering matrix may be described by a 3×3 scattering matrix.

The cascading process of the basic power divider and the two impedance transformers is schematically shown in figure G.1.

The scattering matrix of the basic power divider is denoted S, the scattering matrices of the two impedance transformers are denoted S', respectively S''.

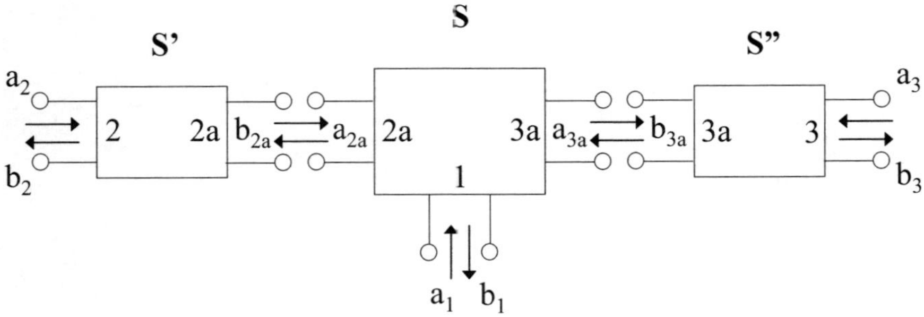

Fig. G.1 Cascading of three-port and two two-ports.

The complex amplitudes of outgoing waves and incoming waves of the basic power divider are related through, see figure G.1

$$b_1 = S_{11}a_1 + S_{12}a_{2a} + S_{13}a_{3a}, \tag{G.1}$$
$$b_{2a} = S_{21}a_1 + S_{22}a_{2a} + S_{23}a_{3a}, \tag{G.2}$$
$$b_{3a} = S_{31}a_1 + S_{32}a_{2a} + S_{33}a_{3a}. \tag{G.3}$$

The complex amplitudes of outgoing waves and incoming waves of the impedance transformer at the left of figure G.1 are related through

$$a_{2a} = S'_{11}b_{2a} + S'_{12}a_2, \tag{G.4}$$
$$b_2 = S'_{21}b_{2a} + S'_{22}a_2, \tag{G.5}$$

and for the impedance transformer at the right of the figure, we find

$$a_{3a} = S''_{11}b_{3a} + S''_{12}a_3, \tag{G.6}$$
$$b_3 = S''_{21}b_{3a} + S''_{22}a_3. \tag{G.7}$$

Substitution of equations (G.4) and (G.6) into equations (G.1), (G.2) and (G.3) results in

$$b_1 = S_{11}a_1 + S_{12}S'_{11}b_{2a} + S_{12}S'_{12}a_2 + S_{13}S''_{11}b_{3a} + S_{13}S''_{12}a_3, \tag{G.8}$$
$$b_{2A} = S_{21}a_1 + S_{22}S'_{11}b_{2a} + S_{22}S'_{12}a_2 + S_{23}S''_{11}b_{3a} + S_{23}S''_{12}a_3, \tag{G.9}$$
$$b_{3A} = S_{31}a_1 + S_{32}S'_{11}b_{2a} + S_{32}S'_{12}a_2 + S_{33}S''_{11}b_{3a} + S_{33}S''_{12}a_3. \tag{G.10}$$

Rearranging the terms in equation (G.10) leads to

$$b_{3a} = \frac{S_{31}}{1 - S_{33}S''_{11}}a_1 + \frac{S_{32}S'_{12}}{1 - S_{33}S''_{11}}a_2 + \frac{S_{33}S'_{12}}{1 - S_{33}S''_{11}}a_3 + \frac{S_{32}S'_{11}}{1 - S_{33}S''_{11}}b_{2a}. \tag{G.11}$$

Substitution of equation (G.11) in equation (G.9) results in an expression for b_{2a} in terms of a_1, a_2 and a_3

$$\begin{aligned} b_{2a} = & \frac{S_{21}(1 - S_{33}S''_{11}) + S_{23}S''_{11}S_{31}}{(1 - S_{22}S'_{11})(1 - S_{33}S''_{11}) - S_{23}S''_{11}S_{32}S'_{11}} a_1 + \\ & \frac{S_{22}S'_{12}(1 - S_{33}S''_{11}) + S_{23}S''_{11}S_{32}S'_{12}}{(1 - S_{22}S'_{11})(1 - S_{33}S''_{11}) - S_{23}S''_{11}S_{32}S'_{11}} a_2 + \\ & \frac{S_{23}S''_{12}(1 - S_{33}S''_{11}) + S_{23}S''_{11}S_{33}S''_{12}}{(1 - S_{22}S'_{11})(1 - S_{33}S''_{11}) - S_{23}S''_{11}S_{32}S'_{11}} a_3. \end{aligned} \tag{G.12}$$

Then, upon substitution of equation (G.12) into equation (G.11) we find for b_{3a}

$$\begin{aligned} b_{3a} = & \left[\frac{S_{31}}{(1 - S_{33}S''_{11})} + \right. \\ & \left. \frac{S_{32}S'_{11}}{(1 - S_{33}S''_{11})} \frac{S_{21}(1 - S_{33}S''_{11}) + S_{23}S''_{11}S_{31}}{(1 - S_{22}S'_{11})(1 - S_{33}S''_{11}) - S_{23}S''_{11}S_{32}S'_{11}} \right] a_1 + \\ & \left[\frac{S_{32}S'_{12}}{(1 - S_{33}S''_{11})} + \right. \\ & \left. \frac{S_{32}S'_{11}}{(1 - S_{33}S''_{11})} \frac{S_{22}S'_{12}(1 - S_{33}S''_{11}) + S_{23}S''_{11}S_{32}S'_{12}}{(1 - S_{22}S'_{11})(1 - S_{33}S''_{11}) - S_{23}S''_{11}S_{32}S'_{11}} \right] a_2 + \\ & \left[\frac{S_{33}S''_{12}}{(1 - S_{33}S''_{11})} + \right. \\ & \left. \frac{S_{32}S'_{11}}{(1 - S_{33}S''_{11})} \frac{S_{23}S''_{12}(1 - S_{33}S''_{11}) + S_{23}S''_{11}S_{33}S''_{12}}{(1 - S_{22}S'_{11})(1 - S_{33}S''_{11}) - S_{23}S''_{11}S_{32}S'_{11}} \right] a_3. \end{aligned} \tag{G.13}$$

Substitution of equations (G.12) and (G.13) in, respectively, equations (G.8), (G.5) and (G.7), bearing in mind that the scattering matrix of the complete cascaded system is given by

$$\begin{pmatrix} b_1 \\ b_2 \\ b_3 \end{pmatrix} = \begin{pmatrix} S^T_{11} & S^T_{12} & S^T_{13} \\ S^T_{21} & S^T_{22} & S^T_{23} \\ S^T_{31} & S^T_{32} & S^T_{33} \end{pmatrix} \begin{pmatrix} a_1 \\ a_2 \\ a_3 \end{pmatrix}, \tag{G.14}$$

finally yields for the scattering coefficients

$$\begin{aligned} S^T_{11} = & S_{11} + S_{12}S'_{11} \frac{S_{21}(1 - S_{33}S''_{11}) + S_{23}S''_{11}S_{31}}{(1 - S_{22}S'_{11})(1 - S_{33}S''_{11}) - S_{23}S''_{11}S_{32}S'_{11}} + \\ & S_{13}S''_{11} \left[\frac{S_{31}}{1 - S_{33}S''_{11}} + \right. \\ & \left. \frac{S_{32}S'_{11}}{1 - S_{33}S''_{11}} \frac{S_{21}(1 - S_{33}S''_{11}) + S_{23}S''_{11}S_{31}}{(1 - S_{22}S'_{11})(1 - S_{33}S''_{11}) - S_{23}S''_{11}S_{32}S'_{11}} \right], \end{aligned} \tag{G.15}$$

$$S^T_{12} = S_{12}S'_{12} + S_{12}S'_{11}\frac{S_{22}S'_{12}(1-S_{33}S''_{11}) + S_{23}S''_{11}S_{32}S'_{12}}{(1-S_{22}S'_{11})(1-S_{33}S''_{11}) - S_{23}S''_{11}S_{32}S'_{11}} + S_{13}S''_{11}\left[\frac{S_{32}S'_{12}}{1-S_{33}S''_{11}} + \frac{S_{32}S'_{11}}{1-S_{33}S''_{11}}\,\frac{S_{22}S'_{12}(1-S_{33}S''_{11}) + S_{23}S''_{11}S_{32}S'_{12}}{(1-S_{22}S'_{11})(1-S_{33}S\text{"}{}_{11}) - S_{23}S''_{11}S_{32}S'_{11}}\right], \tag{G.16}$$

$$S^T_{13} = S_{13}S''_{12} + S_{12}S'_{11}\frac{S_{23}S''_{12}(1-S_{33}S''_{11}) + S_{23}S''_{11}S_{33}S''_{12}}{(1-S_{22}S'_{11})(1-S_{33}S''_{11}) - S_{23}S''_{11}S_{32}S'_{11}} + S_{13}S''_{11}\left[\frac{S_{33}S''_{12}}{1-S_{33}S''_{11}} + \frac{S_{32}S'_{11}}{1-S_{33}S''_{11}}\,\frac{S_{23}S''_{12}(1-S_{33}S''_{11}) + S_{23}S''_{11}S_{33}S''_{12}}{(1-S_{22}S'_{11})(1-S_{33}S''_{11}) - S_{23}S''_{11}S_{32}S'_{11}}\right], \tag{G.17}$$

$$S^T_{21} = S'_{21}\frac{S_{21}(1-S_{33}S''_{11}) + S_{23}S''_{11}S_{31}}{(1-S_{22}S'_{11})(1-S_{33}S''_{11}) - S_{23}S''_{11}S_{32}S'_{11}}, \tag{G.18}$$

$$S^T_{22} = S'_{22} + S'_{21}\frac{S_{22}S'_{12}(1-S_{33}S''_{11}) + S_{23}S''_{11}S_{32}S'_{12}}{(1-S_{22}S'_{11})(1-S_{33}S''_{11}) - S_{23}S''_{11}S_{32}S'_{11}}, \tag{G.19}$$

$$S^T_{23} = S'_{21}\frac{S_{23}S''_{12}(1-S_{33}S''_{11}) + S_{23}S''_{11}S_{33}S''_{12}}{(1-S_{22}S'_{11})(1-S_{33}S''_{11}) - S_{23}S''_{11}S_{32}S'_{11}}, \tag{G.20}$$

$$S^T_{31} = S''_{21}\left[\frac{S_{31}}{1-S_{33}S''_{11}} + \frac{S_{32}S'_{11}}{1-S_{33}S''_{11}}\,\frac{S_{21}(1-S_{33}S''_{11}) + S_{23}S''_{11}S_{31}}{(1-S_{22}S'_{11})(1-S_{33}S''_{11}) - S_{23}S''_{11}S_{32}S'_{11}}\right], \tag{G.21}$$

$$S^T_{32} = S''_{21}\left[\frac{S_{32}S'_{12}}{1-S_{33}S''_{11}} + \frac{S_{32}S'_{11}}{1-S_{33}S''_{11}}\,\frac{S_{22}S'_{12}(1-S_{33}S''_{11}) + S_{23}S''_{11}S_{32}S'_{12}}{(1-S_{22}S'_{11})(1-S_{33}S''_{11}) - S_{23}S''_{11}S_{32}S'_{11}}\right], \tag{G.22}$$

$$S^T_{33} = S''_{22} + S''_{21}\left[\frac{S_{33}S''_{12}}{1-S_{33}S''_{11}} + \frac{S_{32}S'_{11}}{1-S_{33}S''_{11}}\,\frac{S_{23}S''_{12}(1-S_{33}S''_{11}) + S_{23}S''_{11}S_{33}S''_{12}}{(1-S_{22}S'_{11})(1-S_{33}S''_{11}) - S_{23}S''_{11}S_{32}S'_{11}}\right]. \tag{G.23}$$

Grouping common terms results in the following - easier to handle - equations for the elements of the split-T power divider scattering matrix

$$S_{11}^T = S_{11} + S_{31}G + FC, \tag{G.24}$$

$$S_{12}^T = S_{12}S'_{12} + S_{32}S'_{12}G + FD, \tag{G.25}$$

$$S_{13}^T = S_{13}S''_{12} + S_{33}S"_{12}G + FE, \tag{G.26}$$

$$S_{21}^T = S'_{21}C, \tag{G.27}$$

$$S_{22}^T = S'_{22} + S'_{21}D, \tag{G.28}$$

$$S_{23}^T = S'_{21}E, \tag{G.29}$$

$$S_{31}^T = S''_{21}\left[\frac{S_{31}}{A} + HC\right], \tag{G.30}$$

$$S_{32}^T = S''_{21}\left[\frac{S_{32}S'_{12}}{A} + HD\right], \tag{G.31}$$

$$S_{33}^T = S''_{22} + S''_{21}\left[\frac{S_{33}S''_{12}}{A} + HE\right], \tag{G.32}$$

where

$$A = 1 - S_{33}S''_{11}, \tag{G.33}$$

$$B = A(1 - S_{22}S'_{11}) - S_{23}S''_{11}S_{32}S'_{11}, \tag{G.34}$$

$$C = \frac{S_{21}A + S_{23}S''_{11}S_{31}}{B}, \tag{G.35}$$

$$D = \frac{S_{22}S'_{12}A + S_{23}S''_{11}S_{32}S'_{12}}{B}, \tag{G.36}$$

$$E = \frac{S_{23}S''_{12}A + S_{23}S''_{11}S_{33}S''_{12}}{B}, \tag{G.37}$$

$$F = S_{12}S'_{11} + \frac{S_{13}S''_{11}S_{32}S'_{11}}{A}, \tag{G.38}$$

$$G = \frac{S_{13}S''_{11}}{A}, \tag{G.39}$$

$$H = \frac{S_{32}S'_{11}}{A}. \tag{G.40}$$

Index